DIGITAL CIRCUITS WITH WITH MICROPROCESSOR APPLICATIONS

**Matrix Series in
Computer Engineering**

Paul M. Chirlian, *Editor*

*Analysis and Design of Digital Circuits and
Computer Systems, Chirlian*

*Fundamental Principles of Microcomputer
Architecture, Doty*

*Microprocessor Systems Engineering,
Camp, Smay and Triska*

*Digital Circuits with Microprocessor
Applications, Chirlian*

DIGITAL CIRCUITS WITH MICROPROCESSOR APPLICATIONS

Paul M. Chirlian

Professor
Department of Electrical Engineering
and Computer Science
Stevens Institute of Technology

MATRIX PUBLISHERS, INC.
Beaverton, Oregon

To Barbara, Lisa and Peter

© **Copyright, Matrix Publishers, Inc., 1982**
All rights reserved. No part of this book may be reproduced or utilized in any form or by any means, electronic or mechanical, including photocopying, recording or by any information storage and retrieval system, without written permission from the publisher.

10 9 8 7 6 5 4 3 2 1

Library of Congress Cataloging in Publication Data

Chirlian, Paul M.
   Digital circuits with microprocessor applications.

   (Matrix series in computer engineering)
   Bibliography: p.
   Includes index.
   1. Digital electronics.   2. Microprocessors.
I. Title.   II. Series.
TK7868.D5C45   621.3815'3   81-12407
ISBN 0-916460-32-0   AACR2

Production: Patricia Miller
Illustrations: Scientific Illustrators
Printing: Pantagraph Printing
Editor: Merl K. Miller

Matrix Publishers, Inc.
11000 SW 11th St., Suite E
Beaverton, OR 97005

Contents

Preface

This introductory book's objective is to provide the reader with the basic ideas and tools needed to analyze and design digital circuits and computer systems. Since this is a text for a basic course, the fundamental ideas of switching circuits and their design is stressed and covered in great detail. However it is felt that, after completing such a first course, the student should have an understanding of a complete digital computer or a computer system. Therefore, such things as microprocessor computer organization and machine language are discussed. Interfacing the computer is also considered so that the reader develops an understanding of microprocessor programming and interfacing. Of course, this material is not covered in as much detail as it would be in a text devoted to complete computer systems.

The discussion of such things as number systems, change of radix, and elementary binary arithmetic is very detailed. This is done so that, if desired, this material need not be covered in class but may just be assigned for home reading.

The first chapter is an introductory one which is designed to provide the student with a brief picture of digital device usage. Chapter Two considers number systems and simple binary arithmetic operations. The elementary ideas of logic and basic gates are discussed in Chapter Three.

In Chapter Four the basic ideas of Boolean algebra are presented on an axiomatic basis. The necessary Boolean relations are developed. These are applied in Chapter Five where minimization techniques for combinational circuits are studied. The topics covered include basic ideas of switching algebra, minimization using maps, prime implicants, the Quine-McCluskey method and the minimization of multiple-output networks.

In the sixth chapter, we start with a general discussion of sequential circuits. Next we discuss the various flip-flops. Synchronization and triggering are considered in detail. The basic tools needed to analyze synchronous circuits are discussed, followed by a discussion of registers. A detailed discussion of the design of synchronous sequential circuits is presented. Counters are considered here. The basic ideas of asynchronous sequential circuits are discussed, and the problems of hazards and races are considered. The design of asynchronous sequential circuits is then presented.

After developing the basic tools, some of the components of the digital computer are discussed. In Chapter Seven we consider registers and memories. Random access memories are considered in great detail. Next, auxiliary memories are discussed. Finally, read-only memories, decoders, and programmable logic arrays are studied.

In Chapter Eight, the ideas of binary codes are discussed. First we consider BCD codes, then alphnumeric codes. Next, reflecting codes are considered. Finally we study error-detecting and correcting codes.

Chapter Nine covers binary arithmetic. We start with a consideration of modular arithmetic. Next, complementary arithmetic is studied. Floating-point arithmetic is discussed. Precision and accuracy are considered in detail.

Computer arithmetic is presented in Chapter Ten. The actual computer implementations of addition, subtraction, multiplication, and division are presented. BCD conversion is also considered here. Circuits which implement these operations are explained. The chapter concludes with a complete discussion of the arithmetic logic unit.

In Chapter Eleven the complete microprocessor-based digital computer is considered. We start with a general discussion of the components of a computer. This is followed by a discussion of microprocessor architecture and of the operation of a microprocessor-based computer. Timing is discussed in detail. Machine language programming of a computer is then presented. The machine language is not related to any specific microprocessor but is typical of that used in microprocessors in general. Examples of programming are given. Both control and computation examples are discussed. This is followed by a detailed discussion of interfacing the computer with the external world.

Gate circuits are covered in Appendix A so that the reader can gain some insight into the actual circuitry that is used.

Many varied problems are given at the end of each chapter.

This book draws heavily on the author's previous book, *The Analysis and Design of Digital Circuits and Computer Systems*. However, because much of the material has been thoroughly revised, and because of the new emphasis on microprocessor systems, it was decided to change the title rather than to call this book a second edition.

The author would like to express his gratitude to his colleagues Professors Alfred C. Gilmore, Jr., Emil C. Neu, and Otto C. Boelens for the numerous and lengthy discussions held during the writing of the book. They provided much help and inspiration.

Much loving gratitude and heartfelt thanks are again due my wife, Barbara, who not only provided me with continuous encouragement but who also typed parts of the rough draft and final draft of the manuscript. Loving thanks are also given to my daughter, Lisa, who typed much of the revision.

Paul M. Chirlian

1

Introduction

This book discusses the analysis and design of the elements of digital systems with particular attention being paid to the digital computer, although other digital systems will be covered as well. First, the basic principles of switching theory will be considered and then these concepts will be applied to the design of general digital systems. Before commencing the study of switching theory, we shall utilize this first chapter to discuss some general ideas of the digital computer.

1.1 A GENERAL DISCUSSION OF THE DIGITAL COMPUTER

The digital computer can perform many operations. It can, in seconds, solve equations which would take years to calculate by hand. An entire industrial process can be controlled by a digital computer. For instance, special sensors could determine the temperature, pressure, acidity, etc., existing during a chemical process. In an appropriate form, this information would then be supplied to a digital computer where calculations on it are performed. As a result of these calculations, electrical signals controlling the flow of the process would be generated. For instance, in response to these signals, heaters could be turned on or off, or additional chemicals might be added. In such cases, the digital computer could be called a *digital controller*.

A radar signal reflected from a distant planet (or object) is often very weak and obscured by noise and ordinary techniques cannot be used to detect it. However, a digital computer can process the radar signal and render it detectable. We have considered only several of the host of tasks that can be performed by a digital computer. In order to perform such complex operations, the digital computer must contain very many circuits. If fact, each computer is made up of the interconnection of thousands of relatively simple circuits and we must understand the individual circuits before considering how they are intereconnected to form the complete computer. In this book we shall consider both these logic circuits and their interconnections in detail.

Integrated circuit techniques can be used to produce simple digital computers which are very small. Such computers, called *microprocessors*, are used to perform many kinds of tasks. For instance, a microprocessor in an oscilloscope can be used to obtain a reading of the true rms value of a signal and, as we shall see, microprocessors are used for many of the previously discussed control tasks.

The earliest digital computers used relays as their fundamental elements. A relay is simply a switch controlled by an electromagnet. These switches were either completely on or completely off (i.e., either open or closed). In present digital computers the relays have been replaced with electronic devices. These, although much smaller, faster and less expensive than relays, still act as controlled switches which are either on or off. In the next section we shall discuss how these on or off conditions give rise to a signal which has two levels that are called *zero* and *one*. All the signals in a digital system are made up of combinations of such zeros and ones. In contrast, the signals in an *analog* system can vary continuously. For instance, if the number 1.23 is the result of a computation by an analog computer, then 1.23 volts might actually appear at the output. In a digital computer, the number 1.23 would be represented by a sequence of zeros and ones. In the next chapter we shall see how this is accomplished. In addition, the output of an analog system can vary continuously while that of a digital one can vary only in discrete steps. The digital system seems more complex and, indeed, it often is. However, it has many advantages. A digital system can be extremely accurate with many systems providing greater than 16-place accuracy. Such accuracy is impossible with an analog system, where the limit of precision is usually three or four significant figures. In addition, the characteristics of analog systems are often sensitive to time, temperature, etc. (i.e., drift) while digital systems are not. Furthermore, digital systems are often more flexible to use and, thus, in spite of their complexity, digital computers are very widely used for a great variety of operations. In fact, digital systems have replaced analog systems in very many operations. One reason for this is that the use of integrated circuits now allows the extremely large circuits needed for digital computers to be reliably built in small volumes at relatively low cost, often at much lower cost than the corresponding analog system.

1.2 SOME FUNDAMENTALS OF DIGITAL OPERATIONS

Digital systems and computers always manipulate numbers. For instance, if a nonlinear differential equation is to be solved, then the computer is programmed to implement an appropriate numerical analysis procedure. The program instructs the computer to perform many arithmetic operations on a set of input data (i.e., numbers). These operations result in the solution to the equation. Similarly, when a digital system is used to control a chemical process, data must be supplied to the digital controller in numerical form. For instance, there may be a sequence of numbers, each of which represents the temperature of the process at the time at which it is supplied to the controller. There are *transducers* that convert quantities such as temperature or pressure to voltage. These transducers are analog in nature and supply an analog voltage. Special electronic circuits called *analog to digital converters* change the voltage level into a number composed of a sequence of zeros and ones. It is this numeral that is then supplied to the digital controller.

The operations performed by the digital computer are controlled by a series of instructions called a *program*. In a subsequent chapter we shall discuss how such programs direct digital computation. We shall also see how the digital computer controls external devices such as heaters or lights.

Note that the words digital computer, digital system, and digital controller all have essentially the same meaning when the internal structure of the device is considered. In general, a digital system is one whose input consists of numbers consisting of only the digits zero and one. Such a system might more correctly be called a binary system, but the word digital is commonly used.

A digital computer consists of several basic parts. There is a *control unit* which receives the instructions of the program and directs the appropriate computer operations. The *central processor unit* performs the operations of addition, subtraction, multiplication and division plus some additional operations. The *memory* is used to store both data and programs. There are also *input-output* devices such as card readers, terminals and printers which are used by the human operators to supply the computer with the data and instructions and which, in turn, are used by the computer to return the calculated values.

All the data handled by the computer must be in numeric form. If the computer performs nonnumeric procedures such as alphabetizing a list of names then this information must be converted into numerical form using a suitable *code*. This shall also be discussed in a subsequent chapter.

The storage and manipulation of numbers is of fundamental importance in the study of digital computers. For this reason, we shall now study some fundamental ideas of number systems. We all learned and used the *decimal number* system which is based on the 10 digits 0, 1, 2, 3, 4, 5, 6, 7, 8, and 9. Such a number system probably was developed because humans have 10 fingers; thus 10 was used for convenience. When mechanical adding machines and desk calculators were built, the decimal system was easily carried over to them since gears with 10, or multiples of 10, teeth were used. However, electronic devices have replaced almost all the mechanical ones, not only in large computers but also in simple adding machines and desk calculators. It is not necessary or, in fact, desirable, to use the decimal system with such electronic devices.

We shall illustrate our discussion with the simple electronic circuit of Fig. 1-1a. The box represents an idealized electronic device; it could approximate a transistor. A plot of output voltage versus input voltage for this device is shown in Fig. 1-1b. Let us consider this characteristic.

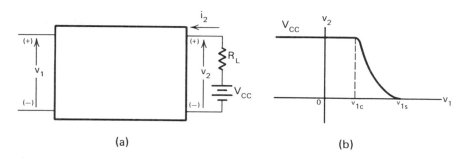

(a)   (b)

Figure 1-1. (a) An idealized electronic circuit; (b) the transfer characteristics of this circuit

If the input voltage v_1 is less than v_{1c}, then the output circuit of the device acts as an open circuit and $v_2 = V_{CC}$. In this condition, the output current $i_2 = 0$ and the device is said to be *cut off*. If we want to be sure that the device is cut off, then we can make v_1 substantially less than v_{1c}.

If v_1 is greater than v_{1s}, then the output circuit of the device acts as a short circuit and $v_2 = 0$. In this case, the device is said to be *saturated*. Again, if we want to be sure that the device is saturated, then we can make v_1 substantially greater than v_{1s}.

The typical digital computer is made up of many devices of this type. Their cut-off and saturation values can be easily established. These voltage levels are relatively far apart so that they are readily distinguishable. For instance, the saturation voltage may range between 0.0 and 0.5 volts while a cut-off voltage could range between 9 and 10 volts. Note that the cut-off values for all the devices need not be the same. They only must fall within some specified range. A similar statement can be made for the saturation values. Thus, the digital computer becomes extremely reliable since a small shift in levels will still result in voltages which lie within the specified ranges and, thus, be interpreted correctly. Hence, there are two levels (ranges of levels) that are used. If the cut-off and saturation levels are represented by numbers, then only two digits are required. As we have seen, these are called *zero* and *one*, and this is called a *binary number system*. Thus, in digital computers, we work with an arithmetic system that uses only the two digits zero and one. These are called *binary digits* or *bits*. Note that this does *not* mean that the data cannot be entered in the usual decimal form, but only that if it is, it must be converted to the binary form within the computer. Thus, we must study the *binary number system*.

Efficiency is another reason for operating the active devices of the system at either saturation or cut off only. At saturation, the current is high but the voltage across the device is extremely small; at cut off, the voltage is large but the current is extremely small. Either of these conditions will usually result in a small power dissipation. This is very important, especially when many electronic components are contained on one small chip. Now that we have discussed the reasons for using a binary number system, we shall consider this number system in detail in the next chapter.

BIBLIOGRAPHY

Booth, T.L., *Digital Networks and Computer Systems*, 2nd Edition, Chap. 1, John Wiley and Sons, Inc., New York, 1978

Hill, F.J. and Peterson, G.R., *Digital Systems Hardware Organization and Design*, Chap. 1, John Wiley and Sons, Inc., New York, 1973

Peatman, J.B., *The Design of Digital Systems*, Chap. 1, McGraw-Hill Book Co., New York, 1972

PROBLEMS

1-1. How many signal levels are used in a digital computer?

1-2. Compare digital and analog systems.

1-3. List the basic elements of a digital computer.

1-4. Why is the binary number system used in a digital computer?

1-5. Why are codes needed in a digital system?

2

Number Systems

Since digital computers work with only two digits, we shall now discuss two-digit number systems. In addition, because it is sometimes convenient to work with other number systems, these too shall be considered in this chapter.

2.1 FUNDAMENTAL IDEAS OF NUMBER SYSTEMS

A number system can be established using any integral number of digits greater than one. The number of *digits* used is called the *radix* or *base* of the system. Thus, we commonly work with a radix 10 or base 10 system, and arithmetic in the digital computer is performed in a radix 2 or base 2 system. At times, people using computers find it convenient to work in the *octal* system (radix 8) and the *hexadecimal* system (radix 16). Reasons for using these other systems will be discussed subsequently. First, we shall compare some numbers as they are written in various systems, as shown in Table 2-1. Note that, since the hexadecimal system requires 16 digits, it is conventional to use the letters A, B, C, D, E and F to represent the digits 10, 11, 12, 13, 14, and 15, respectively.

If a number is to have any meaning, then we must know what radix is being used. It is often obvious. However, if this is not the case, then a subscript will be appended to indicate the radix. For instance, see Table 2-1, we can write

$$8_{10} = 1000_2 = 10_8 = 8_{16} \tag{2-1}$$

The subscript is always written in radix 10. When the radix is obvious, the subscript will be omitted.

Table 2-1 A Comparison of Various Number Systems

Radix 10 Decimal	Radix 2 Binary	Radix 8 Octal	Radix 16 Hexadecimal
0	00000	0	0
1	00001	1	1
2	00010	2	2
3	00011	3	3
4	00100	4	4
5	00101	5	5
6	00110	6	6
7	00111	7	7
8	01000	10	8
9	01001	11	9
10	01010	12	A
11	01011	13	B
12	01100	14	C
13	01101	15	D
14	01110	16	E
15	01111	17	F
16	10000	20	10
17	10001	21	11
18	10010	22	12
19	10011	23	13
20	10100	24	14
21	10101	25	15
22	10110	26	16
23	10111	27	17
24	11000	30	18
25	11001	31	19
26	11010	32	1A
27	11011	33	1B
28	11100	34	1C
29	11101	35	1D
30	11110	36	1E
31	11111	37	1F
32	100000	40	20
33	100001	41	21
34	100010	42	22
35	100011	43	23
36	100100	44	24

Now we shall discuss number systems, starting with radix 10 since this is the most familiar system. We shall then extend the results to an arbitrary radix. After considering whole numbers, we will generalize to numbers with fractional parts. Suppose we write the number 165_{10}. This number actually represents the following:

$$165_{10} = 1(10^2) + 6(10^1) + 5(10^0) \tag{2-2}$$

where all the numbers are written in radix 10. That is, the number 165_{10} represents 1 one hundreds, 6 tens and 5 ones. Let us generalize this. Suppose that we have a number represented by

$$a_n a_{n-1} a_{n-2} \ldots a_2 a_1 a_0 \tag{2-3}$$

where each a_j represents a digit of the base 10 number. For instance, in the previous case, where we considered the number 165,

$$a_0 = 5; \quad a_1 = 6; \quad a_2 = 1$$

We can represent the number of (2-3) by the equation

$$a_n a_{n-1} \ldots a_2 a_1 a_0 = a_n 10^n + a_{n-1} 10^{n-1} + \ldots$$

$$+ a_2 10^2 + a_1 10^1 + a_0 10^0 \tag{2-4}$$

Thus, each digit represents the number of times that the appropriate power of 10 appears in the summation.

Now let us consider numbers with fractional parts. For example

$$256.123_{10} = 2(10^2) + 5(10^1) + 6(10^0) + 1(10^{-1}) + 2(10^{-2}) + 3(10^{-3})$$

The decimal point is used to separate the positive powers of 10 from the negative ones. Therefore, in general, we can write

$$(a_n a_{n-1} \ldots a_2 a_1 a_0 . a_{-1} a_{-2} \ldots a_{-k})_{10}$$

$$= a_n 10^n + a_{n-1} 10^{n-1} + \ldots + a_2 10^2 + a \quad a_1 10^1 + a_0 10^0$$

$$+ a_{-1} 10^{-1} + a_{-2} 10^{-2} + \ldots + a_{-k} 10^{-k} \tag{2-5}$$

In a decimal system, we refer to the point which separates the positive and negative powers of 10 as the decimal point. The general name given to this point is the *radix point*. Therefore in binary systems, the radix point is called the *binary point*; in octal systems, the radix point is called the *octal point*, etc.

Now let us consider the expression of a number in a general radix. We shall express the general radix in terms of numbers in radix 10, since we are most familiar with numbers in this radix. Integers will be considered first. Thus, following Eq. (2-4), we have

$$(a_n a_{n-1} \ldots a_2 a_1 a_0)_r = a_n r^n + a_{n-1} r^{n-1} + \ldots$$

$$+ a_2 r^2 + a_1 r^1 + a_0 r^0 \qquad (2\text{-}6)$$

Note that the numbers to the right of the equals sign are expressed to base 10. We shall discuss some examples of this. (Table 2-1 can be used to verify some of the results.

Examples:

$$37_8 = 3_{10}(8_{10}^1) + 7_{10}(8_{10}^0) = 24_{10} + 7_{10} = 31_{10}$$

$$1E_{16} = 1_{10}(16_{10}^1) + 14_{10}(10_{10}^0) = 30_{10}$$

$$100100_2 = 1(2_{10}^5) + 0(2_{10}^4) + 0(2_{10}^3) + 1(2_{10}^2)$$

$$+ 0(2_{10}^1) + 0(2_{10}^0) = 32 + 4 = 36_{10}$$

Now let us include a radix point in our number which is expressed in a general radix. Following the example of Eq. (2-5), we obtain

$$(a_n a_{n-1} \ldots a_2 a_1 a_0 . a_{-1} a_{-2} \ldots a_{-k})_r = a_n r^n + a_{n-1} r^{n-1} + \ldots$$

$$+ a_2 r^2 + a_1 r^1 + a_0 r^0 + a_{-1} r^{-1} + a_{-2} r^{-2} + \ldots$$

$$+ a_{-k} r^{-k} \qquad (2\text{-}7)$$

where we have expressed the numbers to the right of the equals sign in decimal radix. We shall illustrate this with some examples.

Examples:

$$127.34_8 = 1(8^2) + 2(8^1) + 7(8^0) + 3(8^{-1}) + 4(8^{-2}) = 87.4375_{10}$$

$$101.101_2 = 1(2^2) + 0(2^1) + 1(2^0)$$

$$+ 1(2^{-1}) + 0(2^{-2}) + 1(2^{-3}) = 5.625_{10}$$

$$EA2.AB3_{16} = 14(16^2) + 10(16^1) + 2(16^0)$$

$$+ 10(16^{-1}) + 11(16^{-2}) + 3(16^{-3}) = 3746.668701_{10}$$

The examples that we have just given illustrate the conversion from an arbitrary radix to radix 10. Let us now illustrate the conversion from radix 10 to an arbitrary radix. Suppose that we have the decimal number

$$N_{10} = (d_n d_{n-1} \ldots d_2 d_1 d_0)_{10}$$

and we want to convert it to a binary system number which we represent as

$$N_2 = b_k 2^k + b_{k-1} 2^{k-1} + \ldots + b_2 2^2 + b_1 2^1 + b_0 2^0 \qquad (2\text{-}8)$$

We must solve for all the unknown b_j which, in this case, are equal to either 0 or 1. There are various procedures that can be used here. The technique that we shall discuss is convenient to use. Divide both sides of Eq. (2-8) by 2. This yields

$$N_2/2 = b_k 2^{k-1} + b_{k-1} 2^{k-2} + \ldots + b_2 2 + b_1 + b_0/2 \qquad (2\text{-}9)$$

The quotient of the division is

$$Q_1 = b_k 2^{k-1} + b_{k-1} 2^{k-2} + \ldots + b_2 2 + b_1 \qquad (2\text{-}10)$$

while the remainder $R_1 = b_0$, which is the *rightmost* binary digit. Now divide the quotient Q_1 by 2. This yields

$$Q_1/2 = b_k 2^{k-2} + b_{k-1} 2^{k-3} + b_2 + b_1/2 \qquad (2\text{-}11)$$

The quotient is

$$Q_2 = b_k 2^{k-2} + b_{k-1} 2^{k-3} + b_2 \qquad (2\text{-}12)$$

while the remainder is

$$R_2 = b_1$$

which is the second rightmost binary digit. Proceeding in this way we can use the remainders to obtain all the binary digits. Let us illustrate this with an example.

Example:

Obtain the binary equivalent of 47_{10}.

$47_{10}/2 = 23 + \frac{1}{2}$

Thus, $R_1 = 1$

$Q_1 = 23$

Dividing by 2, we obtain

$Q_1/2 = 11 + \frac{1}{2}$

Therefore,

$Q_2 = 11$

$R_2 = 1$

Proceeding similarly, we have

$Q_2/2 = 5 + \frac{1}{2}$

$R_3 = 1$

$Q_3 = 5$

$Q_3/2 = 2 + \frac{1}{2}$

$R_4 = 1$

$Q_4 = 2$

$Q_4/2 = 1$

$R_5 = 0$

$Q_5 = 1$

$Q_5/2 = \frac{1}{2}$

$Q_6 = 0$

$R_6 = 1$

Thus, the procedure has terminated. We now use all the remainders to obtain the binary digits. Remember that the first remainder yields the rightmost digit. Thus, we have

$47_{10} = 101111_2$

Conversion of Numbers with Fractional Parts

Now let us consider the conversion of numbers with fractional parts. A number can always be considered to be the sum of its whole and fractional parts. Thus, we can use the procedure just discussed to obtain the conversion of the whole part and then add the conversion of the fractional part to it. To convert the portion to the right of the radix point start by writing

$$F_{10} = (0.d_{-1}d_{-2} \ldots d_{-m})_{10} = (0.b_{-1}b_{-2} \ldots)_2 \qquad (2\text{-}13)$$

That is, we have equated the decimal and binary numbers. This represents the equation

$$d_{-1}10^{-1} + d_{-2}10^{-2} + \ldots + d_{-m}10^{-m} = b_{-1}2^{-1} + b_{-2}2^{-2} \ldots \qquad (2\text{-}14)$$

Note that we have represented a decimal fraction of finite length by a binary fraction, which may be of infinite length. We shall consider this subsequently.

Now multiply both sides of Eq. (2-14) by 2. This yields

$$2F_{10} = b_{-1} + b_{-2}2^{-1} + b_{-3}2^{-2} + \ldots \qquad (2\text{-}15)$$

The nonfractional part of this expression represents the first digit to the right of the binary point. The fractional part of Eq. (2-15) is then

$$F_2 = b_{-2}2^{-1} + b_{-3}2^{-2} + \ldots \qquad (2\text{-}16)$$

Multiplying this by two, we have

$$2F_2 = b_{-2} + b_{-3} 2^{-1} + \ldots \tag{2-17}$$

Again, the nonfractional part represents the next digit. We can then proceed in this way to obtain all the succeeding digits. Let us illustrate this with some examples.

Example:

Obtain the binary equivalent of

0.25_{10}

Multiplying by 2, we have

$2(0.25) = 0.5$

There is no whole part. Thus,

$b_{-1} = 0$

Multiplying again, we have

$2(0.5) = 1$

The whole part is 1. Hence,

$b_{-2} = 1$

Removing the whole part, we are left with 0. Hence, the procedure terminates. Then,

$$0.25_{10} = 0.01_2 \tag{2-18}$$

Example:

Obtain the binary equivalent of

0.426_{10}

Multiplication by 2 yields

$2(0.426) = 0.852$

Thus, $b_{-1} = 0$. Proceeding, we have

$2(0.852) = 1.704$

Then,

$b_{-2} = 1$

Removing the whole part and multiplying, we obtain

$2(0.704) = 1.408$

Hence,

$b_{-3} = 1$

Continuing, we have

$2(0.408) = 0.816$

$b_{-4} = 0$

$2(0.816) = 1.632$

$b_{-5} = 1$

$2(0.632) = 1.264$

$b_{-6} = 1$

$2(0.264) = 0.528$

$b_{-7} = 0$

Hence, the result is

$$0.426_{10} = 0.0110110\ldots._2 \qquad (2\text{-}19)$$

Note that the procedure of the last example did not terminate. At times, an infinite number of binary digits must be used to represent a decimal fraction *exactly*. Since humans and computers can only work with a finite number of digits, the conversion of fractions from one radix to another often results in an error. For instance we can write (see the previous example)

$$0.426_{10} = 0.01101_2 + \epsilon \qquad (2\text{-}20)$$

where ϵ represents the error. That is, if we represent 0.426_{10} by 0.01101_2 then there will be an error whose magnitude is equal to ϵ.

Let us estimate the size of this error. In radix 2, we can write the fractional part *exactly* by using the infinite series

$$F_{10} = \sum_{n=1}^{\infty} b_{-n} 2^{-n} \qquad (2\text{-}21)$$

When only a finite number of binary digits are used, we represent this by a finite sum plus an error. Thus,

$$F_{10} = \sum_{n=1}^{N} b_{-n} 2^{-n} + \epsilon \qquad (2\text{-}22)$$

where the error is equal to the terms in the summation which have been left out. Hence,

$$\epsilon = \sum_{n=N+1}^{\infty} b_{-n} 2^{-n} \qquad (2\text{-}23)$$

The b_{-n} are either 0's or 1's. In this case, this series will converge rapidly. In fact, it can be shown that

$$|\epsilon| \leqslant 2^{-N} \qquad (2\text{-}24)$$

Thus, we need only carry the fraction out until the 2^{-N} becomes less than some desired value. For instance, in Eq. (2-20)

$$\epsilon < 2^{-5} = 0.03125_{10}$$

If this is not a sufficiently small error, then additional terms should be taken. For instance, if we write

$0.246_{10} = 0.0110110$

then

$\epsilon < 0.00781250_{10}$

In general, an error will occur when numbers are converted from one radix to another. This is called *roundoff error*. (We shall subsequently discuss other causes of roundoff errors.)

Let us illustrate the conversion of a decimal number containing a whole and a fractional part to binary.

Example:

Express 47.426_{10} in binary form.

We shall use a shorthand form here to represent the divisions. The whole part is 47. Then,

$47/2 = 23 + \frac{1}{2}$

$23/2 = 11 + \frac{1}{2}$

$11/2 = 5 + \frac{1}{2}$

$5/2 = 2 + \frac{1}{2}$

$2/2 = 1 + \frac{0}{2}$

$1/2 = 0 + \frac{1}{2}$

Thus, $47_{10} = 101111_2$. Then, to obtain the fractional part, we perform the following calculations:

$2(0.426) = 0.852$

$2(0.852) = 1.704$

$2(0.704) = 1.408$

$2(0.408) = 0.816$

$2(0.816) = 1.632$

$2(0.632) = 1.264$

$\vdots$

Hence,

$47.426_{10} \approx 101111.011011_2$

We have illustrated conversion from radix 10 to radix 2. The procedures for conversion from base 10 to other bases follow this one. The radix r replaces the 2 in the previous procedures. Let us illustrate this with an example.

Example:

Express 47.426_{10} in octal. Again we work with the whole and fractional parts separately. Thus,

$47/8 = 5 + \frac{7}{8}$

$5/8 = 0 + \frac{5}{8}$

Hence,

$47_{10} = 57_8$

Now consider the fractional part

$8(0.426) = 3.408$

$8(0.408) = 3.264$

$8(0.264) = 2.112$

$8(0.112) = 0.896$

$8(0.896) = 7.168$

$\vdots$

Therefore,

$$47.4026_{10} \approx 57.33207_8$$

The conversion between arbitrary radices follows the same general ideas. It is somewhat more difficult to perform the arithmetic in this case since we are not used to performing computations in other bases, e.g. we do not know the base 6 multiplication table.

Conversion Between Binary and Octal or Hexadecimal

Conversion between binary and octal or hexadecimal is especially easy. Let us write a binary number and show how it can be converted to an octal number.

$$1010111_2 = 1(2^6) + 0(2^5) + 1(2^4) + 0(2^3) + 1(2^2) + 1(2^1) + 1(2^0)$$

Note that the terms to the right of the equals sign are expressed in base 10. Now let us rewrite the binary number as 001010111_2. Note that adding 0's to the left does not change the number. It is done so that the total number of digits in the binary number is evenly divisible by three. We can now write

$$001010111_2 = [0(2^8) + 0(2^7) + 1(2^6)] + [0(2^5) + 1(2^4) + 0(2^3)]$$
$$+ [1(2^2) + 1(2^1) + 1(2^0)]$$

Note that each group of three has a factor of 8^k [$(2^3)^k$] where k is an integer. Thus, we can write

$$001010111_2 = [0(2^2) + 0(2^1) + 1(2^0)]\, 8^2$$
$$+ [0(2^2) + 1(2^1) + 0(2^0)]\, 8^1$$
$$+ [1(2^2) + 1(2^1) + 1(2^0)]\, 8^0$$

The righthand side of the this expression is just the representation of the number in radix 8 where the value of each bracketed term represents an octal digit. Hence, we can write

$$001010111_2 = 1(8^2) + 2(8^1) + 7(8^0)$$

Using the definition of an octal number, we have

$$001010111_2 = 127_8$$

We can use this procedure for any binary-octal conversion of an integer. That is, break up the binary number into groups of three digits, adding 0's at the left if necessary. Start at the binary point and convert each binary group into an octal number. The complete octal number can be easily obtained by inspection.

Example:

Find the octal equivalent of 1010101_2.

$$1010101_2 = 001\ 010\ 101_2 = 125_8$$

Example:

Consider a binary number with a fractional part.

$$0.1011_2 = 1(2^{-1}) + 0(2^{-2}) + 1(2^{-3}) + 1(2^{-4}) + 0(2^{-5}) + 0(2^{-6})$$

Note that we have added 0's to the right so that there are six digits following the binary point. Then, manipulating, we have

$$0.1011_2 = [1(2^2) + 0(2^1) + 1(2^0)]\,8^{-1} + [1(2^2) + 0(2^1) + 0(2^0)]\,8^{-2}$$

Since we have put this in an "octal form," we can write

$$0.1011_2 = 0.54_8$$

Again, we break up the binary digits into groups of three, adding 0's at the right if necessary and then replace each group by its octal equivalent (which will be the same as the decimal equivalent since we are dealing with numbers less than seven). Each group then represents the corresponding octal digit. Note that *there is no roundoff error* when numbers are converted from binary to octal form.

Example:

Convert

$$1101101.10101_2$$

to octal. First add 0's so that there are six digits to the right of the binary point. Then, proceeding as before, we obtain

$$1 \; 101 \; 101. \; 101 \; 010_2 = 155.52_8$$

The converse procedure can be used to convert from octal to binary. That is, each octal digit is replaced by its three-digit binary representation. These are all combined in sequence to obtain the binary number. For instance,

$$155.52_8 = 001 \; 101 \; 101. \; 101 \; 010_2$$

Note that the addition of trailing 0's to a binary number can give a false sense of precision. For instance, consider the number

$$0.1011_2$$

To convert it to octal, we write it as

$$0.101100_2$$

When a number is written in this way, it implies that the last two digits are *known* (e.g., if we write 0.1011, which has four significant figures, then this could actually represent 0.101100 or 0.101101, etc.). Thus,

$$0.101100_2 = 0.54_8$$

implies six digits of precision in the binary number whereas there are actually only four digits of precision, and a false precision has been obtained.

The conversion from binary to hexadecimal is essentially the same as that used for binary to octal, except that groups of four binary digits are used. For instance,

$$1101101.10101_2 = 0110 \; 1101. \; 1010 \; 1000_2 = 6D.A8_{16}$$

Of course, the converse procedure is used to convert the hexadecimal number to binary. Note that no roundoff error occurs when there is a conversion from binary to hexadecimal or vice versa.

After gaining some experience, the conversions from binary to octal or hexadecimal and vice versa can be done by inspection. (The conversion is always simple when the radices involved are powers of

2.) This simplicity of conversion is a reason for the importance of the octal and hexadecimal systems for *humans* working with computers. As we have discussed, the computer works with and stores numbers in binary form so *people* working with the computer must often inspect these binary numbers to study the computer's operation. These binary numbers generally have many more digits than do their octal or hexadecimal equivalents and people find them tedious to use. Therefore, the operator will often use the octal or hexadecimal representations of the binary numbers. Note that, although conversions could be made from the binary to the decimal system, this is more complex and would be subject to roundoff error.

2.2 A FUNDAMENTAL DISCUSSION OF SOME BASIC ARITHMETIC OPERATIONS

In this section we shall consider some fundamental aspects of addition using binary numbers. In a subsequent chapter we shall consider arithmetic operations in greater detail. In general, the rules of arithmetic apply no matter what radix is used. We shall specifically consider how they are applied to binary arithmetic. In addition, we will also discuss some aspects of arithmetic resulting from constraints imposed by digital computer circuitry.

Let us start by considering the addition of two numbers. We shall use whole numbers here. However, the discussion also applies to numbers with fractional parts. We begin with examples in base 10 since these will be familiar to us. Consider the addition

$$
\begin{array}{r}
12734 \\
+ \ 26251 \\
\hline
38985
\end{array}
$$

Each column is added producing the result. In the chosen example, the sum of each column was less than the radix (10). If this were not true, then we would "carry a number" into the next column to the left. For instance, consider the addition

$$
\begin{array}{r}
1\overset{\frown}{4}\overset{\frown}{9}\overset{\frown}{7}8 \\
+ \ 36415 \\
\hline
51393
\end{array}
$$

The first column's addition is $8 + 5 = 13$. The 3 is written down as the answer while the remaining 10 is added to the next column by adding 1 to that column. This procedure is repeated, where necessary, in all successive columns. (Note that 1000 is carried into the fourth column, etc.)

The same procedure is used if the radix is other than 10. For instance, let us consider octal addition,

$$
\begin{array}{r}
1247_8 \\
+\ 2510_8 \\
\hline
3757_8
\end{array}
$$

The reader can verify that this arithmetic will "check" if the octal numbers are converted to decimal numbers. Now let us consider octal addition with carrying. (Table 2-1 will be helpful here.)

$$
\begin{array}{r}
111 \\
1365_8 \\
+\ 4627_8 \\
\hline
6214_8
\end{array}
$$

Consider the rightmost column: $5_8 + 7_8 = 14_8$. The 4 is written in the answer and the 10_8 (8_{10}) is added to the next column. Note that octal addition may seem awkward. However, this in only because we are not as familiar with the "octal addition table" as we are with the base 10 addition table.

Let us discuss binary addition. It follows the previously discussed ideas. As before, we shall start with an example which does not require carrying.

$$
\begin{array}{r}
10101_2 \\
+\ 01000_2 \\
\hline
11101_2
\end{array}
$$

Consider the following example which does require carrying.

$$
\begin{array}{r}
111 \\
10101_2 \\
+\ 00111_2 \\
\hline
11100_2
\end{array}
\qquad (2\text{-}25)
$$

Inspect the rightmost column. Here we have $1_2 + 1_2 = 10_2$. Again, we write the 0 in the answer and carry the 10_2 (2_{10}) to the next column.

Thus, addition in the binary system is essentially the same as addition in the decimal system. Of course, the binary addition table, rather than the decimal addition table, must be used.

Now let us consider the addition of binary numbers with fractional parts. Again, this addition follows the rules for "ordinary" addition. For example

$$\begin{array}{r} 10110.1101_2 \\ + \ 00101.1101_2 \\ \hline 11100.1010_2 \end{array} \qquad (2\text{-}26)$$

Thus, binary addition can be readily accomplished.

When numbers are added in a computer, a problem occurs which does not occur when we add numbers using a pencil and paper. Consider the following addition.

$$\begin{array}{r} 10110_2 \\ + \ 11101_2 \\ \hline 110011_2 \end{array}$$

The two numbers added each have five digits. Each binary digit is called a *bit*. Thus, each number has five bits. However, a carry takes place in the leftmost addition. Thus, the resulting answer has six bits. When humans add, this extra digit presents no problem. However, in computers, numbers are stored in devices called *registers*. When addition is performed, the result is stored in a register called an *accumulator*. Any register or accumulator can store only a fixed number of bits. If the binary number has more bits than can be stored, errors may result. To understand this, we shall discuss some fundamental ideas of storage here.

A register which can store an 8-bit binary number is represented in Fig. 2-1. Note that the rightmost digit is called the *least significant digit* since it has the smallest numerical value. Similarly, the leftmost digit is called the *most significant digit* since it has the largest numerical value. For the time being, we shall restrict ourselves to integer numbers. In a subsequent chapter we shall generalize these results.

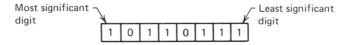

Figure 2-1. A diagrammatic representation of a register storing 10110111_2

In a register, the numbers stored are the least signficant digits. For instance, if we use the register of Fig. 2-1, to store the results of the addition

$$11010110_2$$
$$\underline{11011001_2}$$
$$110101111_2$$

then the number stored in the register would be

$$10101111_2$$

which is equivalent to 175_{10} rather than to 431_{10}. This is a substantial error and is called an *overflow*. Thus, the computer designer and user, to avoid error, must have a thorough understanding of number storage. In a subsequent chapter we shall discuss how these disadvantages can, at times, be used to advantage.

BIBLIOGRAPHY

Hill, F. J. and Peterson, G.R., *Introduction to Switching Theory and Logical Design*, Second Ed., Chap. 2, John Wiley and Sons, Inc., New York, 1974

Mano, M.M., *Digital Logic and Computer Design*, Chap. 1, Prentice Hall, Inc., Englewood Cliffs, NJ, 1979

O'Malley, J., *Introduction to the Digital Computer*, Chap. 1, Holt, Rinehart and Winston, Inc., New York 1970

PROBLEMS

2-1. Express the numbers from 0_{10} to 36_{10} in base 5.

2-2. Convert the following numbers from base 2 to base 10.
101, 101000, 11010110, 1110011101

2-3. Convert the following numbers from base 8 to base 10.
127, 4567, 7772, 1467

2-4. Convert the following numbers from base 16 to base 10.
37, 1000, 1A3, ABCD

2-5. Repeat Prob. 2-2 for
101.101, 10110.11010, 101.0110101

2-6. Repeat Prob. 2-3 for

127.1, 4567.0714, 1775.01245, 11.0121741

2-7. Repeat Prob. 2-4 for

37.A1, 1A3.101AB, BCA101.AC29B

2-8. Convert the following numbers from base 10 to base 2.

26, 98, 147, 1679, 24873

2-9. Repeat Prob. 2-8 but now write the numbers in octal.

2-10. Repeat Prob. 2-8 but now write the numbers in hexadecimal.

2-11. Convert the following numbers from base 10 to binary. Estimate any error.

26.123, 98.476, 147.001, 1246.8012, 24.16793

2-12. Repeat Prob. 2-11 but now write the numbers in octal.

2-13. Repeat Prob. 2-11 but now write the numbers in hexadecimal.

2-14. Convert the following binary numbers to octal.

101.10101, 101.0101, 101.0100101

2-15. Repeat Prob. 2-14 but now write the numbers in hexadecimal.

2-16. Convert the following octal numbers to binary.

127, 1243.107, 721,7461

2-17. Convert the following hexadecimal numbers to binary.

AB1, ABC.01A, 1A7.96BAE

2-18. Perform the following addition. The numbers are in binary.

$$\begin{array}{r} 101101.0101 \\ \underline{101010.0100} \end{array}$$

2-19. Perform the addition A+B where

$A = 7261.427_8$

$B = 674.3165_8$

2-20. Repeat Prob. 2-19 for

$A = CB2.3A_{16}$

$B = AB9.90A_{16}$

2-21. Repeat Prob. 2-19 for

$A = 1011.1101_2$

$B = AB.ACE_{16}$

3

Introduction to Combinational Logic Operations

In a digital computer, we deal with binary numbers where each binary digit can have one (and only one) of two values, 1 or 0. In mathematical logic, we consider propositions that are either *true* or *false*. Thus, we can define logical variables that have either the value "true" or the value "false." Since mathematical logic and digital computers are both two-variable systems, much of the mathematics that has been developed over the years for logical variables can be applied to the analysis and design of digital systems. In this chapter we shall introduce some fundamental ideas of logical operations as they apply to digital computers, and give some examples which illustrate these applications. In the next chapter we shall formalize these ideas.

3.1 BASIC LOGICAL OPERATIONS

To introduce logical operations, let us consider a simple binary addition of two single-digit numbers, A and B. Suppose that we write this in the following way:

$$
\begin{array}{r}
a_1 \\
\underline{b_1} \\
c_1 s_1
\end{array}
\qquad (3\text{-}1)
$$

where a_1, b_1, c_1 and s_1 all represent binary digits (i.e., either 0 or 1). For instance, if $A = 1$, $a_1 = 1$ and if $B = 1$, then $b_1 = 1$, in which case, we have

$$
\begin{array}{r}
1 \\
+\underline{1} \\
10
\end{array}
$$

where $s_1 = 0$ and $c_1 = 1$. We can now make the following statements. If a_1 and b_1 are both 1, then $s_1 = 0$ and $c_1 = 1$. If a_1 and b_1 are both 0, then $s_1 = 0$ and $c_1 = 0$. If either a_1 or b_1, but not both, are 1 then $s_1 = 1$ and $c_1 = 0$. These statements have their analogs in mathematical logic. In that logic there are also two variables, called *true* and *false*. Thus, the above statements could be written in terms of mathematical logic as: if a_1 and b_1 are both true, s_1 is false and c_1 is true; if a_1 and b_1 are both false, then s_1 and c_1 are both false; if either a_1 or b_1, but not both, are true, then s_1 is true and c_1 is false. In books dealing with logic, true and false are written as T and F, respectively. Since we are dealing with computers, 1 and 0 will be used correspondingly.

We have illustrated a simple addition as a logical operation. Actually, we can represent almost all computer operations by logical statements, although some may be considerably more complex than the simple ones we have discussed.

Let us now formalize our previous discussion and introduce the concept of circuits which perform the logical operations electronically. That is, we shall consider operations or circuits with a number of inputs which can be either 1 or 0 (true or false). These inputs produce an output in accordance with logical rules. For instance, we previously discussed a set of logical rules used to obtain simple binary addition. In this section we shall introduce *combinational logic* and the corresponding circuits which are called *combinational circuits*. In these, the present values, or logical inputs, 1's and 0's, produce the output. There is no memory associated with these circuits. In a subsequent chapter, we shall consider *sequential logic* and *sequential circuits* where memory is involved, i.e., the output is a function not only of the present values of the inputs, but also of the past values.

Let us start by defining a *binary variable*. It is simply a variable which can take on the value 1 or 0 or, equivalently, T or F. For instance, in (3-1), a_1, b_1, c_1 and s_1 are binary variables. Now consider a switch which is closed if a particular binary variable is a 1 and open if that binary variable is a 0, as illustrated in Fig. 3-1. Such switches can be implemented electronically using semiconductor devices. In their earliest forms they were *relays*, that is, switches which were opened and closed magnetically (by the current in a solenoid).

Figure 3-1   A switch which is open if $A = 0$ and closed if $A = 1$

We shall now work with the circuit of Fig. 3-2. Here we assume that the output V_o is a 1 if $V_o = V$ and a 0 if $V_o = 0$. Note that, in an actual circuit, the designer can specify the voltage level that corresponds to 0

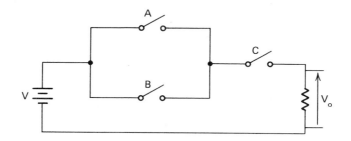

Figure 3-2 A circuit containing three switches

and to 1. They do not have to be zero volts and one volt. In Fig. 3-2, V_o will be a 1 if A or B is 1, and, in addition, if C is also 1. For instance, the following combinations result in $V_o = 1$.

$A = 1, B = 0, C = 1$ (3-2a)

$A = 0, B = 1, C = 1$ (3-2b)

$A = 1, B = 1, C = 1$ (3-2c)

Also, the following combinations of variables result in $V_o = 0$.

$A = 0, B = 0, C = 0$ (3-3a)

$A = 0, B = 0, C = 1$ (3-3b)

$A = 1, B = 0, C = 0$ (3-3c)

$A = 0, B = 1, C = 0$ (3-3d)

$A = 1, B = 1, C = 0$ (3-3e)

Figure 3-2 represents a combinational logic circuit and Eqs. (3-2) and (3-3) characterize it. Since complex logic circuits are characterized by many such equations, it is desirable to write them in a compact form. This is done by constructing a table which is called a *truth table* where each line represents one combination of variables. For instance, the truth table for Fig. 3-2 is

Truth Table for Figure 3-2

A	B	C	V_o
0	0	0	0
0	0	1	0
0	1	0	0
0	1	1	1
1	0	0	0
1	0	1	1
1	1	0	0
1	1	1	1

In mathematical literature, and in some electronics texts, the 1 and 0 of the truth tables will be replaced by T and F. For instance,

Equivalent Truth Table for Figure 3-2

A	B	C	V_o
F	F	F	F
F	F	T	F
F	T	F	F
F	T	T	T
T	F	F	F
T	F	T	T
T	T	F	F
T	T	T	T

As we have noted, in this book 1 and 0 shall be used instead of T and F.

Let us consider the procedure used to write a truth table. We list all the independent binary variables on one side of a vertical line and the dependent binary variable(s) on the other. In this truth table, the first line represents Eq. (3-3a). To do things in an orderly fashion, the independent binary variables are written in order of increasing binary value. That is, if the lefthand side of the truth table is considered to be a binary number, then each row is one more (in value) than the preceding one. In this way, the output for *every possible set of inputs is given*.

Let us consider an additional example. We shall write the truth table for the logic circuit of Fig. 3-3.

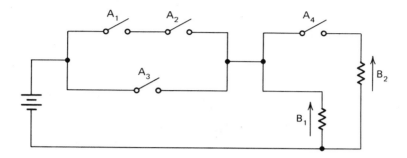

Figure 3-3 A more complex switching circuit

Truth Table for Figure 3-3

A_1	A_2	A_3	A_4	B_1	B_2
0	0	0	0	0	0
0	0	0	1	0	0
0	0	1	0	1	0
0	0	1	1	1	1
0	1	0	0	0	0
0	1	0	1	0	0
0	1	1	0	1	0
0	1	1	1	1	1
1	0	0	0	0	0
1	0	0	1	0	0
1	0	1	0	1	0
1	0	1	1	1	1
1	1	0	0	1	0
1	1	0	1	1	1
1	1	1	0	1	0
1	1	1	1	1	1

Now that we have discussed some elementary ideas of logical operations, we shall formalize these ideas in the next section.

3.2 BASIC COMBINATIONAL LOGIC OPERATIONS — BINARY CONNECTIVES

All of the operations of combinational logic circuits can be expressed in terms of some fundamental logic operations which are called *binary connectives* which ae used to characterize and study logic circuits. Let us now consider these fundamental operations.

The AND Operation

If we have two logical variables A_1 and A_2, the AND operation gives a value of 1 if, and only if, A_1 and A_2 are both 1. It gives a value of 0 otherwise. This is indicated by the switch circuit of Fig. 3-4. Note that

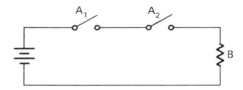

Figure 3-4 A simple AND switch circuit

B will be 1 if A_1 *and* A_2 are 1. It will be 0 otherwise. We indicate the AND operation by a dot ($\cdot$) between the variables. This is the same as an ordinary multiplication sign. Thus, for Fig. 3-4, we have

$$B = A_1 \cdot A_2 \qquad\qquad (3\text{-}4a)$$

The truth table for the AND operation is

Truth Table for Equation 3-4a

A_1	A_2	B
0	0	0
0	1	0
1	0	0
1	1	1

At times, as in the case of ordinary multiplication, the dot is omitted. Thus, we could write

$$B = A_1 A_2 \tag{3-4b}$$

An alternative notation that is used for the AND operation is

$$B = A_1 \wedge A_2 \tag{3-4c}$$

In general, we shall use the notation of Eq. (3-4a) or that of (3-4b).

We have illustrated the AND function with two independent variables. However, there can be more. For instance, in Fig. 3-4, if we had
three switches in series, the relation would be

$$B = A_1 \cdot A_2 \cdot A_3 = A_1 \wedge A_2 \wedge A_3 \tag{3-5}$$

In this case, B would be 1 if, and only if, A_1 *and* A_2 *and* A_3 were *all* 1's.

In modern digital computers, semiconductor circuits called *gates*
perform logical operations. A gate that performs the AND operation is
called an *AND gate*. A great many gates can be incorporated on a single
silicon chip. A computer may incorporate the interconnections of
thousands of gates. To simplify the wiring diagram, special symbols
are used to represent the various gates. The AND gate symbol is illustrated in Fig. 3-5. Two-input and three-input AND gates are shown in
Figs. 3-5a and 3-5b, respectively. Note that there are many symbols
for gates in general use. Another symbol for the AND gate is shown in
Fig. 3-5c, but we shall use the more commonly used forms of Figs.
3-5a and 3-5b in this text.

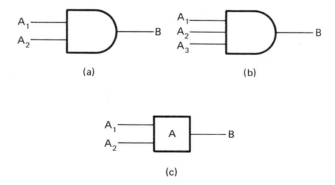

Figure 3-5 Symbols for AND gates. (a) Two input; (b) three input; (c) an
alternative symbol for the AND gate.

The OR Operation

A switch circuit which accomplishes the OR operation is shown in Fig. 3-6. Note that if *either or both* of the independent variables A_1

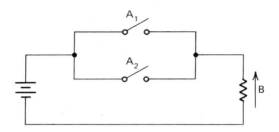

Figure 3-6 A simple OR switch circuit

and A_2 are 1, then $B=1$. Only if both A_1 and A_2 are 0 will $B=0$. The OR operation is indicated by a + sign. Thus, the logical equation for the switch circuit of Fig. 3-6 is

$$B = A_1 + A_2 \tag{3-6a}$$

An alternative symbolic notation for the OR operation is

$$B = A_1 \lor A_2 \tag{3-6b}$$

The truth table for the OR operation is

A_1	A_2	B
0	0	0
0	1	1
1	0	1
1	1	1

Symbols for the OR gate are shown in Fig. 3-7. The logical equation that represents Fig. 3-7b is

$$B = A_1 + A_2 + A_3 + A_4 = A_1 \lor A_2 \lor A_3 \lor A_4 \tag{3-7}$$

In this case, B is 1 if any or all of A_1 or A_2 or A_3 or A_4 are 1.

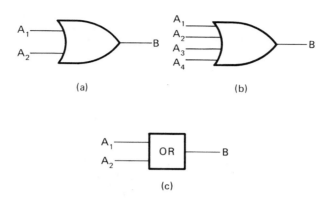

(a)   (b)

(c)

Figure 3-7 Symbols for OR gates. (a) Two input; (b) four input; (c) an alternative symbol for the OR gate.

We can combine gates to form more complicated logical operations. For instance, consider the circuit of Fig. 3-2. We can represent it by the logical expression

$$V_o = (A+B) \cdot C \tag{3-8}$$

That is, V_o is 1 if A or B is 1, and in addition, if C is 1 (i.e., if $(A+B)$ and C are both 1). We can write the truth table for this as

A	B	C	$A+B$	$(A+B) \cdot C$
0	0	0	0	0
0	0	1	0	0
0	1	0	1	0
0	1	1	1	1
1	0	0	1	0
1	0	1	1	1
1	1	0	1	0
1	1	1	1	1

Thus, breaking the expression into parts enables us to obtain the truth table for complex logical expressions. A representation of Fig. 3-2 using gate symbols is shown in Fig. 3-8.

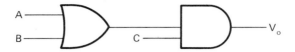

Figure 3-8 An interconnection of gates which results in the logical expression
$V_o = (A + B) \cdot C$

Complement — NOT or Inversion

Before we consider the next gate, let us consider a fundamental operation of mathematical logic. The *complement* of a variable A is defined in the following way. If A is 1, then its complement is 0, and if A is 0, then its complement is 1. We shall indicate the complement by placing a bar over the variable. Thus, the complement of A is $\overline{A}$. (At times, a prime, i.e., A' is used to represent the complement.) Similarly, the complement of $(B+C) \cdot A$ is written as $\overline{(B+C) \cdot A}$. The truth table for the complement is given by

A	$\overline{A}$
0	1
1	0

A gate which performs the mathematical operation of taking the complement is called a NOT gate (or an inversion gate or an inverter). Its symbol is shown in Fig. 3-9. An alternative symbol consisting of a

Figure 3-9 The symbol for a NOT gate

square with an N written in it is sometimes used. The small circle in Fig. 3-9 can also be used by itself to represent a complement operation. For instance, Fig. 3-10 represents

$B = A_1 + \overline{A_2}$

The operation of taking the complement is called a *unary* operation since only one variable is involved. The operations of AND and OR are called *binary* operations since two (or more) variables are involved.

Figure 3-10   A representation of the logic equation $B = A_1 + \overline{A_2}$

The NOR Operation

This consists of taking the complement of the OR operation. That is, the NOR operation is defined by the logical expression

$$B = \overline{A_1 + A_2} \tag{3-10}$$

An alternative name which is sometimes used for the NOR operation is the *Pierce arrow* and an alternative notation is

$$B = A_1 \downarrow A_2 \tag{3-11}$$

The truth table for the NOR operation is

A_1	A_2	B
0	0	1
0	1	0
1	0	0
1	1	0

The symbolic representation for a NOR gate is shown in Fig. 3-11. Note that it consists of the symbol for an OR gate with the "inverting circle" added.

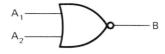

Figure 3-11   A symbol for the NOR gate

The NAND Operation

We just discussed an operation that was the complement of the OR operation. Now we shall consider one which is the complement of the AND operation. It is called the NAND operation and is defined by

$$B = \overline{A_1 \cdot A_2} \qquad (3\text{-}12)$$

The NAND operation has also been called the *Sheffer stroke*. An alternative representation of this operation is

$$B = A_1 \uparrow A_2 \qquad (3\text{-}13)$$

The truth table for the NAND operation is

A_1	A_2	B
0	0	1
0	1	1
1	0	1
1	1	0

The symbolic representation for the NAND gate is shown in Fig. 3-12 where three inputs are shown. The logical expression for this is

$$B = \overline{A_1 \cdot A_2 \cdot A_3} \qquad (3\text{-}14)$$

Note that the symbol for the NAND gate is simply that for the AND gate followed by the "inverting circle." Naturally, the output of the NAND gate is just the complement of the output of an AND gate.

Figure 3-12   The symbol for a three-input NAND gate

The Exclusive OR (XOR) Operation

The XOR is similar to the OR operation except that the output is a 0 if the inputs are either both 1 or both 0. That is, if the inputs are A_1 and A_2 and B is the output, then B is a 1 if either A_1 *or* A_2 is a 1. However, B is 0 if *both* A_1 and A_2 are 0 or if *both* A_1 and A_2 are 1. The symbolic representation for the exclusive OR operation is

$$B = A_1 \oplus A_2 \qquad (3\text{-}15)$$

The truth table for the exclusive OR (XOR) operation is

A_1	A_2	B
0	0	0
0	1	1
1	0	1
1	1	0

The symbolic representation for an XOR gate is shown in Fig. 3-13.

Figure 3-13 An exclusive OR gate (XOR)

Let us discuss XOR operations involving more than two variables. This is somewhat more involved than multiple-variable AND and OR operations. Consider

$$B = A_1 \oplus A_2 \oplus A_3 \tag{3-16a}$$

We can write this as

$$B = (A_1 \oplus A_2) \oplus A_3 \tag{3-16b}$$

That is, we apply the XOR operation to A_1 and A_2 and then apply the XOR operation again to the result of this first operation and A_3. Let us illustrate this with a truth table.

A_1	A_2	A_3	$A_1 \oplus A_2$	$(A_1 \oplus A_2) \oplus A_3$	
0	0	0	0	0	
0	0	1	0	1	
0	1	0	1	1	
0	1	1	1	0	(3-17)
1	0	0	1	1	
1	0	1	1	0	
1	1	0	0	0	
1	1	1	0	1	

Note that if $A_1 = 1$, $A_2 = 1$, and $A_3 = 1$, then $A_1 \oplus A_2 \oplus A_3 = 1$, and not 0.

We need not place the parentheses as in Eq. (3-16b). For instance, all the following are equivalent.

$$A_1 \oplus A_2 \oplus A_3 = (A_1 \oplus A_2) \oplus A_3 = A_1 \oplus (A_2 \oplus A_3) = (A_1 \oplus A_3) \oplus A_2 \quad (3\text{-}18)$$

This can be demonstrated by writing the truth table for each of these. The details of this will be left to the reader. Although the placement of parentheses was not critical in this case, it *is* critical in many logical equations.

The If and Only If Connective

The binary connectives that we have used thus far (i.e. AND, OR, NOT, NAND, NOR, XOR) are realized using gates. We shall now consider other binary connectives that are not directly realized with gates but which are nonetheless important. The first is *if and only if*. It is symbolically represented by $\equiv$. Let us illustrate this by considering the AND operation. $A_1 \cdot A_2$ represents A_1 AND A_2. The result of this operation is 1 if A_1 and A_2 are both 1 and is 0 otherwise. Similarly,

$$A_1 \equiv A_2$$

is 1 if A_1 is 1 *when A_2* is 1 and A_1 is 0 when A_2 is 0. That is, $A_1 \equiv A_2$ is 1 only when A_1 and A_2 are identical. For instance, if we write

$$B = (A_1 \equiv A_2) \quad\quad\quad\quad\quad\quad\quad\quad\quad\quad\quad\quad\quad\quad (3\text{-}19)$$

then B is described by the following truth table.

A_1	A_2	B
0	0	1
0	1	0
1	0	0
1	1	1

This can be implemented by the gate circuit of Fig. 3-14.

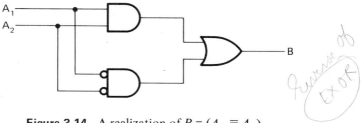

Figure 3-14 A realization of $B = (A_1 \equiv A_2)$

The If-Then Connective

Another useful relation is the if-then connective which is written as $\supset$. It is defined by

$$A \supset B$$

where $(A \supset B)$ is 1 if B is 1 whenever A is 1. It is also 1 if A is 0 no matter what value B has. To illustrate this, consider

$$C = (A_1 \supset A_2)$$

The truth table is:

A_1	A_2	C
0	0	1
0	1	1
1	0	0
1	1	1

Inspection of the truth table yields the following: C is 1 whenever A_1 is 0 or when A_2 is 1. Hence, this can be implemented by a single OR gate whose inputs are A_1 and $\overline{A}_2$, see Fig. 3-15.

Figure 3-15 An implementation of $C = (A_1 \supset A_2)$

Let us consider an additional example.

Example:

Construct the truth table for

$C = (\overline{A} + A) \supset B$

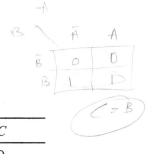

A	B	$\overline{A}$	$\overline{A} + A$	C
0	0	1	1	0
0	1	1	1	1
1	0	0	1	0
1	1	0	1	1

The column for C is identical to that for B. Hence, we have

$C = B$

3.3 REPRESENTATION OF GENERAL COMBINATIONAL LOGIC EXPRESSIONS — FUNCTIONALLY COMPLETE SETS

In general, it can be shown that we can represent any combinational logic expression using only the operations discussed in the last section. Alternatively, we can say that any combinational logic circuit can be built from an interconnection of the gates presented there. Note that the $\equiv$ and $\supset$ operations can be represented by Figs. 3-14 and 3-15, respectively. In general, we can represent the behavior of a complex interconnection of gates by relations such as Eq. (3-8). Such logical expressions are also called *switching functions*. If a set of operations is such that it can be used to express *any* switching function, then that set is said to be *functionally complete*. Actually we do not need all the operations presented here to obtain a functionally complete set. In fact, we can show that many of the operations can be expressed in terms of the others. For instance, if we have a set consisting of AND, OR and NOT, we can obtain NAND, NOR and XOR. NAND and NOR represent the complement of the AND and OR operations, respectively. Thus, these operations can be obtained by cascading a NOT gate with an AND or with an OR gate as shown in Figs. 3-16a and b. We can

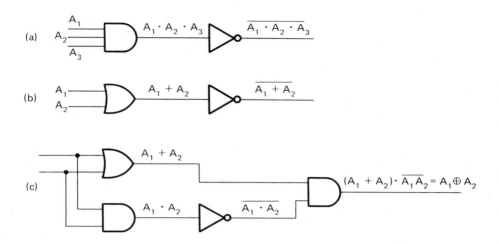

Figure 3-16 Interconnection of AND, OR and NOT gates to obtain other logical operations. (a) NAND; (b) NOR; (c) XOR.

demonstrate the equivalence by comparing truth tables. For instance, consider Fig. 3-16b, which is equivalent to the NOR circuit. Let us write its truth table.

A_1	A_2	$A_1 + A_2$	$\overline{A_1 + A_2}$
0	0	0	1
0	1	1	0
1	0	1	0
1	1	1	0

If this is compared with the truth table for the NOR gate, it is seen that the results are the same.

Now let us demonstrate that we can replace an exclusive OR gate by an interconnection of AND, OR and NOT gates. Consider the interconnection of gates shown in Fig. 3-16c. That this is equivalent to the XOR operation can be demonstrated by constructing the truth table for $(A_1 + A_2) \cdot \overline{A_1 \cdot A_2}$ and showing that it is equivalent to that for the XOR. Alternatively, we can use the following reasoning. $(A_1 + A_2) \cdot \overline{A_1 \cdot A_2}$ represents $A_1 + A_2$ AND $\overline{A_1 \cdot A_2}$. Now $A_1 + A_2$ will be 1 when A_1 or A_2 is 1. In addition, $\overline{A_1 \cdot A_2}$ will be 1 except when A_1 and A_2 are both 1. Thus, $(A_1 + A_2) \cdot \overline{A_1 \cdot A_2}$ will be 1 only when A_1 or A_2, but not both, are 1. Thus, we have achieved the exclusive OR. We have now shown that the AND, OR and NOT operations are functionally complete, at least as far as NAND, NOR and XOR are concerned. It can be shown that the set of AND, OR and NOT is functionally complete for all possible switching functions. We shall assume that this is true without proof.

There are other functionally complete sets of operations. For instance, the NOR operation is functionally complete by itself. We can demonstrate this by showing that we can construct the NOT, OR and AND gates just from interconnections of NOR gates. Since we have just "shown" that the set of NOT, OR and AND operations is functionally complete, this will demonstrate that the NOR operation is itself functionally complete.

Consider Fig. 3-17a. This is a NOR gate with both inputs connected together. The truth table for this circuit can be obtained from the previous truth table given for the NOR operation for those cases where $A_1 = A_2$. Using this we obtain the truth table

A	B
0	1
1	0

This is the same as the truth table for the NOT operation.

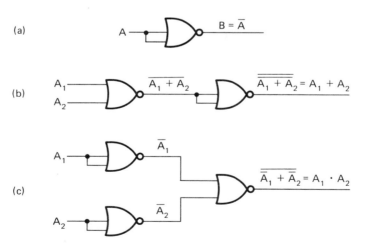

Figure 3-17 Interconnections of NOR gates to achieve other logical operations. (a) NOT; (b) OR; (c) AND.

Now consider Fig. 3-17b. The output of the first NOR gate is $\overline{A_1 + A_2}$ (by definition). We have just shown that the second gate merely takes the complement. The complement of a complement is simply the variable itself. That is

$$\overline{\overline{A}} = A \tag{3-22}$$

This can be verified by considering the truth table. Thus, the output of the second gate is $\overline{\overline{A_1 + A_2}} = A_1 + A_2$. Hence, Fig. 3-17b represents an OR gate.

Finally, consider Fig. 3-17c. The output is $\overline{\overline{A}_1 + \overline{A}_2}$. This is equivalent to the AND operation. We shall demonstrate this by constructing the truth table.

A_1	A_2	$\overline{A}_1$	$\overline{A}_2$	$\overline{A}_1 + \overline{A}_2$	$\overline{\overline{A}_1 + \overline{A}_2}$	$A_1 \cdot A_2$
0	0	1	1	1	0	0
0	1	1	0	1	0	0
1	0	0	1	1	0	0
1	1	0	0	0	1	1

The last two columns are identical. This demonstrates that

$$A_1 \cdot A_2 = \overline{\overline{A}_1 + \overline{A}_2} \tag{3-23}$$

Therefore, Fig. 3-17c represents an AND gate. Hence, we have demonstrated that the NOR operation is functionally complete in itself.

In a similar way, we can show that the NAND operation is also functionally complete. In Fig. 3-18 we illustrate the NOT, AND and OR operations realized using NAND gates. In Fig. 3-18a we obtain a NOT gate. Note that $\overline{A \cdot A} = \overline{A}$. This can be verified by writing the truth table.

In Fig. 3-18b we obtain the AND operation. Note that the output of the NAND gate is the complement of the AND operation. Thus, following a NAND gate by a NOT gate produces the desired result.

The output of Fig. 3-18c is $\overline{A}_1 \cdot \overline{A}_2$. This is equivalent to the OR operation. The following truth table verifies this.

A_1	A_2	$\bar{A}_1$	$\bar{A}_2$	$\bar{A}_1 \cdot \bar{A}_2$	$\overline{\bar{A}_1 \cdot \bar{A}_2}$	$A_1 + A_2$
0	0	1	1	1	0	0
0	1	1	0	0	1	1
1	0	0	1	0	1	1
1	1	0	0	0	1	1

Thus, Fig. 3-18c is an OR gate. Hence, we have demonstrated that the NAND operation is functionally complete in itself.

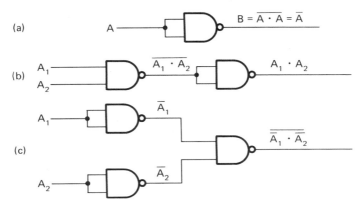

Figure 3-18 Interconnections of NAND gates to achieve other logical operations. (a) NOT; (b) AND; (c) OR.

We have demonstrated functional completeness by realizing NOT, AND and OR operations. Actually, we need only consider certain pairs of these. For instance, NOT and AND are functionally complete. This is true since NAND is functionally complete and NAND can be realized by cascading AND and NOT gates. Simlarly, using the NOR operations, we can demonstrate that the set of NOT and OR is functionally complete.

When computers are built, many complex logical expressions must be implemented using gate circuits. It is necessary to work with a functionally complete set of gates so that the designer can implement any required logical operation no matter how complex, using standard gates. Each of the functionally complete sets we have discussed (i.e. AND and NOT, or OR and NOT, or NAND, or NOR) can be used for this purpose. Actually, since NAND and NOR gates can be simply constructed using semiconductors, they are often used for this purpose.

3.4 SOME EXAMPLES OF COMBINATIONAL LOGIC CIRCUITS — ELEMENTARY COMBINATIONAL LOGIC DESIGN

Let us now see how logic circuits can be interconnected to result in some practical operations. First we shall consider an adder circuit which is a circuit whose input is two binary numbers and whose output is a binary number which is the sum of these input numbers. Again, this example will be very elementary. We shall consider actual adder circuits in much greater detail in a subsequent chapter.

We shall start by discussing a circuit called a *half adder* which adds two one-bit binary numbers, a_1 and b_1 [see (3-1)]. The truth table for the half adder is

a_1	b_1	c_1	s_1	
0	0	0	0	
0	1	0	1	(3-24)
1	0	0	1	
1	1	1	0	

Note that we have

$$\frac{\begin{array}{l} a_1 \\ b_1 \end{array}}{c_1 s_1}$$

It is conventional to call s_1 the sum digit and c_1 the carry digit. From the truth table we see that the logical expressions for c_1 and s_1 are

$$s_1 = a_1 \oplus b_1 \tag{3-25a}$$

$$c_1 = a_1 \cdot b_1 \tag{3-25b}$$

We will use AND, OR and NOT gates to construct the half adder. [Note (see Sec. 3-3) that we could implement this using NOR or NAND gates, etc. This is often done in actual computers.] Figure 3-13 can be used for the XOR gate for s_1 while a single AND gate can be used for c_1. The resulting half adder is shown in Fig. 3-19a. Note that there are two AND circuits that both have the same inputs, a_1 and b_1. Thus, their outputs are the same and only one AND circuit is needed. This is

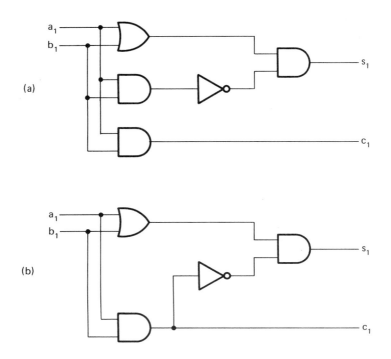

Figure 3-19 (a) A half adder; (b) an implementation that uses one less gate.

illustrated in Fig. 3-19b, where there is one less gate. Here, the leftmost AND circuit not only supplies c_1 but also supplies the input to the inverter. Thus, Fig. 3-19b represents a simpler or reduced form of Fig. 3-19a.

This is an example of a multilevel logic circuit in that a signal must pass through several gates. (This is true for s_1, but not for c_1.) There is a disadvantage to this. Practical gates take a finite amount of time to respond. For instance, if a signal must pass through two gates in succession, the second gate cannot respond until the first one has produced its output. For example, if it takes 10^{-8} seconds for each gate to respond and the signal must pass through two gates, then the response time is 2×10^{-8} seconds. In a computer, an extremely large number of such operations are performed when even a simple computer program is executed. Thus, the seemingly small response times of each component can result in minutes, or hours of actual computer time when a lengthy program is run.

When logic circuits are designed, they are usually *minimized*. That is, the number of components is reduced. An example of an extremely simple minimization has been illustrated in Fig. 3-19. Minimization is important not only because it reduces the number of components (this can be substantial since there are very many components in a computer) but also because it increases reliability. Reduction in the number of levels will also reduce response time.

The half adder we have studied can be used to add two binary digits. If the number to be added has more than one bit, it would seem as though we need only use a half adder for each digit. This would be true if there were no carrying. However, carrying does occur and the half adder has no provision for the carry bit.

A more complex circuit called the *full adder* resolves this problem. Remember that the adder circuit works with only one bit. Each full adder must work with single bits of A and B and also with a carry bit from the preceding digit. Note that the half adder produces the sum plus a carry bit. The full adder must do this also. Let us consider a simple addition

$$
\begin{array}{r}
1111 \\
0\overset{\frown}{1}\overset{\frown}{0}\overset{\frown}{1}\overset{\frown}{1} \\
+\ 00111 \\
\hline
10010
\end{array}
$$

Here, each addition after the first digit must accept a carry digit. The inputs to the adder which adds the "j^{th}" bit will be a_j, b_j, and c_{j-1} where a_j and b_j represent the jth digit of A and B respectively and c_{j-1} represents the carry from the $j-1$ column addition. Thus, we have

$$
\begin{array}{r}
c_{j-1} \\
a_j \\
b_j \\
\hline
c_j s_j
\end{array}
\qquad (3\text{-}26)
$$

Note that in Eqs. (3-22) and in the preceding comments, we used the notation a_1, b_1, etc. Here we use the more generalized notation a_j, b_j, etc.

Let us write the desired truth table, which is obtained by simply performing the binary addition. For instance, if $c_{j-1}=1$, $a_j=1$ and $b_j=1$, then the addition yields $c_j=1$, $s_j=1$. Proceeding similarly, we obtain the following truth table:

a_j	b_j	c_{j-1}	s_j	c_j
0	0	0	0	0
0	0	1	1	0
0	1	0	1	0
0	1	1	0	1
1	0	0	1	0
1	0	1	0	1
1	1	0	0	1
1	1	1	1	1

$$(3\text{-}27)$$

In a complex case, it is desirable to obtain a logical expression or switching function (see Sec. 3-3) which characterizes each of the outputs. From this, the gate circuit can be designed. Let us discuss how this can be done.

The value of the logical expression should be a 1 corresponding to each time that a 1 appears in the "output column" of the truth table. Let us consider the first line where $s_j = 1$. In this case, $a_j = 0$, $b_j = 0$ and $c_j = 1$. When $a_j = 0$, then $\bar{a}_j = 1$. Thus, we can write this in compact form as: $s_j = 1$ if

$$\bar{a}_j \cdot \bar{b}_j \cdot c_{j-1} = 1$$

There are a number of possibilities that result in $s_j = 1$. Since the occurrence of *any* one of them causes $s_j = 1$, we must combine all these possibilities using the OR relation. Thus, for the truth table (3-27), we have

$$s_j = \bar{a}_j \cdot \bar{b}_j \cdot c_{j-1} + \bar{a}_j \cdot b_j \cdot \bar{c}_{j-1} + a_j \cdot \bar{b}_j \cdot \bar{c}_{j-1} + a_j \cdot b_j \cdot c_{j-1} \qquad (3\text{-}28)$$

Note that any other possible combination of a_j, b_j and c_{j-1} results in a 0 for s_j. Thus, none of these terms appears in Eq. (3-28). Proceeding similarly, we can obtain the expression for c_j. It is

$$c_j = \bar{a}_j \cdot b_j \cdot c_{j-1} + a_j \cdot \bar{b}_j \cdot c_{j-1} + a_j \cdot b_j \cdot \bar{c}_{j-1} + a_j \cdot b_j \cdot c_{j-1} \qquad (3\text{-}29)$$

We can use these logical expressions to accomplish the design of a simple switching circuit. For instance, Eq. (3-28) indicates that we should have an OR gate with four inputs, $\bar{a}_j \cdot \bar{b}_j \cdot c_{j-1}$, $\bar{a}_j \cdot b_j \cdot \bar{c}_{j-1}$, $a_j \cdot \bar{b}_j \cdot \bar{c}_{j-1}$, and $a_j \cdot b_j \cdot c_{j-1}$. Each of these inputs can be obtained from a separate AND gate. The input to the AND circuits are a_j, b_j and c_{j-1}, or their complements (which can be obtained from a NOT gate). This is shown in the upper portion of Fig. 3-20 where s_j is obtained. In a similar way, the lower portion of Fig. 3-20 implements Eq. (3-29) where c_j is obtained. Note that Eqs. (3-28) and (3-29) have one term in common, $a_j \cdot b_j \cdot c_{j-1}$. Thus, one of the AND gates is used by both parts of the diagram. This represents a saving of one gate.

Let us generalize the procedure that we have used. First, form the truth table. Then, for each of the output variables, determine which entries produce 1's. Each set of inputs which produces a 1 represents a set of inputs to an AND gate. The outputs of all the AND gates are supplied to an OR gate. Thus, if n 1's appear in the output column, then n AND gates, one OR gate, and the necessary NOT gates are used.

Although this method is simple, it often results in an excessive number of gates. Equations such as (3-28) and (3-29) can usually be manipulated so as to minimize the number of gates required. The general techniques for minimization will be discussed in detail in the next chapter. In this section we shall just consider some examples that illustrate that these minimizations can take place.

We shall start by illustrating a minimization of the full adder that was just discussed. We can write s_j in terms of the XOR operation. That is

$$s_j = a_j \oplus b_j \oplus c_{j-1} \tag{3-30}$$

This can be seen by comparing truth table (3-17) for the XOR operation applied to three variables and truth table (3-27) for s_j.

Now let us consider the expression for c_j. We can write this as

$$c_j = a_j \cdot b_j + b_j \cdot c_{j-1} + c_{j-1} \cdot a_j \tag{3-31}$$

This can be verified by constructing the truth table for Eq. (3-31) and comparing it with truth table (3-27) as it applies to c_j. If this is done, it is seen that they are the same. Thus, we can characterize the full adder by

$$s_j = a_j \oplus b_j \oplus c_{j-1} \tag{3-32a}$$

$$c_j = a_j \cdot b_j + b_j \cdot c_{j-1} + c_{j-1} \cdot a_j \tag{3-32b}$$

Now let us consider the half adder. From Eqs. (3-25), we have

$$s = a \oplus b \tag{3-33a}$$

$$c = a \cdot b \tag{3-33b}$$

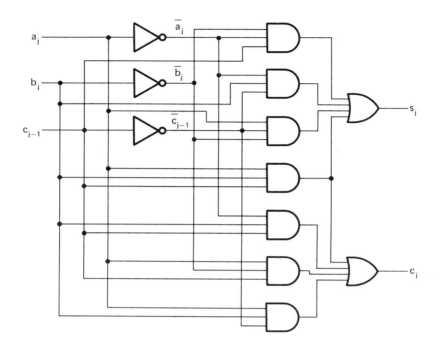

Figure 3-20   A full adder based upon Eqs. (3-28) and (3-29). This is an unreduced form.

The half adder performs an XOR and an AND operation. We can use two half adders and an OR gate to obtain a full adder. Let us demonstrate this. We shall use some simplified diagrams here. In Fig. 3-21a, we show a *block diagram* of a half adder. The inputs and outputs are shown. However, the internal connections of the gates are omitted. This is a "shorthand" type of drawing. Figure 3-19b could represent the internal wiring of the box.

In Fig. 3-21b we have shown the interconnection of two half adders and an OR gate to produce a full adder. Since the s output of a half adder produces an XOR, then s_j is of the form of Eq. (3-32a), which is correct. Let us consider c_j. The c output of the second half adder is $(a_j \oplus b_j) \cdot c_{j-1}$. Thus, the output of the OR gate is $(a_j \oplus b_j) \cdot c_{j-1} + a_j \cdot b_j$. Let us show that this is equivalent to $a_j \cdot b_j + b_j \cdot c_{j-1} + c_{j-1} \cdot a_j$. We shall use the truth table to do this.

a_j	b_j	c_{j-1}	$a_j \oplus b_j$	$(a_j \oplus b_j)\cdot c_{j-1} + a_j\cdot b_j$	$a_j\cdot c_{j-1} + b_j\cdot c_{j-1} + a_j\cdot b_j$
0	0	0	0	0	0
0	0	1	0	0	0
0	1	0	1	0	0
0	1	1	1	1	1
1	0	0	1	0	0
1	0	1	1	1	1
1	1	0	0	1	1
1	1	1	0	1	1

Comparing the last two columns, we have

$$(a_j \oplus b_j)\cdot c_{j-1} + a_j\cdot b_j = a_j\cdot c_{j-1} + b_j\cdot c_{j-1} + a_j\cdot b_j \qquad (3\text{-}34)$$

Thus, the proper value for c_j is obtained, see Eq. (3-32b).

The half adder of Fig. 3-19b uses four gates. Thus, the full adder of Fig. 3-21b requires a total of nine gates. The full adder of Fig. 3-20 uses 12 gates. Thus, Fig. 3-21b represents a substantial reduction in the number of gates (25%). Digital computers contain an extremely large number of gates. Thus, if we can substantially reduce this number, a considerable saving takes place. In this case, we simply appeared to obtain the reduced form through luck. Again, systematic procedures do exist for reducing the number of gates and they will be considered in subsequent chapters.

Each full adder adds two one-bit numbers plus a carry term. Thus, if we wish to add numbers which have a total of 16 bits, 16 adders are required. The interconnection of these will be considered subsequently.

Let us consider one further example of a reduction. Figure 3-22 is a block diagram of a device called a *multiplexer*. This particular one has four inputs a_0, a_1, a_2 and a_3, one output b, and two control inputs c_0 and c_1. The multiplexer is a controlled switch which connects the output b to *one* of the inputs in accordance with the digital number supplied by the control. For instance, if $c_0 = 1$ and $c_1 = 1$ ($11_2 = 3_{10}$) then b will be effectively connected to a_3. That is, the value of b will be equal to that of a_3. Multiplexers find extensive use in digital systems. For instance, suppose that we are working with a digital communication system that has four different output channels and that we want to be able to select one of these channels at a time. If each channel is

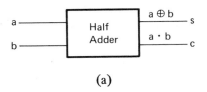

(a)

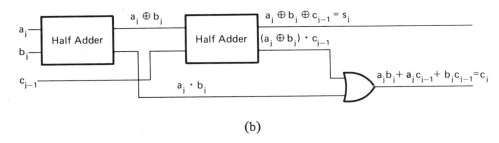

(b)

Figure 3-21 (a) A half adder; (b) a full adder composed of two half adders and an OR gate.

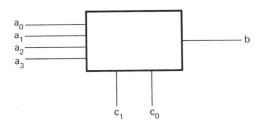

Figure 3-22 The block diagram for a multiplexer

connected to one input of a multiplexer, then we can electronically select the desired signal by placing the appropriate signal on the control leads.

In digital computers, we often want to vary the interconnections in accordance with the operation being performed. A multiplexer can be used here as a high speed, easily controlled, electronic switch. This greatly adds to the versatility of the computer.

The truth table for this device has 64 entries. For purposes of brevity, we shall only consider part of it.

c_1	c_0	a_0	a_1	a_2	a_3	b
0	0	0	0	0	0	0
0	0	0	0	0	1	0
0	0	0	0	1	0	0
0	0	0	0	1	1	0
0	0	0	1	0	0	0
0	0	0	1	0	1	0
0	0	0	1	1	0	0
0	0	0	1	1	1	0
0	0	1	0	0	0	1
0	0	1	0	0	1	1
0	0	1	0	1	0	1
0	0	1	0	1	1	1
0	0	1	1	0	0	1
0	0	1	1	0	1	1
0	0	1	1	1	0	1
0	0	1	1	1	1	1

In the total table, half (or 32) of the entries will be 1's. (The details of constructing the remainder of the table are similar to our previous discussion and will be omitted.) Thus, if we used the formal procedure discussed earlier in this section, we would require 32 AND gates, 4 NOT gates, and an OR gate. The OR gate would require 32 inputs and each of the AND gates would require six inputs. That is, a 32-term switching function can be obtained from the truth table to characterize b.

We shall now show that b can be expressed in terms of a four-term switching function. We shall just state this without derivation. Again, the purpose of presenting it here is to show that minimization techniques can be very powerful. We can write b as

$$b = \bar{c}_1 \cdot \bar{c}_0 \cdot a_0 + \bar{c}_1 \cdot c_0 \cdot a_1 + c_1 \cdot \bar{c}_0 \cdot a_2 + c_1 \cdot c_0 \cdot a_3 \qquad (3\text{-}35)$$

This equation represents four AND gates. If any of the inputs to an AND gate is 0, then its output will be 0. The control variables or their complements in each term represent the subscript of the input which is to be "connected" to the output. For instance, if $\bar{c}_1 \cdot \bar{c}_0 = 1$, then $c_1 = 0$ and $c_0 = 0$ ($00_2 = 0_{10}$) and $b = a_0$. Similarly, if $c_1 \cdot \bar{c}_0 = 1$, then

$c_1 = 1$, $c_0 = 0$ ($10_2 = 2_{10}$) and $b = a_2$. Thus, the only AND gate whose output can possibly be 1 is properly selected by the control values. That is, if the number $(c_1 c_0)_2 = d_{10}$, then the output will be given by

$$b = a_d \tag{3-36}$$

Thus, we have achieved the desired multiplex operation. However, only 4 AND gates rather than 32 are required. The gate diagram for this circuit is shown in Fig. 3-23.

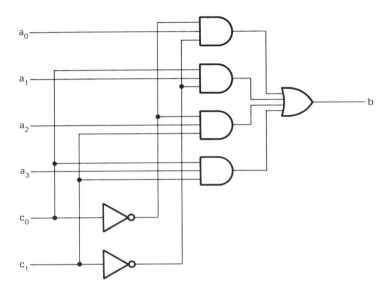

Figure 3-23 A multiplexer

3.5 POSITIVE AND NEGATIVE LOGIC

In an actual circuit, different voltage levels are used to represent 0's and 1's (e.g. a 1 could be represented by +5 volts and a 0 by 0 volts). The voltages representing 0's and 1's can both be positive, both be negative, or one can be positive and the other negative. If the voltage for a 1 is more positive than the voltage for a 0, then we say that *positive logic* is being used. Conversely, if the voltage for a 0 is more positive than that for a 1, we say that *negative logic* is being used.

Suppose that we have a circuit whose output voltage v_o is either 5 volts or 0.1 volt. If positive logic is used, then the 5 volt signal represents a 1 while the 0.1 volt signal represents a 0. On the other hand, if negative logic is used, then the 5 volt signal would represent a 0 while the 0.1 volt signal would represent a 1.

Now consider a particular circuit which has two inputs v_A and v_B and an output v_o. Table 3-1 expresses v_o as a function of v_A and v_B.

Table 3-1 v_o as a Function of v_A and v_B

v_A	v_B	v_o
0.1	0.1	0.1
0.1	5	0.1
5	0.1	0.1
5	5	5

Suppose that positive logic is used. Then, the circuit of Table 3-1 yields the following truth table:

Table 3-2 Truth Table for the Circuit of Table 3-1 when Positive Logic is Used

A	B	v_o
0	0	0
0	1	0
1	0	0
1	1	1

Thus, this circuit represents an AND gate. On the other hand, suppose that negative logic is used. Then, the truth table becomes

Table 3-3 Truth Table for the Circuit of Table 3-1 when Negative Logic is Used

A	B	v_o
0	0	0
0	1	1
1	0	1
1	1	1

Thus, this circuit now represents an OR gate. In general, a positive logic AND gate is a negative logic OR gate and vice versa.

BIBLIOGRAPHY

Booth, T.L., *Digital Networks and Computer Systems*, 2nd Ed., Chap. 1, John Wiley and Sons, Inc., New York 1978

Mano, M.M., *Digital Computer Logic and Design*, Chap. 1, Prentice Hall, Inc., Englewood Cliffs, N.J. 1979

O'Malley, J., *Introduction to the Digital Computer*, Chap. 2, Holt, Rinehart and Winston, Inc., New York 1972

Hill, J.H. and Peterson, G.R., *Introduction to Switching Theory and Logical Design*, Chap. 3, John Wiley and Sons, Inc., New York 1974

Peatman, J.B., *The Design of Digital Systems*, Chap. 2, McGraw-Hill Book Co., New York 1972

PROBLEMS

3-1. Write a truth table for the switch circuit of Fig. 3-24.

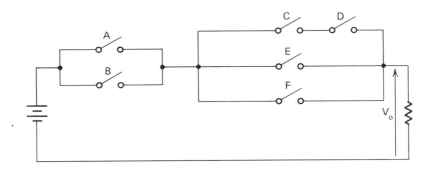

Figure 3-24

3-2. Write a truth table for the gate circuit of Fig. 3-25.

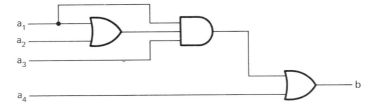

Figure 3-25

3-3.   Write a logical expression for the gate circuit of Fig. 3-25.

3-4.   Repeat Prob. 3-2 for Fig. 3-26.

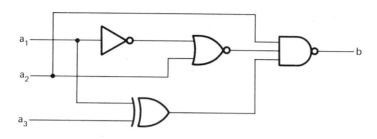

Figure 3-26

3-5.   Repeat Prob. 3-3 for Fig. 3-26.

3-6.   Show that the NAND operation is functionally complete.

3-7.   Redraw the circuit of Fig. 3-25 using NOR gates only.

3-8.   Repeat Prob. 3-7 using NAND gates.

3-9.   Obtain a truth table for a half subtractor (i.e., one that does not provide for borrowing). The output should consist of a borrow bit and a difference bit.

3-10. Write a switching function for the truth table of Prob. 3-9.

3-11. Obtain a gate circuit using AND, OR and NOT gates for the truth table of Prob. 3-9.

3-12. Obtain a truth table for a full subtractor, which does not have provision for borrowing.

3-13. Obtain a switching function for the truth table of Prob. 3-12.

3-14. Obtain a gate circuit for the truth table of Prob. 3-12.

3-15. Redraw Fig. 3-19b using only NOR gates.

3-16. Redraw Fig. 3-19b using only NAND gates.

3-17. Obtain a switching function for a multiplexer with eight inputs.

3-18. Draw a gate circuit for the multiplexer of Prob. 3-17.

4

Boolean Algebra — Switching Circuits

In Chapter 3 we introduced you to switching algebra and demonstrated the desirability of minimizing the switching functions. In this chapter we shall develop the fundamental mathematical tools used to minimize switching functions and also in the design of switching circuits. In the followin chapters, we shall apply the methods developed here to actual minimizations and shall discuss actual design procedures

4.1 BOOLEAN ALGEBRA

The mathematics of switching circuits is essentially the same as that encountered when dealing with many topics of mathematical logic. Mathematical loic deals with two values, T and F (true and false), while switching circuits deal with the values 1 and 0. The manipulations involving T and F are often the same as those which deal with the values 1 and 0. In this section we shall see how a logical algebra can be developed and then applied to switching circuits. We shall refer to this as *switching algebra.*

The mathematical treatment of logic was introduced in the 19th century by George Boole, and is called *Boolean algebra.* Subsequently, E. B. Huntington introduced a set of postulates that formalized the algebra. We shall use the ideas of Boole and Huntington to develop an algebra that can be used for switching circuits.

A rather formal presentation shall be used in this section so that the reader will see how such an algebra can be developed. We shall use symbols such as · and +. At first, these will represent general operations but when we discuss switching algebra, later in the chapter, they will refer to AND and OR, respectively.

We start by giving some basic definitions and then some postulates that apply to switching theory will be stated. Let us begin by considering some basic ideas of set theory.

A *set* of elements is a collection of objects that have some property in common. For instance,

$$A = [1, 2, 3, 4, 5, 6, 7, 8, 9,] \tag{4-1}$$

is the set of all positive integers less than 10.

The symbol $\in$ implies that an element is a member of a set. For instance, regarding (4-1), we can write

$$3 \in A$$

The symbol $\notin$ implies that an element is not a member of a set. Accordingly, for (4-1),

$$15 \notin A$$

A *binary operator* defines a manipulation of two members of a set, and it must result in another member of the same set. For instance, suppose that we have the set of all positive integers:

$$I = [1, 2, 3, 4, \ldots] \tag{4-2}$$

An example of a binary operator is $\times$ (multiplication). Note that the product of any positive integers will itself be a positive integer. Therefore, multiplication fits our definition of a binary operation. In this section we shall use the symbol * to denote a general binary operator. For instance, * might represent either + or $\cdot$.

If we are to have a useful and consistant mathematics, then an appropriate set of postulates must be developed. Although branches of mathematics which are not applicable to practical problems can be developed, if we are to have a mathematics that can be used with switching circuits, then the postulates that are used must be consistent with the ideas of switching circuits. If we follow this standard, then the resulting algebra will be practical.

We shall now work with postulates. These will be discussed in terms of a general set S. We shall first state the postulates in terms of the general binary operator * and then in terms of $\cdot$ and + operators. Although these will become the AND and OR operators, for the time being, we shall not define them as such. That is, we shall now use · and + to represent two arbitrary, but different, binary operations.

Postulates

1. Equivalence

Two elements of a set are equivalent if one can be substituted for the other in all relations. For instance, consider the set of integers [1,2,3,4,3]. The third and fifth elements are equivalent.

2. Closure

A set is said to be closed with respect to a given binary operator * if the application of that binary operator to any two members of the set results in a unique member of the same set. For instance, suppose that, with the set of all positive integers as defined in (4-2), we consider the binary operation of multiplication, ×. Then, I is closed with respect to ×. Note that the product of any two integers results in a unique third positive integer. However, I is not closed with respect to the operation of subtraction (-). For instance, $3 - 5$ results in a negative integer which is not a member of the set.

Let us consider the specific binary operators · (AND) and + (OR).

(2a) · is closed with respect to I (4-3a)

(2b) + is closed with respect to I (4-3b)

Although we have postulated that these two operators exist, we have not, as yet, defined them. Later we shall define them in terms of · representing AND, and + representing OR. It may help to think of these operators in this way. Note that, in this case, if the set S contains 0 and 1, then the set will be closed with respect to · and + since the result of any AND or OR operation is either a 0 or a 1. Note that we are *not* proving a postulate but merely demonstrating that it is reasonable to use it as a basis for switching algebra.

3. Identity

A set S has an identity element i with respect to a binary operation * if, for every $x \in S$ (every element x that is in S),

$$x*i = i*x = x \tag{4-4}$$

Now assuming that the set has two elements called 0 and 1, we postulate:

(3a) 1 is the identity element of S with respect to $\cdot$.

$$1 \cdot x = x \cdot 1 = x \tag{4-5a}$$

(3b) 0 is the identity element of S with respect to +.

$$0 + x = x + 0 = x \tag{4-5b}$$

Again we note that these postulates conform to our ideas of switching algebra.

4. Associative Law

A binary operation on S is associative if:

$$(x * y) * z = x * (y * z) \text{ for all } x, y, z \in S \tag{4-6}$$

Specifically we have:

(4a) $(x \cdot y) \cdot z = x \cdot (y \cdot z)$ $\hspace{4cm}$ (4-7a)

(4b) $(x + y) + z = x + (y + z)$ $\hspace{4cm}$ (4-7b)

5. Distributive law

Two binary operators * and $\circledast$ are distributive on S if:

$$x * (y \circledast z) = (x * y) \circledast (x * z) \text{ for all } x, y, z \in S \tag{4-8}$$

We specifically write the postulates

(5a) $x \cdot (y + z) = (x \cdot y) + (x \cdot z)$ $\hspace{3.5cm}$ (4-9a)

(5b) $x + (y \cdot z) = (x + y) \cdot (x + z)$ $\hspace{3.5cm}$ (4-9b)

Note that although these relations are not valid for ordinary algebra, they are valid for switching algebra. (Note that $\circledast$ is simply the symbol for another general binary operator.)

6. Commutative law

A binary operator * is said to be commutative on S if

$$x * y = y * x \text{ for all } x, y \in S \tag{4-10}$$

Specifically, we write the postulates

$$(6a) \quad x \cdot y = y \cdot x \tag{4-11a}$$

$$(6b) \quad x + y = y + x \tag{4-11b}$$

7. Complement (we shall not generalize this rule)

For every $x \in S$ there is another element $\bar{x}$ such that:

$$x \cdot \bar{x} = 0 \tag{4-12a}$$

$$x + \bar{x} = 1 \tag{4-12b}$$

where $\bar{x}$ is called the *complement* of x.

8. There are at least two elements in S (x and y) such that:

$$x \neq y \tag{4-13}$$

Duality

Note that all of the postulates that relate to $\cdot$ and $+$ are presented in pairs. One postulate can be obtained from the other simply by replacing $+$ with $\cdot$, $\cdot$ with $+$, 0 with 1 and 1 with 0. The postulates of each pair are said to be *duals*. Duality can be very helpful. If we develop a relation, then its dual will also be true. For instance, if we develop a relation using postulates 4a, 5b, and 3b then, by using postulates 4b, 5a, and 3a we can develop the dual relation.

We have stated the postulates here in a very general way. Now let us become more specific and rewrite them so that they apply directly to switching algebra.

4.2 TWO-VALUED BOOLEAN ALGEBRA—SWITCHING ALGEBRA

We will now restrict the variables with which we deal to be those of switching functions. That is, suppose that we have a set of variables:

$$S = [a, b, c, d, e, 0, 1] \tag{4-14}$$

Each variable element of the set (a, b, c, d, and e) can take on either the value 0 or the value 1, each of which is also a member of the set. We shall also restate our postulates more specifically than we did in Sec. 4-1. However, they will not apply directly to switching algebra. In addition, $\cdot$ and $+$ are now defined as AND and OR, respectively, (see Sec. 3-2).

Postulates

1. Equivalence

Two elements of the set S are equivalent if they are both 1 or both 0. For example if, whenever $a=1$, $b=1$ and whenever $a=0$, $b=0$, then $a=b$. That is, a and b are said to be equivalent.

2. Closure

(2a) S is closed with respect to $\cdot$. $\tag{4-15a}$

(2b) S is closed with respect to $+$. $\tag{4-15b}$

Since 0 and 1 are in S, the result of $+$ or $\cdot$ must also be in S. Again note that we are not proving the postulates but are simply demonstrating that they are reasonable.

3. Identity

For all $x \in S$

(3a) $x \cdot 1 = x$ $\tag{4-16a}$

(3b) $x + 0 = x$ $\tag{4-16b}$

4. Associative law

(4a) $(x \cdot y) \cdot z = x \cdot (y \cdot z)$ (4-17a)

(4b) $(x + y) + z = x + (y + z)$ (4-17b)

5. Distributive law

(5a) $x \cdot (y + z) = (x \cdot y) + (x \cdot z)$ (4-18a)

(5b) $x + (y \cdot z) = (x + y) \cdot (x + z)$ (4-18b)

6. Commutative law

(6a) $x \cdot y = y \cdot x$ (4-19a)

(6b) $x + y = y + x$ (4-19b)

7. Complement

(7a) $x \cdot \bar{x} = 0$ (4-20a)

(7b) $x + \bar{x} = 1$ (4-20b)

8. Since 0 and 1 are in S, there are at least two nonequal elements in S.

Proof by Perfect Induction

We can and shall prove theorems using these postulates, just as we do in ordinary algebra. However, since in switching algebra the variables can take on only one of two values, we can also demonstrate that a theorem is true by showing that it is valid for all possible combinations of the variables. For instance, see Eq. (3-23), we demonstrate the validity of the statement,

$$A_1 \cdot A_2 = \overline{\overline{A}_1 + \overline{A}_2}$$

by letting A_1 and A_2 take on all possible combinations of 0 and 1. For *every* possible combination $A_1 \cdot A_2$ and $\overline{\overline{A}_1 + \overline{A}_2}$ were equal. Thus we proved that $A_1 \cdot A_2$ does equal $\overline{\overline{A}_1 + \overline{A}_2}$. This is called a proof by *perfect induction.* Note that we cannot use perfect induction in ordinary algebra since an infinite number of combinations of variables would have to be checked.

Note that we can "prove" many of these postulates by perfect induction. For instance, consider the distributive law of Eq. (4-18b). We can write the following truth table:

x	y	z	$x+y$	$x+z$	$(x+y)\cdot(x+z)$	$y\cdot z$	$x+(y\cdot z)$
0	0	0	0	0	0	0	0
0	0	1	0	1	0	0	0
0	1	0	1	0	0	0	0
0	1	1	1	1	1	1	1
1	0	0	1	1	1	0	1
1	0	1	1	1	1	0	1
1	1	0	1	1	1	0	1
1	1	1	1	1	1	1	1

Comparing the sixth and eighth columns we see that (14-18b) is valid. We could prove many of the postulates of switching algebra in this way. Then they could be stated as theorems. However, since they are stated as postulates in the more general case of Boolean algebra, it is conventional to do this for switching algebra. The "proofs" then demonstrate that the postulates are well-suited for use in switching algebra.

We shall now prove some relations that are very helpful when dealing with switching algebra.

Indempotence Laws

$$x + x = x \tag{4-21a}$$

$$x \cdot x = x \tag{4-21b}$$

We prove (4-21a) in the following way:

$x + x = (x + x) \cdot 1$	by Postulate 3a
$x + x = (x + x) \cdot (x + \bar{x})$	by Postulate 7b
$x + x = x + x \cdot \bar{x}$	by Postulate 5b

$x + x = x + 0$	by Postulate 7a
$x + x = x$	by Postulate 3b

Thus Eq. (4-21a) is proven. Perfect induction could also have been used here. Note that (4-21b) is the dual of (4-21a). Hence, (4-21b) is true by duality.

0 and 1 Associated with a Variable

$$x + 1 = 1 \qquad (4\text{-}22a)$$

$$x \cdot 0 = 0 \qquad (4\text{-}22b)$$

Let us prove Eq. (4-22a).

$x + 1 = 1 \cdot (x + 1)$	by Postulates 3a and 3b
$x + 1 = (x + \bar{x}) \cdot (x + 1)$	by Postulate 7b
$x + 1 = x + \bar{x} \cdot 1$	by Postulate 5b
$x + 1 = x + \bar{x}$	by Postulate 3a
$x + 1 = 1$	by Postulate 7b

This can also be simply proven by perfect induction and we can prove that Eq. (4-22b) is true by duality.

Absorption

The next two relations are very valuable when switching functions are minimized.

$$x + x \cdot y = x \qquad (4\text{-}23a)$$

$$x \cdot (x + y) = x \qquad (4\text{-}23b)$$

Let us prove Eq. (4-23a)

$x + x \cdot y = x \cdot 1 + x \cdot y$	by Postulate 3a
$x + x \cdot y = x \cdot (1 + y)$	by Postulate 4a
$x + x \cdot y = x \cdot 1$	by Eq. (4-22a)

$x + x \cdot y = x$ by Postulate 3a

Again, Eq. (4-23) can be proven by duality.
There are two other forms of absorption.

$x + \bar{x} \cdot y = x + y$ (4-24a)

$x \cdot (\bar{x} + y) = x \cdot y$ (4-24b)

Let us prove Eq. (4-24a).

$x + \bar{x} \cdot y = (x + \bar{x}) \cdot (x + y)$ by Postulate 5b
$x + \bar{x} \cdot y = 1 \cdot (x + y)$ by Postulate 7b
$x + \bar{x} \cdot y = x + y$ by Postulate 3a

Equation (4-24b) is proven by duality.
Let us demonstrate how these equations can be used to reduce switching functions.

Example:

Reduce the switching function

$$f(x,y,z) = x + x \cdot y + x \cdot z + x \cdot (y \cdot \bar{z} + \bar{x})$$

Using Eq. (4-23a) we can write $x + x \cdot y$ as x. Hence,

$$f(x,y,z) = x + x \cdot z + x \cdot (y \cdot \bar{z} + \bar{x})$$

Applying Eq. (4-23a) again, we have

$$f(x,y,z) = x + x \cdot (y \cdot \bar{z} + \bar{x})$$

Using Eq. (4-23a) again and considering that $y \cdot \bar{z} + \bar{x}$ can be thought of as a single variable, we obtain

$$f(x,y,z) = x$$

This represents a considerable reduction in the function.

DeMorgan's Theorem

The following relations are widely used.

$$\overline{x + y} = \bar{x} \cdot \bar{y} \tag{4-25a}$$

$$\overline{x \cdot y} = \bar{x} + \bar{y} \tag{4-25b}$$

Note that when the bar extends over the entire function, it means that the function must first be evaluated and then the complement taken. These results can be proved as before. However, let us take this opportunity to illustrate a proof using perfect induction. We shall prove Eq. (4-25a).

x	y	$\bar{x}$	$\bar{y}$	$x \cdot y$	$\overline{x \cdot y}$	$\bar{x} + \bar{y}$
0	0	1	1	0	1	1
0	1	1	0	0	1	1
1	0	0	1	0	1	1
1	1	0	0	1	0	0

The last two columns of the truth table are identical. Thus, Eq. (4-25a) is proven and Eq. (4-25b) can be proven by duality.

Equations (4-25a) and (4-25b), which are called *DeMorgan's theorems*, have been stated for two variables. For instance, consider $\overline{x + y + z}$. Let $y + z = a$. Thus, we have $\overline{x + a} = \bar{x} \cdot \bar{a} = \bar{x} \cdot \overline{y + z}$. Again, using Eq. (4-25a) we obtain

$$\overline{x + y + z} = \bar{x} \cdot \bar{y} \cdot \bar{z} \tag{4-26a}$$

Using duality we have

$$\overline{x \cdot y \cdot z} = \bar{x} + \bar{y} + \bar{z} \tag{4-26b}$$

We can use these theorems with mixed AND and OR equations. For instance, suppose that we want the complement of $x + y \cdot z$. Let $y \cdot z = a$. Thus,

$$\overline{x + (y \cdot z)} = \overline{x + a} = \bar{x} \cdot \bar{a} = \bar{x} \cdot \overline{(y \cdot z)}$$

Using Eq. (4-25b), we have

$$\overline{x + (y \cdot z)} = \bar{x} \cdot (\bar{y} + \bar{z}) \tag{4-27}$$

If we continue in this fashion, we obtain a general expression for DeMorgan's theorem which states: *The complement of any switching function can be obtained by replacing each variable by its complement, replacing each AND operation by an OR operation and each OR operation by an AND operation. When they are present, 0's are replaced by 1's and vice versa.* Equation (4-27) is one example of this. Let us consider another example.

$$\overline{(x + \bar{y}) \cdot (z \cdot a)} = (\bar{x} \cdot y) + (\bar{z} + \bar{a}) \tag{4-28}$$

Now let us illustrate how a simple reduction utilizes DeMorgan's theorems.

Example:

Reduce the following

$$f(x,y,z) = x \cdot \left\{ [x + (y \cdot z)] + [\bar{x} \cdot (\bar{y} + \bar{z})] \right\} \tag{4-29}$$

Consider the terms within the braces. Using DeMorgan's theorems, we have

$$\overline{x + (y \cdot z)} = \bar{x} \cdot (\bar{y} + \bar{z})$$

Thus

$$f(x,y,z) = x \cdot \left\{ [x + (y \cdot z)] + \overline{[x + (y \cdot z)]} \right\}$$

Then, substituting Postulate 7b, we have

$$f(x,y,z) = x \cdot 1$$

Substituting Postulate 3a, we obtain

$$f(x,y,z) = x$$

This is considerably simpler than Eq. (4-25).

Shannon's Theorems

There are two other general relations, known as *Shannon's theorems*, that also prove helpful. They are

$$f(x,y,z, \ldots) = x \cdot f(1,y,z, \ldots) + \bar{x} \cdot f(0,y,z, \ldots) \qquad (4\text{-}30a)$$

and

$$f(x,y,z, \ldots) = [x + f(0,y,z, \ldots)] \cdot [\bar{x} + f(1,y,z, \ldots)] \qquad (4\text{-}30b)$$

Let us clarify the notation. $f(x,y,z, \ldots)$ is a switching function containing the variables x, y and z among others. $f(1,y,z, \ldots)$ is the same function where every x is replaced by a 1.

Perfect induction can be used to demonstrate the validity of Eqs. (4-30). For instance, in Eq. (4-30a), if $x = 1$, we have

$$f(1,y,z, \ldots) = 1 \cdot f(1,y,z, \ldots) + 0 \cdot f(0,y,z, \ldots)$$

Now applying Postulate 3a and Eq. (4-22b) we obtain

$$f(1,y,z, \ldots) = f(1,y,z, \ldots)$$

Now consider that $x = 0$. Then

$$f(0,y,z, \ldots) = 0 \cdot f(1,y,z, \ldots) + 1 \cdot f(0,y,z, \ldots)$$

Proceeding as before, we have

$$f(0,y,z, \ldots) = f(0,y,z, \ldots)$$

We have demonstrated that Eq. (4-30a) is valid for $x = 1$ and all possible $y,z, \ldots$ and that it is valid for $x = 0$ and all possible $y,z, \ldots$. Thus, the relation is always valid. We can prove Eq. (4-28) using duality.

Involution

The complement of a complement gives the variable itself.

$$\bar{\bar{x}} = x \qquad (4\text{-}31)$$

To prove this, we can use perfect induction.

x	$\bar{x}$	$\bar{\bar{x}}$
0	1	0
1	0	1

Hence, $\bar{\bar{x}} = x$.

The results that we have developed in this section are very useful. For convenience, we shall tabulate them.

Table 4-1 Important Boolean Algebra Relations

	Postulate		Equation
$x \cdot 1 = x$	Identity	3a	(4-16a)
$x + 0 = x$	Identity	3b	(4-16b)
$x + 1 = 1$			(4-22a)
$x \cdot 0 = 0$			(4-22b)
$(x \cdot y) \cdot z = x \cdot (y \cdot z)$	Associative	4a	(4-17a)
$(x + y) + z = x + (y + z)$	Associative	4b	(4-17b)
$x \cdot (y + z) = (x \cdot y) + (x \cdot z)$	Distributive	5a	(4-18a)
$x + (y \cdot z) = (x + y) \cdot (x + z)$	Distributive	5b	(4-18b)
$x \cdot y = y \cdot x$	Commutative	6a	(4-19a)
$x + y = y + x$	Commutative	6b	(4-19b)
$x \cdot \bar{x} = 0$	Complement	7a	(4-20a)
$x + \bar{x} = 1$	Compelment	7b	(4-20b)
$x + x = x$	Indempotence		(4-21a)
$x \cdot x = x$	Indempotence		(4-21b)
$x + x \cdot y = x$	Absorption		(4-23a)
$x \cdot (x + y) = x$	Absorption		(4-23b)
$x + \bar{x} \cdot y = x + y$	Absorption		(4-24a)
$x \cdot (\bar{x} + y) = x \cdot y$	Absorption		(4-24b)
$\overline{x + y} = \bar{x} \cdot \bar{y}$	DeMorgan's Theorem		(4-25a)
$\overline{x \cdot y} = \bar{x} + \bar{y}$	DeMorgan's Theorem		(4-25b)
$f(x,y,z,\ldots) = x \cdot f(1,y,z,\ldots) + \bar{x} \cdot f(0,y,z,\ldots)$			
	Shannon's Theorem		(4-30a)
$f(x,y,z,\ldots) = [x + f(0,y,z,\ldots)] \cdot [\bar{x} + f(1,y,z,\ldots)]$			
	Shannon's Theorem		(4-30b)
$\bar{\bar{x}} = x$	Involution		(4-31)

4.3 VENN DIAGRAMS

Venn diagrams are utilized in set theory to provide a pictorial representation of the relations among sets. They can also be used in switching theory to represent and simplify functions. We shall define Venn diagrams here and show how they can be used.

An elementary Venn diagram is shown in Fig. 4-1. Consider the circle labeled x. In set theory it would represent the set S. If an element is in x, then it lies within the circle. If an element is not in x, then it lies outside the circle. In switching theory, x represents a variable. The interior of the circle represents $x = 1$ while the exterior of the circle represents $x = 0$ or $\bar{x} = 1$. When we work with Venn diagrams, the area that is 1 represents the function. For instance, in Fig. 4-1, x is represented by the interior of the circle while $\bar{x}$ is represented by the region which is exterior to the circle. Note that the location of the region in itself has no significance. We could have drawn the circle at any point. Actually, the circle is merely a convenient shape. We could use any closed curve here.

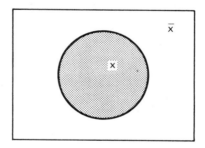

Figure 4-1 An elementary Venn diagram

Now let us use a Venn diagram to represent the function

$$f = x \cdot y$$

This is shown in Fig. 4-2. The circles represent x and y as labeled. x is 1 within the x circle and y is 1 within the y circle. (Remember that this is simply a symbolic representation and that the coordinates of the diagrams themselves have no significance.) The function $x \cdot y$ is 1 within the area enclosed by *both* circles. Thus, that region is labeled $x \cdot y$. Various other functions are labeled on the diagram. For instance, consider $\bar{x} \cdot \bar{y}$. This is 1 when $x = 0$ AND when $y = 0$. Hence, this is represented by the area outside of both circles. Similarly, $x \cdot \bar{y}$ is the region within the x circle which is *not* part of the y circle.

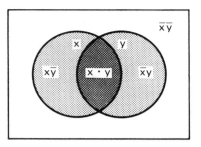

Figure 4-2   Venn diagram for $f = x \cdot y$

Let us demonstrate how Venn diagrams can be used to evaluate or simplify switching functions. We shall use the Venn diagram of Fig. 4-3 to prove Eq. (4-23a) which we repeat here

$$x + x \cdot y = x \qquad\qquad (4\text{-}32)$$

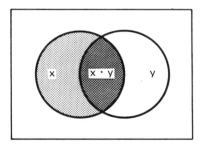

Figure 4-3   Demonstration that $x + x \cdot y = x$

The area within the x circle represents $x = 1$. The area common to both the x and y circles represents $x \cdot y = 1$. Hence, the sum of these two areas represents $x + x \cdot y = 1$. But the sum of the two areas is simply the x circle, and hence Eq. (4-32) is valid.

As a final example, let us use the Venn diagram of Fig. 4-4 to prove one of DeMorgan's relations.

$$\overline{x + y} = \bar{x} \cdot \bar{y} \qquad\qquad (4\text{-}33)$$

In Fig. 4-4a we show x and y. The area outside the x circle represents $\bar{x} = 1$ and the area outside of the y circle represents $\bar{y} = 1$. Thus, the area outside of *both* circles represents $\bar{x} \cdot \bar{y}$. Now consider Fig. 4-4b. It is the same representation as Fig. 4-4a. The total area enclosed by both circles represents $x + y$. Hence, the area external to the circles represents $\overline{x + y}$. Since Figs. 4-4a and 4-4b are the same, we have shown that Eq. (4-33) is valid.

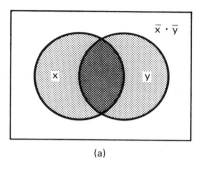
(a)

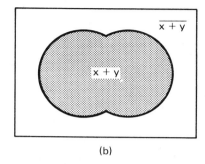
(b)

Figure 4-4 Demonstration that $\overline{x + y} = \bar{x} \cdot \bar{y}$

BIBLIOGRAPHY

Givone, D. D., *Introduction to Switching Circuit Theory*, Chap III, McGraw-Hill Book Co., New York 1970

Hill, F. J. and G. R. Peterson, *Introduction to Switching Theory and Logical Design*, 2nd Ed., Chap. 4, John Wiley and Sons, New York 1974

Mano, M. M. , *Digital Logic and Computer Design*, Chap. 2, Prentice Hall, Inc., Englewood Cliffs, NJ 1979

PROBLEMS

4-1. Discuss the choice of a set of postulates.

4-2. Show that the postulates of Secs. 4-1 and 4-2 are valid for a switching algebra.

4-3. Discuss the concept of duality.

4-4. Discuss proofs by perfect induction and by deduction. Also tell why proofs by perfect induction are used with switching algebra but never with ordinary algebra.

4-5. Which of the following switching functions are equivalent to $(x + y) \cdot x$?

$x \cdot y \cdot z + y \cdot z + x \cdot y \cdot z + \bar{y} \cdot z \cdot x$

$(x + z) \cdot (y + \bar{z})$

$(x + z) \cdot (z + x \cdot y)$

4-6. Use perfect induction to prove Eq. (4-23a).

4-7. Use the postulates to prove Eq. (4-23b).

4-8. Repeat Prob. 4-7 but now use perfect induction.

4-9. Use the postulates to prove Eq. (4-24b).

4-10. Repeat Prob. 4-9 but use perfect induction.

4-11. Use the postulates to prove Eq. (4-25a).

4-12. Prove Eq. (4-24b) using perfect induction.

4-13. Simplify the following expression.

$f(x,y,z) = x + x \cdot y + x \cdot y + x \cdot x$

4-14. Simplify the following expression

$f(x,y,z) = x \cdot (y + \bar{y}) + y \cdot x \cdot \bar{x}$

4-15. Repeat Prob. 4-14 for

$f(x,y,z) = \bar{x} \cdot (\bar{y} + y) \cdot (x + x)$

4-16. Simplify the following switching function

$f(w,x,y,z) = x + y + x \cdot z + y \cdot z + x \cdot w + y \cdot w$

4-17. Repeat Prob. 4-16 for

$f(w,x,y,z) = w \cdot x + (\bar{w} + \bar{x})(y + z)$

4-18. Find the complement of the following switching function.

$f(x,y,z) = (x + y) \cdot (z + x) + y \cdot z$

4-19. Repeat Prob. 4-18 for

$f(w,x,y,z) = w \cdot (x + y \cdot \bar{z}) + x \cdot \bar{y}(\bar{y} + z)$

4-20. Repeat Prob. 4-13 but now use two Venn diagrams.

4-21. Repeat Prob. 4-14 but now use a Venn diagram.

Minimization of Switching Functions

We are now in a position to discuss the minimization of switching functions. However, before we do this, we shall discuss some standard procedures for writing them.

5.1 CANONICAL FORMS – STANDARD FORMS

There are several general forms that can be used to express any switching function. These are called *canonical forms*. These generalized forms are usually not in minimal form, so we shall discuss procedures whereby they can be reduced. The study of canonical forms requires two fundamental definitions. We shall now consider the first of these. Suppose that we have a switching function of three variables such as $f(x, y, z)$. If we consider the three variables and their complements, we have a total of six variables. Now set up all the possible combinations of these variables using the AND operation. Note that a variable and its complement cannot both appear in the same function. There are eight such functions which are:

$$\bar{x} \cdot \bar{y} \cdot \bar{z}$$
$$\bar{x} \cdot \bar{y} \cdot z$$
$$\bar{x} \cdot y \cdot \bar{z}$$
$$\bar{x} \cdot y \cdot z$$
$$x \cdot \bar{y} \cdot \bar{z}$$
$$x \cdot \bar{y} \cdot z$$
$$x \cdot y \cdot \bar{z}$$
$$x \cdot y \cdot z$$

$$(5\text{-}1)$$

These terms are called *minterms* or *canonical products*. Note that if there are three binary variables, then there are eight possible minterms. In general, if there are n binary variables, there are 2^n possible minterms.

If we call the AND operation a *logical product* or simply a *product*, then the minterms are the products of each variable or its complement with all the other variables or their complements. They are called minterms since each minterm will be equal to 0 except for one combination of the binary variables. For instance, $x \cdot y \cdot \bar{z}$ is 0 except if $x = 1$ AND $y = 1$, AND $z = 0$. The other seven combinations of x, y and z all result in $x \cdot y \cdot \bar{z} = 0$. In Sec. 3-4 we demonstrated a procedure for obtaining a switching function from a truth table. Note that the procedure discussed there which generated a sum of minterms is also general. For instance, any switching function composed of three variables can be expressed as an OR combination of minterms such as those in (5-1). In fact, Eqs. (3-28) and (3-29) represent a function as an OR combination of minterms.

Before proceeding, let us discuss some convenient terminology. If two logical expressions are related by an OR operation, then they are said to be a *logical sum* or simply a *sum*. For instance, $A + B$ is the sum of A and B. Similarly, the AND operation is said to be a logical product or simply a product. For instance, $A \cdot B$ is said to be the product of A and B. Thus, we can state that any logical function can be expressed as a sum of minterms. Actually, see Sec. 3-4, we have demonstrated this by showing that a logical expression could be obtained from a truth table by summing the minterms which correspond to a 1 for the value of the function. Let us illustrate this procedure by obtaining a switching function that represents the following truth table.

x	y	z	A
0	0	0	1
0	0	1	0
0	1	0	1
0	1	1	0
1	0	0	1
1	0	1	1
1	1	0	0
1	1	1	1

(5-2)

The logical expression for A is then

$$A = \bar{x} \cdot \bar{y} \cdot \bar{z} + \bar{x} \cdot y \cdot \bar{z} + x \cdot \bar{y} \cdot \bar{z} + x \cdot \bar{y} \cdot z + x \cdot y \cdot z \qquad (5\text{-}3)$$

That is, A is the sum of all the minterms that correspond to 1's in the A-column of the truth table.

Since *any* switching function can be expressed as a sum of minterms, such general expressions are *canonical* forms. In particular, such an expression is called a *canonical sum of products form* or a *disjunctive canonical form* or a *standard sum of products*.

Let us discuss some notation which will be of help in describing canonical forms. A *product term* is the product of variables (either complemented or uncomplemented). A *normal sum* or a *disjunctive normal form* is the sum of product terms. Thus, a canonical sum is a normal sum. However, the converse need not be true. For instance,

$$F(x,y,z) = x \cdot y + \bar{x} \cdot y \cdot z + x \cdot \bar{z}$$

is a normal sum, but it is *not* a canonical sum since $x \cdot y$ and $x \cdot \bar{z}$ are not minterms. Note that $x \cdot y$ and $x \cdot \bar{z}$ do not contain all three variables.

Now let us define another quantity called a *maxterm*, or *canonical sum*. Again, we will illustrate this with a logical expression of three variables $f(x,y,z)$. Consider logical expressions which are the sum (OR operations) of each of the three variables or their compelements. These are (for three variables):

$$x + y + z$$
$$x + y + \bar{z}$$
$$x + \bar{y} + z$$
$$x + \bar{y} + \bar{z} \qquad\qquad\qquad (5\text{-}4)$$
$$\bar{x} + y + z$$
$$\bar{x} + y + \bar{z}$$
$$\bar{x} + \bar{y} + z$$
$$\bar{x} + \bar{y} + \bar{z}$$

These are called *maxterms* since each will be 1 for all but one possible combinations of x, y and z. For instance, $x + \bar{y} + z$ is 1 except when $x=0$, $y=1$ and $z=0$.

We shall now proceed with a development which will show how to express any switching function as a product of maxterms. We start by expressing the complement of the variable ($\bar{A}$) in terms of minterms. This can be obtained by adding all the minterms for which $A=0$ (i.e., $\bar{A}=1$). For instance, using the truth table (5-2), we obtain

$$\bar{A} = \bar{x} \cdot \bar{y} \cdot z + \bar{x} \cdot y \cdot z + x \cdot y \cdot \bar{z} \tag{5-5}$$

Now take the complement of both sides of the expression. Using DeMorgan's theorem, we obtain

$$A = (x + y + \bar{z}) \cdot (x + \bar{y} + \bar{z}) \cdot (\bar{x} + \bar{y} + z) \tag{5-6}$$

Note that $\bar{\bar{A}} = A$. Thus, we have expressed A as a product of maxterms.

This procedure can be used as a general method. We can always express the complement of the output as a sum of minterms, and taking the complement of a minterm, we obtain a maxterm. Thus, since any switching function can be expressed as a product of maxterms, the product of maxterms is also a canonical form, which is called a *canonical product of sums form* or a *conjunctive canonical form* or a *standard product of sums*.

In Sec. 3-4 we demonstrated that a switching function consisting of the sum of n minterms could be realized by the interconnection of n AND gates and an OR gate, with a suitable number of inverters. In a similar way we see that a switching function consisting of the product of m maxterms can be realized using m OR gates and one AND gate. These two realizations do not necessarily use the same number of gates. For instance, the realization of A using Eq. (5-3) requires four AND gates and one OR gate, while the realization using Eq. (5-6) requires three OR gates and one AND gate. In general, if there are more 1's than 0's in the output column of the truth table, then the realization using maxterms will require fewer gates than the realization using minterms. Similarly, if there are more 0's than 1's, the realization using minterms will have fewer gates. However, there may often be other realizations which use far fewer gates than either of these.

Minterms and maxterms are often listed in terms of a shorthand notation. In Table 5-1, we shall illustrate this notation assuming three independent variables.

Table 5-1 Minterms and Maxterms for Three Switching Variables

x_1	x_2	x_3	*Minterm*	*Designation*	*Maxterm*	*Designation*
0	0	0	$\bar{x}_1 \cdot \bar{x}_2 \cdot \bar{x}_3$	m_0	$x_1 + x_2 + x_3$	M_0
0	0	1	$\bar{x}_1 \cdot \bar{x}_2 \cdot x_3$	m_1	$x_1 + x_2 + \bar{x}_3$	M_1
0	1	0	$\bar{x}_1 \cdot x_2 \cdot \bar{x}_3$	m_2	$x_1 + \bar{x}_2 + x_3$	M_2
0	1	1	$\bar{x}_1 \cdot x_2 \cdot x_3$	m_3	$x_1 + \bar{x}_2 + \bar{x}_3$	M_3
1	0	0	$x_1 \cdot \bar{x}_2 \cdot \bar{x}_3$	m_4	$\bar{x}_1 + x_2 + x_3$	M_4

x_1	x_2	x_3	*Minterm*	*Designation*	*Maxterm*	*Designation*
1	0	1	$x_1 \cdot \bar{x}_2 \cdot x_3$	m_5	$\bar{x}_1 + x_2 + \bar{x}_3$	M_5
1	1	0	$x_1 \cdot x_2 \cdot \bar{x}_3$	m_6	$\bar{x}_1 + \bar{x}_2 + x_3$	M_6
1	1	1	$x_1 \cdot x_2 \cdot x_3$	m_7	$\bar{x}_1 + \bar{x}_2 + \bar{x}_3$	M_7

The switching variables are listed in an orderly manner, just as when we write a truth table. In the minterm column we write the minterm which is 1 when the switching variables on the corresponding row of the table have the designated values. The minterm is designated by a subscripted lower case m. The subscript is in base 10 and is equal to the base 2 representation of a binary number whose digits are the values of those variables that make the minterm 1. For instance, $\bar{x}_1 \cdot x_2 \cdot x_3$ is designated m_3 since $3_{10} = 001_2$ and this minterm is 1 when $x_1 = 0$, $x_2 = 1$ and $x_3 = 1$. Similarly, m_4 corresponds to $x_1 \cdot \bar{x}_2 \cdot \bar{x}_3$ since this is 1 when $x_1 = 1$, $x_2 = 0$ and $x_3 = 0$ ($4_{10} = 100_2$). Note that there is some ambiguity here. We arbitrarily listed the independent variables in the order x_1, x_2, x_3. Other orders could have been used (e.g. x_3, x_2, x_1). Then the minterm representing m_1 would be $\bar{x}_3 \cdot \bar{x}_2 \cdot x_1$, etc.

A minterm is 1 for only one of the 2^n possible combinations of the independent variables. Thus, the designation of the minterms is based on their being 1. Conversely, a maxterm is 0 for only one of the 2^n combinations of the binary variables. Hence, we designate the maxterms on the basis of their being 0. For instance, M_5 designated $\bar{x}_1 + x_2 + \bar{x}_3$ because $5_{10} = 101_2$ and $\bar{x}_1 + x_2 + \bar{x}_3 = 0$ when $x_1 = 1$, $x_2 = 0$ and $x_3 = 1$. Note that the corresponding minterms and maxterms are complements. That is

$$m_j = \bar{M}_j \qquad (5\text{-}7)$$

Suppose that we have a switching function

$$f(x_1, x_2, x_3) = \bar{x}_1 \cdot x_2 \cdot \bar{x}_3 + x_1 \cdot \bar{x}_2 \cdot \bar{x}_3 + x_1 \cdot x_2 \cdot x_3 \qquad (5\text{-}8a)$$

We can write this as

$$f(x_1, x_2, x_3) = m_2 + m_4 + m_7 \qquad (5\text{-}8b)$$

A shorthand notation is sometimes adopted to make such expressions more compact. For instance,

$$f(x_1, x_2, x_3) = \Pi m(0,2,7) \tag{5-8c}$$

The symbol Πm indicates that the numbers which follow designate minterms, and that their sum is to be taken.

A similar notation can be used for products of maxterms, for instance:

$$g(x_1, x_2, x_3) = (x_1 + x_2 + x_3) \cdot (\bar{x}_1 + x_2 + \bar{x}_3) \cdot (\bar{x}_1 + \bar{x}_2 + \bar{x}_3) \tag{5-9a}$$

We can write this as

$$f(x_1, x_2, x_3) = M_0, M_2, M_7 \tag{5-9b}$$

This can also be expressed in shorthand form. We use the notation

$$g(x_1, x_2, x_3) = \Pi M(0,2,7) \tag{5-9c}$$

Note that the symbol ΠM indicates that the numbers which follow designate maxterms, and that their product is to be taken.

Expression of an Arbitrary Switching Function without Using the Truth Table

We have seen how an arbitrary switching function can be expressed as a sum of minterms or as a product of maxterms. These forms can be easily generated from the truth table. Thus, if we are given an arbitrary switching function and we want to express it in canonical form, we need only construct its truth table. However, there are other procedures which can be used to obtain a canonical form without writing the truth table. Let us illustrate this with an example. We shall express the function

$$A(x,y,z) = x + \bar{y} \cdot z \tag{5-10a}$$

as a sum of minterms. Each term in the sum should be a product of each of the three independent variables x, y and z, or their complements. Consider each term. The first is x. Since $x \cdot 1 = x$ and $y + \bar{y} = 1$, see Eqs. (4-16a) and (4-20b), we can write

$$x = x \cdot (y + \bar{y}) = x \cdot y + x \cdot \bar{y}$$

Now repeat this procedure with z. Then,

$$x = (x \cdot y + x \cdot \bar{y}) \cdot (z + \bar{z}) = x \cdot y \cdot z + x \cdot y \cdot \bar{z} + x \cdot \bar{y} \cdot z + x \cdot \bar{y} \cdot \bar{z}$$

Hence, we have expressed x as a sum of minterms. The second term in Eq. (5-10a) has no x term. Thus, we write

$$\bar{y} \cdot z = \bar{y} \cdot z \cdot (x + \bar{x}) = \bar{y} \cdot z \cdot x + \bar{y} \cdot z \cdot \bar{x} = x \cdot \bar{y} \cdot z + \bar{x} \cdot \bar{y} \cdot z$$

Combining terms, we have

$$A(x,y,z) = x \cdot y \cdot z + x \cdot y \cdot \bar{z} + x \cdot \bar{y} \cdot z + x \cdot \bar{y} \cdot \bar{z} + \bar{x} \cdot \bar{y} \cdot z \tag{5-10b}$$

There are several terms which are repeated. Using Eq. (4-21a), we obtain

$$A(x,y,z) = x \cdot y \cdot z + x \cdot y \cdot \bar{z} + x \cdot \bar{y} \cdot z + x \cdot \bar{y} \cdot \bar{z} + \bar{x} \cdot \bar{y} \cdot z \tag{5-10c}$$

This type of procedure can be used as a general method. That is, the switching function is expressed as a sum of terms which are products of variables. Next, the "missing" variables are included by means of the above procedure and, thus, the switching function can be expressed as a sum of minterms.

If we want to express a function as a product of maxterms, we can obtain the complement of the given function, express it as a product of minterms and, finally, take the complement of the resulting expression. For example, let us express Eq. (5-10a) as a sum of maxterms. Using DeMorgan's theorem to obtain the complement yields

$$\bar{A} = \bar{x} \cdot (y + \bar{z}) = \bar{x} \cdot y + \bar{x} \cdot \bar{z}$$

Now we use the previously discussed procedure to express this as a sum of minterms.

$$\bar{A} = \bar{x} \cdot y \cdot (z + \bar{z}) + \bar{x} \cdot \bar{z} \cdot (y + \bar{y})$$

Manipulation yields

$$\bar{A} = \bar{x} \cdot y \cdot z + \bar{x} \cdot y \cdot \bar{z} + \bar{x} \cdot y \cdot \bar{z} + \bar{x} \cdot \bar{y} \cdot \bar{z}$$

Elimination of the repeated terms yields

$$\bar{A} = \bar{x} \cdot y \cdot z + \bar{x} \cdot y \cdot \bar{z} + \bar{x} \cdot \bar{y} \cdot \bar{z}$$

This is now a sum of minterms. Taking the complement of both sides of this expression, we obtain

$$A = (x+\bar{y}+\bar{z}) \cdot (x+\bar{y}+z) \cdot (x+y+z) \tag{5-11}$$

Thus we have the desired expression.

Uniqueness of Canonical Forms

The canonical form representing a switching function is unique. This means that, if a function is represented as a sum of minterms, then *any* sum of minterms which represents this function would contain *exactly* the same minterms. We can prove this by contradiction. Suppose that the same switching function could be represented by two different sums of minterms. Each minterm is 1 for *one, and only one*, of the 2^n possible combinations of the independent binary variables. Thus, there must be one set of binary variables which causes one sum of minterms to be 1 and the other sum to be 0. However, a function cannot be 0 and 1 simultaneously. Thus, the assumption that sums of different minterms could represent the same function is false. This demonstrates the uniqueness. A similar proof can be used to demonstrate the uniqueness of the product of maxterms.

We have used the ($\cdot$) to indicate the AND operation. However, it is both conventional and convenient to omit it if no ambiguity is present. In the remainder of this book we shall do so. Thus, we shall write

$$x \cdot y = xy \tag{5-12}$$

Logical Expressions from Design Specifications

In Chapter 3 we obtained logical expressions for switching functions that were based upon design specifications. For intance, the switching functions for adders and a multiplexer were obtained there. In subsequent chapters we shall discuss additional examples of this procedure and shall consider formal procedures for obtaining the switching functions.

At the present time let us consider a simple example that illustrates how a switching function can be obtained from design specifications that have nothing to do with digital systems. Suppose that we want to obtain a logical circuit that can be used to indicate if a student should be admitted to an honors program.

Whether or not the student will be admitted to the honors program will be determined by the passing or failing of three tests. The first two, T_1 and T_2, cover specialized areas and the student is only required to know one of the subjects. The third is general and all students are required to pass it. Thus, to enter the honors program, the student must pass either T_1 or T_2, or both, and T_3. Thus, the truth table for entrance to the honors program is:

T_1	T_2	T_3	P	
0	0	0	0	
0	0	1	0	
0	1	0	0	
0	1	1	1	(5-13)
1	0	0	0	
1	0	1	1	
1	1	0	0	
1	1	1	1	

where, for T_1, T_2 and T_3, 1 represents pass and 0 represents fail. If P is 1, the student enters the program, while if it is 0, he does not. We want to design a simple gate circuit which will determine if a student enters the program. Using the truth table, we can write P as a sum of minterms. This is

$$P = \bar{T}_1 T_2 T_3 + T_1 \bar{T}_2 T_3 + T_1 T_2 T_3 \tag{5-14a}$$

This can be realized using three AND gates, one OR gate and two NOT gates. However, we can simplify the relation. The following steps can be used:

$$P = (T_1 + \bar{T}_1) T_2 T_3 + T_1 \bar{T}_2 T_3$$

$$P = T_2 T_3 + T_1 \bar{T}_2 T_3 \tag{5-14b}$$

$$P = T_3 (T_2 + T_1 \bar{T}_2)$$

Using Eq. (4-24a) we have

$$P = T_3 (T_1 + T_2) \tag{5-14c}$$

Now we need use only one OR gate and one AND gate. No NOT gates are required. This realization is shown in Fig. 5-1.

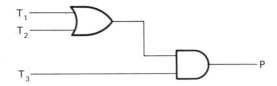

Figure 5-1 The gate circuit realization of Eq. (5-14)

5.2 KARNAUGH MAPS—MINIMIZATION OF SWITCHING FUNCTIONS

In Sec. 4-2, we illustrated several minimizations of switching functions. However, the unsystematic procedures used there become very tedious when relatively large logical expressions are encountered. In this section we shall discuss a symbolic representation of the truth table called a *Karnaugh map* that will provide a systematic means of reducing the function. The Karnaugh map is a graphical visualization of the truth table which enables us to easily use Eqs. (4-20) and (4-23) to minimize the function in an organized way.

We shall start by considering the example of a switching function of two variables, for instance,

$$f(x,y) = x\bar{y} + xy + \bar{x}y \tag{5-15}$$

The Karnaugh map of this function is shown in Fig. 5-2. A Karnaugh map is a pictorial representation of a canonical sum of products. It consists of a number of boxes called *cells*. There is one cell in the Karnaugh map for every *possible* minterm. If a minterm is actually present in the canonical sum of products, then a 1 is placed in the cell corresponding to that minterm. The Karnaugh map of Fig. 5-2 is a pictorial representation of Eq. (5-15). To clarify this, consider Fig. 5-3 where we illustrate a Karnaugh map and mark each cell with its minterm designation. The designated minterm becomes a 1 when the values of x and y correspond to those indicated on the axes of the map. The numbers along the top correspond to x while those along the side correspond to y. For instance, $m_2 = 1$ when $y = 0, x = 1$.

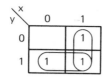

Figure 5-2 A two-variable Karnaugh map for the switching function of Eq. (5-15)

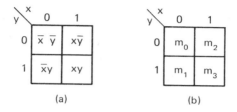

Figure 5-3 Karnaugh maps with cells marked in accordance with minterms

Now consider the actual Karnaugh map of Fig. 5-2. The cells marked with 1's are called 1-cells. The 1-cells represent the actual minterms of the function. All other cells are called 0-cells and could actually be marked with 0's. However, this is usually not done to avoid cluttering the map.

Now let us see how the Karnaugh map can be used to simplify a switching expression. We shall make use of the following equation,

$$yf + \bar{y}f = (y + \bar{y})f = f \tag{5-16}$$

where f is an arbitrary switching function. That is, if we have two terms and one is a function f multiplied by y and the other is the same function multiplied by $\bar{y}$, then the terms can be combined and the variable y can be eliminated.

Now consider the Karnaugh map. If we move vertically from one cell to an adjacent cell then one, and only one, of the two variables of the corresponding minterms differs. The variable appears in one minterm while its complement appears in the other. For instance, the two righthand boxes correspond to $x\bar{y}$ and xy. In general, *any* two adjacent boxes will be of this form. (We shall see how this extends to more than two independent variables subsequently.) Now, for Eq. (5-16) we have

$$x\bar{y} + xy = x(y+\bar{y}) = x \tag{5-17}$$

Thus, if 1's appear in adjacent vertical boxes, the *two minterms can be combined into a single, simpler term.* This is indicated on the Karnaugh map by encircling the variables, see Fig. 5-2.

In a similar way, if two 1's are adjacent horizontally, then only one variable of the corresponding minterms will change (to its complement). For instance, the bottom row of the Karnaugh map of Fig. 5-2 signifies

$$\bar{x}y + xy = (\bar{x}+x)y = y \qquad (5\text{-}18)$$

Thus, if two 1's are adjacent horizontally, the corresponding minterms can be combined into a simpler expression. We indicate this also by encircling the appropriate 1-cells. (Note that we can combine 1-cells that are adjacent horizontally or vertically, but *not* diagonally.)

Proceeding with the simplification, we can write

$$f(x,y) = x + \bar{x}y = y + x\bar{y} \qquad (5\text{-}19a)$$

or

$$f(x,y) = y + x\bar{y} \qquad (5\text{-}19b)$$

Using Eq. (4-23a) we can write either of these as

$$f(x,y) = x + y \qquad (5\text{-}20)$$

Now let us use the Karnaugh map to interpret this. One circled pair represents x; the other circled pair represents y, see Eqs. (5-16) and (5-17). Their combination is the sum of the two, $x+y$. Thus, we can use the Karnaugh map to systematically reduce the expression. We shall subsequently demonstrate that the simplified expression can be obtained by a simple inspection of the Karnaugh map.

We have illustrated some simple ideas using a two-variable Karnaugh map. Now let us extend these to additional variables. The Karnaugh map is always two-dimensional. Thus, each cell may have to represent more than two variables. For instance, a three-variable Karnaugh map is shown in Fig. 5-4. The upper 0's and 1's represent the values of x and y. The leftmost number of each pair represents x and the other number represents y, since x and y are written in this order above the map. Note that the order of the numbers along the top of the map is *not* in ascending binary order, but is such that *adjacent numbers differ by only one binary digit.* The Karnaugh map is useful for mini-

z \ xy	00	01	11	10
0	m_0	m_2	m_6	m_4
1	m_1	m_3	m_7	m_5

Figure 5-4 Three-variable Karnaugh map which is used to represent $f(x,y,z)$

mization of functions because cells which are adjacent (horizontally or vertically) represent a change of just one variable to its complement. Thus, as we have seen, minterms represented by 1's in adjacent boxes can be combined using Eq. (5-16) to yield a single term in the switching function which has one less variable than the original minterm. Ordering the binary numbers along the top edge of the map so that adjacent numbers differ by only one binary digit accomplishes this.

In a subsequent chapter we shall discuss general procedures for encoding numbers so that successive numbers differ by only one binary digit. Such encoding yields a *Gray code* or a *reflected code*. Here, we need not consider the general details of Gray coding. The appropriate ordering of the binary digits of the Karnaugh map will simply be given in the text.

Let us illustrate minimization using the Karnaugh map of Fig. 5-5 as an example. The switching function that corresponds to this map is

$$f(x,y,z) = \bar{x}y\bar{z} + xy\bar{z} + \bar{x}yz \tag{5-21}$$

The two 1's in the second column of Fig. 5-5 represent

$$\bar{x}y\bar{z} + \bar{x}yz = \bar{x}y(z+\bar{z}) = \bar{x}y \tag{5-22a}$$

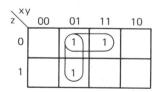

Figure 5-5 A three-variable Karnaugh map for the switching function of Eq. (5-21)

As in the case of the two-variable Karnaugh map, a pair of vertically adjacent 1's results in the elimination of a variable and the combination of two terms. Now consider the two adjacent horizontal 1's. They represent

$$\bar{x}y\bar{z} + xy\bar{z} = (\bar{x}+x)(y\bar{z}) = y\bar{z} \qquad (5\text{-}22b)$$

Thus, adjacent 1-cells can be combined, eliminating one variable. Combining Eqs. (5-22a) and (5-22b), we have

$$f(x,y,z) = \bar{x}y + y\bar{z} \qquad (5\text{-}23)$$

Note that the value of the function is the sum of the values of the encircled pairs. This is because the sum of all the encircled pairs is just the sum of all the minterms. Remember that a minterm can be included more than once in such a sum, see Eq. (4-21a).

Let us consider some additional examples. The switching function for the Karnaugh map of Fig. 5-6 is

$$f(x,y,z) = \bar{x}y\bar{z} + \bar{x}yz + xy\bar{z} + xyz \qquad (5\text{-}24a)$$

We can simplify this as

$$f(x,y,z) = (\bar{x}+x)y\bar{z} + (\bar{x}+x)yz = y\bar{z} + yz = y(\bar{z}+z)$$

$$\qquad (5\text{-}24b)$$

$$f(x,yz) = y$$

Now consider the circle drawn around all four 1-cells. We could break this up into four loops, each enclosing a pair of horizontal or vertical 1-cells. From what we have seen previously, a horizontal or vertical loop indicates a pair of minterms which can be combined into a single term since they differ only in the complementation of one variable. Because of the Gray coding of the variable locations, we would expect

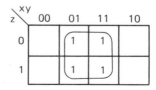

Figure 5-6 A three-variable Karnaugh map for the switching function of Eq. (5-24a)

any group of variables enclosed in a group (such as a square) to contain similarly related terms. Thus, such groupings should contain candidates for reduction. This is illustrated by Eqs. (5-24). For instance, consider each pair of vertical subcubes. They represent

$$xy\bar{z} + xyz = xy \tag{5-25a}$$

and

$$\bar{x}y\bar{z} + \bar{x}yz = \bar{x}y \tag{5-25b}$$

Hence,

$$f(x,y,z) = xy + \bar{x}y = (x+\bar{x})y \tag{5-26a}$$

$$f(x,y,z) = y \tag{5-26b}$$

Note that the two adjacent *pairs* of subcubes of (5-25a) and (5-25b) themselves represent functions that differ in only one variable which is complemented in one function and uncomplemented in the other. Hence, when these are logically summed, a reduction results. This seems to indicate that pairs of adjacent two-cube subcubes can themselves be combined into a single switching function with one less variable. Actually this is the case. In fact, pairs of two-cell subcubes can be combined into a single four-cell subcube which can be represented by a single product of variables. There will be two less variables in this product than in the products that form the original minterms. The grouping on the Karnaugh map systematically yields this minimization. We shall consider several additional examples and then obtain a procedure for determining the reduced function in a very simple way.

Let us consider the Karnaugh map of Fig. 5-7. The switching function for this map is

$$f(x,y,z) = \bar{x}\bar{y}z + \bar{x}\bar{y}\bar{z} + x\bar{y}z + x\bar{y}\bar{z} \tag{5-27}$$

Consider the terms in the upper corners of the map.

$$\bar{x}\bar{y}\bar{z} + x\bar{y}\bar{z} = \bar{y}\bar{z}(\bar{x}+x) = \bar{y}\bar{z} \tag{5-28}$$

This indicates that the Karnaugh map locations which correspond to these minterms should be adjacent. However, they are on opposite

$$
\begin{array}{c|c|c|c|c}
z \diagdown \! ^{xy} & 00 & 01 & 11 & 10 \\
\hline
0 & 1 & & & 1 \\
\hline
1 & 1 & & & 1 \\
\end{array}
$$

Figure 5-7 A three-variable Karnaugh map for the switching function of Eq. (5-27)

sides of the map in the upper righthand and upper lefthand corners, respectively. Thus, they are not physically adjacent. To resolve this difficulty, we consider that the right and left edges of the map are in contact. That is, consider the map to be a cylinder, not a plane. This resolves the difficulty since these squares are now adjacent horizontally. Note that the minterms that are now considered to be adjacent differ by only one variable that is complemented in one minterm and uncomplemented in the other. Proceeding, we can now encircle the square formed by four 1-cells.

Let us see if Eq. (5-27) can be reduced as we might expect. We can write it as

$$f(x,y,z) = \bar{x}\bar{y}(z+\bar{z}) + x\bar{y}(z+\bar{z}) = \bar{x}\bar{y} + x\bar{y}$$

Hence,

$$f(x,y,z) = \bar{y} \tag{5-29}$$

Now let us consider a four-variable Karnaugh map. One showing the minterms that correspond to each cube is shown in Fig. 5-8a. The shorthand notation shown in Fig. 5-8b is sometimes used.

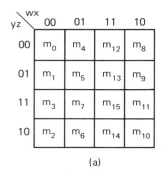

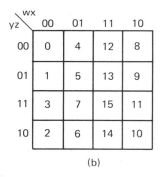

(a)   (b)

Figure 5-8 (a) A four-variable Karnaugh map showing a designation in terms of minterms; (b) an alternative form of this representation

To illustrate the use of the four-variable Karnaugh map, let us consider the switching function

$$f(w,x,y,z) =$$

$$\bar{w}x\bar{y}\bar{z}+wx\bar{y}\bar{z}+\bar{w}x\bar{y}z+wx\bar{y}z+\bar{w}\bar{x}y\bar{z}+\bar{w}xy\bar{z}+wxy\bar{z}+w\bar{x}y\bar{z} \qquad (5\text{-}30)$$

The Karnaugh map is given in Fig. 5-9. Again notice that any pair of 1-cells that are adjacent either horizontally or vertically results in a pair of minterms that differ only in one variable which is complemented in one and uncomplements in the other. Thus, we should

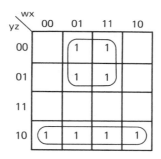

Figure 5-9 A four-variable Karnaugh map for the switching function of Eq. (5-30)

expect the same type of minimization here. Now consider the two groupings isolated in the map. First consider the upper square. This represents the minterm sum

$$f_1 = \bar{w}x\bar{y}\bar{z} + wx\bar{y}\bar{z} + \bar{w}x\bar{y}z + wx\bar{y}z = x\bar{y}\bar{z} + x\bar{y}z = x\bar{y} \qquad (5\text{-}31)$$

Thus, we again see that the groupings indicate where considerable reduction can occur. Now consider the sum of minterms represented by the encircled lower rectangle. This is

$$f_2 = \bar{w}\bar{x}y\bar{z} + \bar{w}xy\bar{z} + wxy\bar{z} + w\bar{x}y\bar{z} = y\bar{z} \qquad (5\text{-}32)$$

Again, a grouping on the Karnaugh map leads to a considerable reduction. Note that these reductions occur because of the appearance of a variable or its complement in each of the products. Thus, we can reduce Eq. (5-30) to

$$f(w,x,y,z) = f_1 + f_2 = x\bar{y} + y\bar{z} \qquad (5\text{-}33)$$

Let us now review some terminology. As we have discussed, each square, representing a single minterm of the Karnaugh map, is called a *cell*. In general, if two cells represent minterms which differ by only one variable (i.e., it is complemented in one and not in the other), then they are said to be adjacent. Thus, if two cells containing 1's (1-cells) are adjacent, a variable can be eliminated by combining that pair of minterms. A grouping of adjacent 1-cells is defined as a *subcube*. The subcube is said to *cover* the cells containing the 1's. For instance, Fig. 5-9 has two subcubes, each of which covers four cells.

Let us consider the reduction of Eq. (5-30) again. Specifically let us discuss f_1, see Eq. (5-31). This is represented by the square, four-cell subcube of Fig. 5-9. This subcube can be considered to be the combination of two adjacent subcubes, one representing

$$f_a = \bar{w}x\bar{y}\bar{z} + wx\bar{y}\bar{z} = x\bar{y}\bar{z}$$

and the other representing

$$f_b = \bar{w}x\bar{y}z + wx\bar{y}z = x\bar{y}z$$

The two-cell subcube representing f_a is adjacent to the two-cell subcube representing f_b. Note that f_a and f_b each represents a product of the same variables. The only difference between f_a and f_b is that z is complemented in *one* and uncomplemented in the other. Thus, $f_a + f_b$ can be combined into a single product with one less variable (i.e., z is eliminated).

$$f_a + f_b = x\bar{y}(z+\bar{z}) = x\bar{y}$$

In general, adjacent two-cell subcubes can, in this way, be combined into a four-cell subcube, and this results in a simplification of the switching function, since a variable has been eliminated.

Now we shall consider some further reductions of four-variable maps. Let us work with the map of Fig. 5-10. It represents the switching function

$$f(w,x,y,z) =$$

$$\bar{w}x\bar{y}\bar{z} + wx\bar{y}\bar{z} + \bar{w}xy\bar{z} + wxy\bar{z} + \bar{w}\bar{x}\bar{y}z + w\bar{x}\bar{y}z + \bar{w}\bar{x}yz + w\bar{x}yz \quad (5\text{-}34)$$

Terms on the left and right edges of the map are adjacent since their minterms only differ in a single variable which is complemented in one

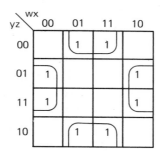

Figure 5-10 A four-variable Karnaugh map for the switching function of Eq. (5-34)

and uncomplemented in the other. This corresponds to the three-variable map. In a similar way, the corresponding cells on the top and bottom edges are also adjacent. Thus, we have encircled two subcubes of four cells each. Consider the subcube which consists of the upper and lower sections, i.e. $m_4 + m_{12} + m_6 + m_{14}$, (see Fig. 5-8). Let us see how it reduces. The minterms which correspond to it are

$$f_1 = \bar{w}x\bar{y}\bar{z} + wx\bar{y}\bar{z} + \bar{w}xy\bar{z} + wxy\bar{z} \tag{5-35a}$$

This can be simplified in the following way:

$$f_1 = x\bar{y}\bar{z} + xy\bar{z} = x\bar{z} \tag{5-35b}$$

Similarly, the minterms which correspond to the remaining subcube, i.e. $m_1 + m_3 + m_9 + m_{11}$, are

$$f_2 = \bar{w}\bar{x}\bar{y}z + w\bar{x}\bar{y}z + \bar{w}\bar{x}yz + w\bar{x}yz \tag{5-36a}$$

This can be simplified in the following way:

$$f_2 = \bar{x}\bar{y}z + \bar{x}yz = \bar{x}z \tag{5-36b}$$

Thus, Eq. (5-34) can be expressed as

$$f(w,x,y,z) = x\bar{z} + \bar{x}z \tag{5-37}$$

Of course, this is the sum of the "functions" corresponding to each of the subcubes.

Before proceeding with some more complex functions, let us make some general statements based upon what we have observed. Each subcube of 1-cells can be expressed as a single product of the independent variables. It is conventional to refer to variables which are represented by letters (e.g., x, y, $\bar{z}$, etc.) as *literals. Thus, each subcube can be represented by a single product of literals.* When a subcube consists of two adjacent cells, one variable can be eliminated from the sum of the corresponding minterms (i.e., the one which changes to its complement). In the examples, we have seen (e.g. Figs. 5-6, 5-7, 5-9 and 5-10) that if a subcube covers four cells, then two variables can be eliminated. This is because the four-cell subcube is composed of two adjacent two-cell subcubes which differ in only one variable which will be complemented in one and uncomplemented in the other. This actually can be generalized. If a map has n variables, then *a subcube covering 2^k adjacent cells can be represented by a single product of $n-k$ literals.* Thus, for a four-variable Karnaugh map, we can write the following table.

Table 5-2 Number of Literals in a Single Product Needed to Represent the 1-Cells Covered by a Subcube in a Four-Variable Karnaugh Map

Number of 1-Cells Covered by a Subcube	*Number of Literals in Corresponding Term*
1	4
2	3
4	2
8	1
16	0

Note that if the entire Karnaugh map consists of 1's, then the function is a 1 and no literals are needed. Thus, it is desirable to use subcubes which are as large as possible. We can then write the product for each subcube and simplify the expression. Actually, we save this step. The minimized expression for any subcube can be obtained simply by inspection. The expression for the subcube is just the *product of all literals whose values do not change in the subcube.* If the value of the literal is always 1, then the literal appears in the product. If the value of the literal is always 0, then the complement of the literal appears in the product. This comes about because the literals which change appear both complenented and uncomplemented. When the function is reduced, these literals can be eliminated. Let us illustrate this with some examples. In Fig. 5-6, the only variable in the subcube that does not change is y. Thus, this subcube (and complete Karnaugh map) can be represented by y. This is verified by Eq. (5-24b).

In Fig. 5-10, there are two subcubes. The first is m_4, m_{12}, m_6 and m_{14}. The two variables that do not change are x and $\bar{z}$. (In the subcube, the value of x is always 1 and that of z is always 0.) Thus, this subcube represents $x\bar{z}$. The second subcube consists of minterms m_1, m_3, m_9 and m_{11}. Again, the variables that do not change are x and z. However, in this case, we have $x=0$ and $z=1$. Thus, the subcube can be represented by $\bar{x}z$. The complete function is given by $x\bar{z}+\bar{x}z$. This is verified by Eq. (5-37). Thus, once the subcubes are identified, the minimized function can be written by inspection.

Let us consider an example of one more minimization. Then we shall formalize some rules. In the Karnaugh map of Fig. 5-11, we have

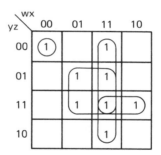

Figure 5-11 A four-variable Karnaugh map for the switching function of Eq. (5-38)

indicated four subcubes. (One has only one cell.) We shall write the switching function in terms of the subcubes. It is

$$
\begin{aligned}
f(w,x,y,z) = \ & \bar{w}\bar{x}\bar{y}\bar{z} \\
& + \bar{w}x\bar{y}z + wx\bar{y}z + \bar{w}xyz + wxyz \\
& + wx\bar{y}\bar{z} + wx\bar{y}z + wxyz + wxy\bar{z} \\
& + wxyz + w\bar{x}yz
\end{aligned}
\tag{5-38}
$$

Note that we have written the minterms corresponding to each subcube on a separate line. We have repeated terms. For instance, the minterm $wxyz$ appears three times. This is allowable since $A+A=A$, see Eq.

(4-21a). It may seem that doing this is contrary to the idea of minimization. Actually, it helps minimization, since increasing the size of the subcube results in its being represented by a simpler function. Also, consider Figs. 5-5 and 5-6. Figure 5-6 has one more minterm but it reduces to a much simpler expression. That is, Eq. (5-24b) is simpler than Eq. (5-23). It is for reasons such as these that we repeat minterms. That is, we form subcubes which are as large as possible. Of course, we do not use more subcubes than necessary.

Now consider the reduction of Eq. (5-38). Let us write

$$f(w,x,y,z) = f_1 + f_2 + f_3 + f_4 \tag{5-39}$$

where f_1, f_2, f_3, and f_4 represent the sum of minterms for the four subcubes. Thus,

$$f_1 = \bar{w}\bar{x}\bar{y}\bar{z} \tag{5-40}$$

This is a one-term subcube and cannot be reduced further.

$$f_2 = \bar{w}x\bar{y}z + wx\bar{y}z + \bar{w}xyz + wxyz \tag{5-41a}$$

Using the rule for obtaining the value of a subcube (or just reducing), Eq. (5-41a) results in

$$f_2 = xz \tag{5-41b}$$

Two variables have been eliminated. Similarly,

$$f_3 = wx\bar{y}\bar{z} + wx\bar{y}z + wxyz + wxy\bar{z} = wx \tag{5-42}$$

$$f_4 = wxyz + w\bar{x}yz = wyz \tag{5-43}$$

Note that each subcube has reduced to a single product term and that the number of terms in the subcube decreases with the size of the subcube, see Table 5-2. Again, this substantiates our observations.

The general procedure is to form subcubes which are as large as possible. Let us state some rules which will help the formulation. The basic idea of these rules is to eliminate those cells that cannot be used in the larger subcubes. The remaining cells can then be used in larger subcubes. The following steps are used:

1. Find all single 1-cells that *cannot be made part of a larger subcube.* Encircle each one. These are single-cell subcubes. For instance, in Fig. 5-11, the cell m_0 is such a subcube.

2. Next find all cells with 1's that can be made part of a two-cell, but no larger, subcube. Encircle each pair. For instance, in Fig. 5-11, the cells m_{15} and m_{11} form such a subcube.

3. Next find all cells with 1's that can be made part of a four-, but no larger, cell subcube. Encircle each group of four. For instance, in Fig. 5-11, the cells m_5, m_{13}, m_7 and m_{15} constitute one such set and m_{12}, m_{13}, m_{15} and m_{14} constitute another set. A subcube should not be picked if *all* of its cells have been covered in an earlier step.

4. Repeat this procedure using groups of eight. Then, repeat using groups of 16. If the Karnaugh map has more than four variables, keep repeating this procedure, doubling the size of the subcubes in each step.

5. When all possible subcubes are covered, stop. At this point, all 1's on the Karnaugh map should be in subcubes. Now examine the Karnaugh map to see if any subcubes can be removed and still have all the 1's covered by subcubes. If this can be done, then these "superfluous" subcubes should removed. Now the procedure is completed.

Remember that no subcube is picked if it can be completely made part of a larger subcube. However, individual cells of one subcube may be part of larger subcubes. Thus, this procedure attempts to cover all the 1-cells with the largest possible subcubes, and the smallest number of subcubes. An illustration of this procedure as applied to the Karnaugh map of Fig. 5-11 is shown in Fig. 5-12.

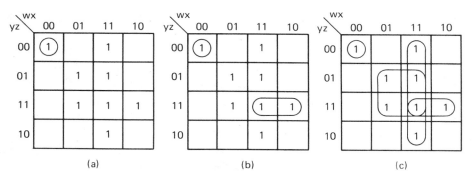

Figure 5-12 An illustration of the minimization procedure used in Fig. 5-11. (a) Karnaugh map after step 1; (b) Karnaugh map after step 2; (c) Karnaugh map after step 3.

At times, this procedure will lead to a unique minimization. That is, only one set of subcubes will be generated. However, this is not always the case. In Fig. 5-13 we illustrate an example where two different minimizations are possible. The subcubes indicated by the dashed lines are not part of the minimization. All of the subcubes, including those indicated by the dashed lines would be generated by the procedure. (The subcubes of Fig. 5-13a and b are the same when all subcubes are considered.) *However, there is one excess subcube in each figure.* In Figs. 5-13a and b, the minimization is completed without using the sub-cube indicated by dashed lines. That is, a different excess subcube is removed in each figure so that two different minimizations result.

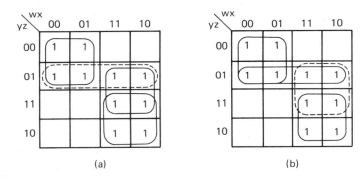

Figure 5-13   Two different minimizations of the same Karnaugh map

Note that all possible subcubes need not be used. For instance, in Fig. 5-13, four subcubes (each with four cells) were picked. However, only three subcubes were used in each of the two minimizations.

5.3 KARNAUGH MAPS WITH MORE THAN FOUR VARIABLES

The size of a Karnaugh map increases as 2^n where n is the number of literals. Thus, for large values of n, these devices become unwieldy to use. In fact, Karnaugh maps are rarely used if the number of variables exceeds seven. Let us illustrate the use of a Karnaugh map if there are six variables. Consider Fig. 5-14 which shows a map with minterm designations.

Because of the Gray coding, all of the previous discussions apply. For instance, suppose that we had a subcube consisting of m_{25}, m_{17}, m_{27} and m_{19}, see Fig. 5-14. This could be represented by the term $\bar{u}v\bar{x}z$. Note that these are the unchanging literals in the subcube. Although we have stated this without proof, the direct application of Eq. (5-16) will

xyz \ uvw	000	001	011	010	110	111	101	100
000	m_0	m_8	m_{24}	m_{16}	m_{48}	m_{56}	m_{40}	m_{32}
001	m_1	m_9	m_{25}	m_{17}	m_{49}	m_{57}	m_{41}	m_{33}
011	m_3	m_{11}	m_{27}	m_{19}	m_{51}	m_{59}	m_{43}	m_{35}
010	m_2	m_{10}	m_{26}	m_{18}	m_{50}	m_{58}	m_{42}	m_{34}
110	m_6	m_{14}	m_{30}	m_{22}	m_{54}	m_{62}	m_{46}	m_{38}
111	m_7	m_{15}	m_{31}	m_{23}	m_{55}	m_{63}	m_{47}	m_{39}
101	m_5	m_{13}	m_{29}	m_{21}	m_{53}	m_{61}	m_{45}	m_{37}
100	m_4	m_{12}	m_{28}	m_{20}	m_{52}	m_{60}	m_{44}	m_{36}

Figure 5-14 A six-variable Karnaugh map. The minterms corresponding to each subcube are illustrated.

demonstrate the validity of this statement. In a similar way, we can simplify any subcube.

The following should be noted. There are subcubes, whose cells are not physically adjacent to each other, which represent products where there is only one change of variable which is complemented in one and uncomplemented in the other. For instance, consider

$$m_{26} + m_{58} = \bar{u}vw\bar{x}y\bar{z} + uvw\bar{x}y\bar{z} \tag{5-44}$$

These cells are said to be adjacent even though they are not physically next to each other. Thus, we cannot rely on cells' being physically adjacent to define subcubes in the higher-ordered Karnaugh map. However, there is a simplification here. Observe the heavy lines drawn in the Karnaugh map. Cells which have mirror symmetry about this line will differ in only one variable. Thus, these cells will also be adjacent cells. This can be verified by studying the Gray code along the top and side of the Karnaugh map. For instance, consider $m_{26} + m_{58}$. As an additional example, consider a subcube which covers four cells. It consists of $m_{27} + m_{59} + m_{31} + m_{63}$. The same type of reflected symmetry existed in the other Karnaugh maps that we studied. For instance, Figs. 5-7 and 5-10 indicate adjacent cells that were not physically adjacent. In the case of four or fewer variables on Karnaugh maps, it was convenient to think of the edges as touching. However, the mirror symmetry of the type discussed here could also be used. If we are dealing with only five variables, then the Karnaugh map is similar to the upper half of the six-variable map.

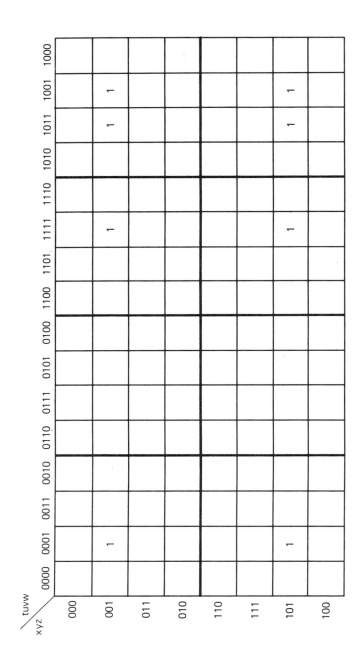

Figure 5-15 A seven-variable Karnaugh map that represents the switching function of Eq. (5-45)

A seven-variable Karnaugh map is indicated in Fig. 5-15. Note that there are now three vertical "reflecting lines" as well as one horizontal one. Again, the validity of these reflecting lines can be verified by studying the minterms corresponding to the subcubes with mirror symmetry about a reflecting line. Subcubes cover not only physically adjacent cells, but also cells which have symmetry about the reflecting lines. Two such four-cell subcubes are shown. Using the previously developed rules, we can express the function represented by this as

$$f(t,u,v,w,x,y,z) = \bar{u}\bar{v}w\bar{y}z + tvw\bar{y}z \qquad (5\text{-}45)$$

Note that a simplified function can easily be obtained once the subcubes are identified.

5.4 SOME STATEMENTS CONCERNING THE FORMALITY OF KARNAUGH MAP MANIPULATIONS

We have made numerous statements in the preceding two sections about the simplification of switching functions using Karnaugh maps. Actually, we have not formally defined what is meant by a minimum function and of course, we have not proved that the Karnaugh map procedures that we have discussed do actually yield this minimum. In addition, we have not explicitly proven all the assertions made about Karnaugh maps.

Let us consider a definition for a minimum function. Before doing this we must specify the form of the function. We shall assume that the functions will all be in a sum-of-products form. Note that these products need not be minterms (i.e., they need not have all the variables). A sum-of-products form is minimized: (1) if there is no other expression for the switching function in a sum-of-products form that has fewer products; and (2) if there is no other sum-of-products form that has the same number of products but fewer literals in its products.

Although we shall not actually consider a logical proof here to demonstrate that the Karnaugh map procedures that we have considered do yield such a minimal function, a consideration of the rules for choosing subcubes given in Sec. 5-2 should convince you that the sum of products obtained using the Karnaugh map will be minimal.

We have also made statements about the Karnaugh maps themselves. For instance, one such statement is that the value of a subcube is equal to the product of those literals of the subcube whose values do not change. This and the other statements made about Karnaugh maps can be proven in every case, for each Karnaugh map, by perfect induction. That is, all possible cases can be studied. Often, this is not necessary and the statements can be verified by studying the Gray coding along the edge and top of the Karnaugh map.

5.5 DON'T CARE CONDITIONS—INCOMPLETELY SPECIFIED FUNCTIONS

We have thus far assumed that the switching function is defined for every possible combination of the literals. However, this may not always be the case and the switching functions may only be specified for certain combinations of the literals. This is called an *incompletely specified function*.

Suppose that we want to represent the alphabet by a code of five binary digits. There are 26 letters. However, five binary digits can represent 32 different quantities. Thus, there must be seven combinations of binary digits which are never used. That is, in a properly working system, those values would never be generated. Thus, we "don't care" what the function is when such combinations occur (since they never occur). There are called *don't care conditions*.

Don't care conditions also arise if a function is not actually a function of one particular variable but, for convenience, or to conform with the other functions in the system, we list it as a function of that variable. For instance, consider Eqs. (5-27) and (5-29). Here we have

$$f(x,y,z) = \bar{x}\bar{y}z + \bar{x}\bar{y}\bar{z} + x\bar{y}z + x\bar{y}\bar{z} = \bar{y} \qquad (5\text{-}46)$$

Here we don't care about the values of x and z. We can indicate this in the following way:

$$f(x,y,z) = f(d,y,d) \qquad (5\text{-}47)$$

When we write a truth table for a function with don't care minterms, a d is used in place of a 0 or 1. For instance, consider the following truth table.

w	x	y	z	f
0	0	0	0	0
0	0	0	1	0
0	0	1	0	d
0	0	1	1	0
0	1	0	0	0
0	1	0	1	1
0	1	1	0	0
0	1	1	1	d
1	0	0	0	d
1	0	0	1	0
1	0	1	0	0
1	0	1	1	0
1	1	0	0	0
1	1	0	1	d
1	1	1	0	0
1	1	1	1	1

$$(5\text{-}48)$$

The designer can choose the ds to be either 0 or 1, whichever simplifies the switching function. This can be illustrated with the Karnaugh map of Fig. 5-16.

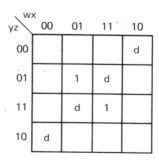

Figure 5-16 A four-variable Karnaugh map indicating a don't care condition

A don't care minterm can be assumed to be either a 0 or a 1 on the Karnaugh map. The value is chosen so that there is a minimium number of subcubes, each of which is as large as possible. For instance, suppose that we have the Karnaugh map of Fig. 5-16. If we consider that every $d=0$, then there are two, one-cell subcubes and the switching function is

$$f(w,x,y,z) = \bar{w}x\bar{y}z + wxyz \qquad (5\text{-}49)$$

However, if the ds corresponding to m_7 and m_{13} are considered to be 1's and the ds corresponding to m_2 and m_8 are considered to be 0's, then we have

$$f(w,x,y,z) = xz \qquad (5\text{-}50)$$

Note that for all cells but the "don't care" ones, both Eqs. (5-49) and (5-50) yield the same results. However, the functions of Eqs. (5-49) and (5-50) are different since they will yield different values for the don't care conditions.

5.6 KARNAUGH MAPS USING MAXTERMS

We have based the previous development on simplification of a sum of minterms. Actually we could proceed in a dual way and work with the product of maxterms and obtain the minimized form as a product of sums. In the latter case, we work with the 0's on the Karnaugh map. Now, it is convenient to mark the 0-cells and leave the 1-cells blank. Note that the Karnaugh map for a function is unique. It does not change if we work with maxterms rather than minterms.

Let us consider an example. We shall use the Karnaugh map of Fig. 5-17. This is *exactly* the same as the Karnaugh map of Fig. 5-9, except

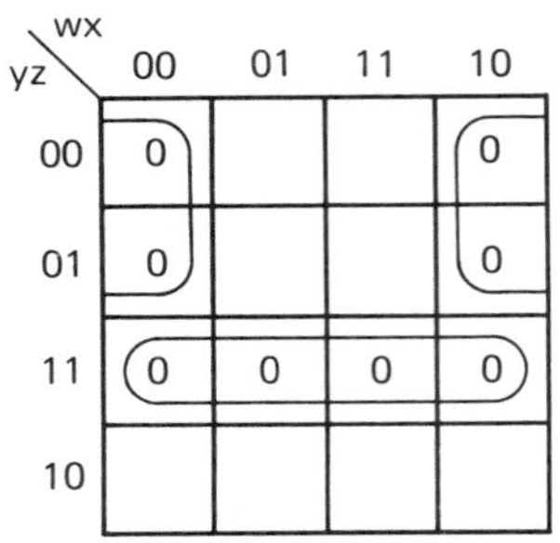

Figure 5-17 A four-variable Karnaugh map which represents the same switching function as Fig. 5-9

that here we have indicated the 0's instead of the 1's. Actually, a complete Karnaugh map could show both 0's and 1's. However, this would be too cluttered. We can easily obtain the function as a product of maxterms by pretending that the 0's are 1's and proceeding as before. This then results in the complement of the function which can then be easily obtained in minimal form as a sum of products. In other words, if we take the complement of the sum, the actual function will be realized as a product of sums. Let us illustrate this procedure with the Karnaugh map of Fig. 5-17.

$$\bar{f}(w,x,y,z) = \bar{w}\bar{x}\bar{y}\bar{z} + \bar{w}\bar{x}\bar{y}z + w\bar{x}\bar{y}\bar{z} + w\bar{x}\bar{y}z$$
$$+ \bar{w}\bar{x}yz + \bar{w}xyz + wxyz + w\bar{x}yz \tag{5-51}$$

This is an unminimized form. If we take the complement we obtain an unminimized product of maxterms.

$$f(w,x,y,z) = (w+x+y+z)(w+x+y+\bar{z})(\bar{w}+x+y+z)(\bar{w}+x+y+\bar{z})$$
$$(w+x+\bar{y}+\bar{z})(w+\bar{x}+\bar{y}+\bar{z})(\bar{w}+\bar{x}+\bar{y}+\bar{z})(\bar{w}+x+\bar{y}+\bar{z}) \tag{5-52}$$

Actually we should have minimized $\bar{f}$ before we took the complement. Remember that when we obtain $\bar{f}$, we pretend that the 0's are 1's. Thus, we form subcubes as before. These are shown in Fig. 5-17. Thus,

$$\bar{f}(w,x,y,z) = \bar{x}\bar{y} + yz \tag{5-53}$$

Now, taking the complement yields a minimized product of sums form.

$$f(w,x,y,z) = (x+y)(\bar{y}+\bar{z}) \tag{5-54}$$

After a little experience is gained with this procedure, it will not be necessary to go through the steps of taking the complement. The following procedure can be used. The subcubes are determined using the same rules except that the 0's take the place of the 1's in the previous development. Each subcube represents a sum of literals. As before, the literals which appear in each sum are those whose values do not change in the cells of the subcube. However, *the complemented literals of the sums are now represented by 1's and the uncomplemented literals are represented by 0's.* Finally, the product of all the sums (each one representing a subcube) is taken and the final minimized form is obtained.

5.7 PRIME IMPLICANTS

The Karnaugh map minimization technique is very easy to use if the number of literals is four or less. However, if there are more than seven literals, it becomes unwieldy. In addition, it does not lend itself to computer implementation. We shall discuss another procedure which is easier to use when the number of literals is large and which can be implemented on a computer. Before discussing the procedure, we must consider some fundamental definitions.

A switching function f is said to *imply* another switching function g if g is 1 whenever f is 1. Let us consider an example of this. Suppose

$$a = xy \tag{5-55a}$$

$$b = xy + \bar{x}z + yz \tag{5-55b}$$

Whenever a is 1, b will also be 1. Note that a and b are not equivalent since b may be 1 when a is 0. a is said to be an *implicant* of b.

If f is an implicant of g, then g is said to *cover* f. That is, g covers f if g is 1 whenever f is 1. For instance, see Eqs. (5-55), b covers a. (Note that b may be 1 when a is 0.)

A *prime implicant* is a special form of implicant. It must be a product of literals and, also, it must be such that if *any* literal is deleted from the product, the resulting term is no longer an implicant. For instance, consider Eqs. (5-55), a is a prime implicant of b since (1) a is an implicant of b; (2) it is a product of literals and; (3) if any literal is deleted from a, it will not longer be an implicant. Consider another example

$$a_1 = xyz \tag{5-56}$$

a_1 is an implicant of b, since whenever $a_1 = 1$, $b = 1$. However, we can delete the z from a_1 and still have an implicant of b. Thus, a_1 is not a prime implicant of b.

If f is an implicant of g, then when $f = 1$, $g = 1$. However, if $f = 0$, g may be either 0 or 1. Suppose that $f_1, f_2, \ldots, f_n$ are all prime implicants of g. If we combine the sums of enough of these implicants such as

$$A = f_1 + f_2 + \ldots + f_n \tag{5-57}$$

then we may have the situation that $A = 1$ whenever $g = 1$. If $g = 0$, then all its implicants must be 0. Thus, $A = 0$ whenever $g = 0$. Hence, we can state

$$g = A \qquad\qquad (5\text{-}58)$$

In this case, we have expressed g as a sum of its implicants. It is always possible to find such a sum. For example, a switching function can always be expressed as a sum of minterms and each minterm is an implicant. If we are interested in minimizing the function, then each implicant used in the expansion should be a prime implicant, since it does not have any unneeded literals. Simply expanding a function in terms of prime implicants does not necessarily produce a minimal function. If, however, the appropriate prime implicants are used, the function will be minimal. We shall now demonstrate this and discuss some ideas which will enable us to obtain minimal functions.

We can use Karnaugh maps to determine prime implicants and to expand the given function in terms of these implicants. We shall do this since it will provide some insight into the process. However, we shall also develop techniques that do not require Karnaugh maps. Consider the minimal functions obtained from the subcubes of the Karnaugh map. These were sums of products, each product representing a single subcube. Each product is a prime implicant. Note that when each product in the sum was 1, the function was 1, and that no variable can be omitted from a product since (we assume that) the minterms covered by each subcube have been expressed in the minimal form discussed in Sec. 5.2. For example, consider Fig. 5-18. Each subcube represents a prime implicant. These are

$\bar{w}\bar{x}\bar{z}$, wxz, xy and $y\bar{z}$

Of course, the sum of these prime implicants gives the function.

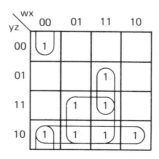

Figure 5-18 A four-variable Karnaugh map showing an optimum set of prime implicants

If a prime implicant covers *at least* one minterm which is *not covered by any other prime implicant*, it is called an *essential prime implicant*. All of the prime implicants shown in Fig. 5-18 are essential prime implicants. However, we shall not always work only with essential prime implicants. We shall illustrate this subsequently.

Now consider the Karnaugh map of Fig. 5-19. Note that four prime implicants are shown here. However, one consisting of $m_7 + m_{15} + m_6 + m_{14}$ contains only minterms which have been included in the other

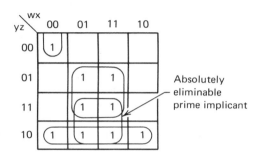

Figure 5-19 A four-variable Karnaugh map

prime implicants. Thus, this prime implicant need not be included in the sum for the function. Such a prime implicant, *all* of whose minterms are contained in essential prime implicants, is called an *absolutely eliminable prime implicant*. Now let us develop a procedure for minimizing switching functions without using Karnaugh maps.

5.8 THE QUINE—McCLUSKEY METHOD

We shall now discuss a procedure using prime implicants that will enable us to minimize switching functions without using Karnaugh maps. The procedure is called the *Quine-McCluskey method*. It is based on obtaining the switching function as a minimum sum of products. Then the following is done:

1. Determine the prime implicants.
2. Determine the essential prime implicants.

If *all* the minterms are covered by the essential prime implicants, then the function has been minimized. If all the minterms are not covered by the essential prime implicants then:

3. Add additional prime implicants to the sum of essential prime implicants, in some minimal way, until all the minterms are covered.

These rules will yield a reduced function. However, it is not obvious how to implement them when the function is complex. We shall develop a procedure which will enable us to do this. For instance, we must determine the essential prime implicants for step 2 and we must choose any additional prime implicants in step 3 in a minimal way. The Karnaugh map provides a procedure for doing this. Since we do not want to use maps, a tabular procedure which is suitable for a large number of variables, and which is also suitable for computer implementation, will be developed.

The first step is to identify all the prime implicants. We assume that we start with a canonical sum of products form (i.e., a sum of minterms). If we used a Karnaugh map, we would identify subcubes. Let us consider what is done there so that we can develop a tabular procedure. Adjacent terms on the Karnaugh map differ in only one literal and that literal is complemented in one term and uncomplemented in the other. Thus, we can consider that a two-cell subcube is formed by eliminating these literals. A four-cell subcube can be considered to result from the elimination of a literal from two adjacent two-cell subcubes. This procedure can be repeated for higher-ordered subcubes.

Let us now discuss a tabular procedure that can be used to implement the process of forming prime implicants or of finding subcubes. We start with the function as a canonical sum of minterms. For example, for the Karnaugh map of Fig. 5-18, we have

$$f(w,x,y,z) = \bar{w}\bar{x}\bar{y}\bar{z} + \bar{w}\bar{x}y\bar{z} + wx\bar{y}z + wxyz$$
$$+ \bar{w}xyz + \bar{w}xy\bar{z} + wxy\bar{z} + w\bar{x}y\bar{z} \tag{5-59}$$

We want to determine, in some systematic way, if the minterms are adjacent to each other. We start by writing each minterm using the following binary representation. Replace each uncomplemented variable with a 1 and each complemented variable with a 0. Thus, we have

$\bar{w}\bar{x}\bar{y}\bar{z} \rightarrow 0000$

$\bar{w}x\bar{y}z \rightarrow 0101$

$\bar{w}xyz \rightarrow 0111$

etc.

The variables must always be listed in the same order. Note that this code is essentially that used in the Karnaugh map. We now form a table containing the binary representation of all the minterms. The table is arranged in groups. The first group contains those minterms with no 1's. The second group contains those minterms with *exactly* one 1; the third group contains those minterms with *exactly* two 1's, etc. Thus, for Eq. (5-59), that table is

Table 5-3 Coded Representation of the Minterms of Eq. (5-59)

Minterm	Binary Representation	Number of 1's
$\bar{w}\bar{x}\bar{y}\bar{z}$	0000	0
$\bar{w}\bar{x}y\bar{z}$	0010	1
$\bar{w}xy\bar{z}$	0110	2
$w\bar{x}y\bar{z}$	1010	2
$\bar{w}xyz$	0111	3
$wx\bar{y}z$	1101	3
$wxy\bar{z}$	1110	3
$wxyz$	1111	4

In Table 5-3 we have written the minterms. This is usually not done; it was just done for illustrative purposes here.

Now suppose that we want to combine two minterms. They must be identical except for one variable, which must be complemented in one and uncomplemented in the other. Thus, if we are to combine two minterms, their coded representations must be identical except for one column. There must be a 0 in one and a 1 in the other. Thus, if two minterms are to be combined into a two-cell subcube, then the number of 1's in their binary representation must differ by exactly unity. We have grouped the minterms in order of· the number of 1's. Thus, starting with the first group (no 1's), if we want to see if a minterm can combine with another, we need only compare it with each minterm in the *next* group. Let us consider this and rewrite Table 5-3 to illustrate the procedure.

Table 5-4 Prime Implicants for Eq. (5-59)

$w\,x\;y\;z$	First Reduction $w\,x\;y\;z$	Second Reduction $w\,x\;y\;z$
0 0 0 0 $\sqrt{}$	0 0 – 0	
0 0 1 0 $\sqrt{}$	0 – 1 0 $\sqrt{}$	– – 1 0
0 1 1 0 $\sqrt{}$	– 0 1 0 $\sqrt{}$	– 1 1 –
1 0 1 0 $\sqrt{}$	0 1 1 – $\sqrt{}$	
0 1 1 1 $\sqrt{}$	– 1 1 0 $\sqrt{}$	
1 1 0 1 $\sqrt{}$	1 – 1 0 $\sqrt{}$	
1 1 1 0 $\sqrt{}$	– 1 1 1 $\sqrt{}$	
1 1 1 1 $\sqrt{}$	1 1 – 1	
	1 1 1 – $\sqrt{}$	

Consider 0000 and 0010. They differ only in the third column. Thus, their sum can be represented by 00–0 which is written in the second column. That is, $\bar{w}\bar{x}\bar{y}\bar{z} + \bar{w}\bar{x}y\bar{z} = \bar{w}\bar{x}\bar{z}$. The dash indicates which variable is missing. It is important to keep track of this. For instance, 0–00 → $\bar{w}\bar{y}\bar{z}$ is different from 00–0 → $\bar{w}\bar{x}\bar{z}$.

Let us consider the table. The groups are arranged so that the number of 1's in a group increases as you proceed down the table so that each minterm in any group has one less 1 than each minterm in the group below it. If we are to combine two minterms to form a two-cell subcube, then those minterms must be in adjacent groups in the table. In addition, all the 1's from the binary representation of the minterm in the upper group must exactly coincide with the 1's in the binary represntation of the minterms in the adjacent lower group. Of course, the binary representation of the minterms in the lower group will have one additional 1. This represents the variable that is eliminated when the two minterms are combined.

When two rows are combined we put check marks on them. This means that the minterms have been combined. Thus, they are not prime implicants and are not needed in the expression. Note that even if a row has been checked off (because of combination with a term from a group below it) it is still compared with the row below it. This is because a minterm can be in more than one subcube. Remember that when two terms are combined, the resulting reduced term is written in the next column.

After all the eliminations have been completed, the column entitled "first reduction" is obtained. It is possible that there can be further reduction of those terms. (Note that Fig. 5-18 shows this because there are subcubes covering four cells and the first reduction only combines pairs of cells. Thus, we would now expect further reduction involving these pairs.)

Now we repeat the procedure using the second column. Note that, because of the ordering of the groups, when a combination occurs, the number of 1's in the reduced element will be the same as the number of 1's that were in the original upper element. Thus, the second column will also be ordered in accordance with the number of 1's in the binary representation.

If two variables are to combine, they must differ in only one column and this difference must be a 0 in the upper row and a 1 in the lower row. Note that this implies that any dashes (–) must coincide. Actually, the dashes are easy to check visually and their presence speeds the process. For instance, 0–10 + 1–10 equals ––10 or, equivalently, $\bar{w}y\bar{z} + wy\bar{z} = y\bar{z}$.

Examine the two terms in the second column. Each has been obtained from two different reductions. For instance, 0–10 + 1–10 = ––10 and also, –010 + –110 = ––10.

After the second reduction is performed, the process is repeated, if possible. In this example, we must stop after the second reduction.

When the process is completed, all the unchecked terms are prime implicants and can be summed to obtain the function. These terms are 00–0, 11–1, ––10, and –11–. Thus, we have

$$f(w,x,y,z) = \bar{w}\bar{x}\bar{z} + wxz + y\bar{z} + xy \tag{5-60}$$

The sum of prime implicants that we obtain from the tabular minimization just discussed may not be in minimum form. The tabular minimization makes use of Eq. (4-20b) to eliminate literals and combine terms. However, too many terms may be generated by this procedure. That is, some prime implicants might be eliminated from the sum of prime implicants generated by this procedure. This is analogous to the case illustrated in Fig. 5-13 where subcubes could be eliminated from the Karnaugh map. In this case we do not have a map so that it is not obvious that there are superfluous prime implicants that must be eliminated. We shall again use a tabular type of procedure in order to be systematic. Now a two-dimensional chart listing the minterms along one axis and the prime implicants along the other shall be used. In this way we can see which prime implicants can be eliminated. Let us illustrate this by continuing the example. In Fig. 5-20 we have drawn the prime implicant choice chart. The minterms which are

Prime implicants \ Minterms	0000	0010	0110	1010	0111	1101	1110	1111
√ 00−0	√	√						
√ 11−1						√		√
√ −−10		√	√	√			√	
√ −11−			√		√		√	√

Figure 5-20 A chart used to determine the optimum set of prime implicants from Table 5-4

covered by a prime implicant can be determined by considering all possible values of the blank. For instance, if we have the prime implicant 00–0, then it covers the minterms 0010 and 0000. That is, $\bar{w}\bar{x}\bar{z} = \bar{w}\bar{x}yz + \bar{w}\bar{x}\bar{y}\bar{z}$. In a similar way, −−10 covers the minterms 0010 and 1010 and 0110 and 1110. That is

$$y\bar{z} = \bar{w}\bar{x}y\bar{z} + w\bar{x}y\bar{z} + \bar{w}xy\bar{z} + wxy\bar{z}$$

We put check marks on the table to indicate which minterms are covered by the corresponding prime implicants.

Next we determine the essential prime implicants, and put a check mark next to them. This is done in the following way: Examine each column. If a column has only one check then that minterm is covered by only one prime implicant. Thus, this must be an essential prime implicant.

After all the essential prime implicants are determined, we examine the columns to determine if there are any minterms that are not covered by the essential prime implicants. If there are such minterms, then these must be covered by some, or all, of the remaining prime implicants. This may involve a trial-and-error procedure to determine a minimal realization. Let us consider some ideas which can be used to help in this choice.

If *all* of one prime implicant's checked columns also have checks from the essential prime implicants, then there is no reason to ever use that prime implicant. In fact, it is an absolutely eliminable prime implicant. In Fig. 5-20 there are no such prime implicants.

The number of dashes in the prime implicant representation is analogous to the size of a subcube. If there are no dashes, then that prime implicant will cover one minterm. If there is one dash, then that prime implicant will cover two minterms. If there are two dashes, then that prime implicant will cover four minterms. In general, we can state that if there are k dashes and n variables, then the prime implicant will cover 2^k minterms and $n-k$ literals will be required to express the prime implicants. Hence, the most desirable prime implicants are those with the most dashes.

The above ideas can aid in the selection of the prime implicants needed to cover the minterms which remain after the essential prime implicants have been chosen.

In the examples we considered, the essential prime implicants cover all the minterms. Indeed, all the prime implicants were essential. Thus, using the checked prime implicants in Fig. 5-20, we obtain

$$f(w,x,y,z) = \bar{w}\bar{x}\bar{z} + wxz + y\bar{z} + xy \qquad (5\text{-}61)$$

In this case we did not reduce the function using the chart.

More Complex Switching Functions

The previous examples considered minimization where the minimization chart was simple in that all the prime implicants were essential ones. However, there are cases where the chart may require study, even after the essential prime implicants are identified. Let us consider such an example.

Suppose that we have the function

$$f(w,x,y,z) = \bar{w}\bar{x}\bar{y}\bar{z} + \bar{w}x\bar{y}\bar{z} + w\bar{x}\bar{y}\bar{z} + \bar{w}xy\bar{z} + w\bar{x}\bar{y}z + wx\bar{y}\bar{z}$$
$$+ \bar{w}xyz + w\bar{x}yz + wxy\bar{z} + wxyz \qquad (5\text{-}62)$$

Its prime implicant determining table is determined as before. Thus, we have

Table 5-5 Prime Implicants for Eq. (5-62)

wx y z	First Reduction wx y z	Second Reduction wx y z
0 0 0 0 √	0 – 0 0 √	– – 0 0
0 1 0 0 √	– 0 0 0 √	– 1 – 0
1 0 0 0 √	0 1 – 0 √	– 1 1 –
0 1 1 0 √	– 1 0 0 √	
1 0 0 1 √	1 0 0 –	
1 1 0 0 √	1 – 0 0 √	
0 1 1 1 √	0 1 1 – √	
1 0 1 1 √	– 1 1 0 √	
1 1 1 0 √	1 0 – 1	
1 1 1 1 √	1 1 – 0 √	
	– 1 1 1 √	
	1 – 1 1	
	1 1 1 – √	

Now form the prime implicant chart. This is indicated in Fig. 5-21. In this more complex case, we proceed in the following way. As before, we determine the essential prime implicants and check them. Next consider the minterms covered by each essential prime implicant and

Prime implicants \ Minterms	0000 √	0100 √	1000 √	0110 √	1001	1100 √	0111 √	1011	1110 √	1111 √
100 –			√		√					
10 – 1					√			√		
1 – 11								√		√
√ – – 00	√	√	√			√				
X – 1 – 0		√		√		√			√	
√ – 11 –				√			√		√	√

Figure 5-21 A chart used in the first step to obtain the optimum prime implicants from Table 5-6

check them. Place these checks along the top of the chart. Now examine the unchecked prime implicants. If any of them cover only checked minterms, they are absolutely eliminable prime implicants. Place a cross next to them. The prime implicant –1–0 is absolutely eliminable.

The remaining prime implicants which are neither checked nor crossed must cover the unchecked minterms in an optimum way. It is often not convenient to use the original chart to determine this since it is very cluttered. Thus, we redraw the chart using only the unchecked minterms and those prime implicants which are neither checked nor crossed. This is shown in Fig. 5-22. From this we see that the two minterms will be covered either by 100– AND 1–11 or by 10–1. The

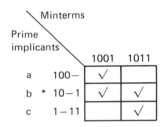

Figure 5-22   Reduced chart used in the second step to obtain the prime implicants from Table 5-6

second set 10–1 is the minimum one. Thus, we place an asterisk (*) next to it. Then, the reduced function consists of the sum of the checked minterms of Fig. 5-21 and the minterm(s) marked with an asterisk in Fig. 5-22. Hence,

$$f(w,x,y,z) = \bar{y}\bar{z} + xy + w\bar{x}z \qquad (5\text{-}63)$$

In examining Fig. 5-22, we obtained two sets of prime implicants which could be used to cover the minterms. If the chart is large, it may be tedious to identify such sets. The following procedure can be used. Assume that each prime implicant is a variable. These are labeled a, b and c in Fig. 5-22. Now consider each column. The term corresponding to the first column will be 1 if $a+b$ is 1. Similarly, the term corresponding to the second column will be 1 if $b+c$ is 1. Then, we can write a function F which will be 1 if both columns are 1. The function is

$$F = (a+b)(b+c) \qquad (5\text{-}64)$$

This represents a function of the prime implicants which cover the minterms. To form the function, we examine each column and form a sum consisting of those variables which have a check in that column. Next we take the product of all such sums. Thus, the minterms are all covered. Now let us simplify the resulting expression. Using Eq. (4-18b) we obtain

$$F = b + ac \qquad (5\text{-}65)$$

(Note that any of the minimization procedures that we have discussed could be used here.) Each of the terms in the sum correspond to a set of prime implicants which will cover the desired minterms. Thus, the sets of prime implicants correspond to b OR ac. That is

10–1

OR

100– AND 1–11

Of course, these are the same as the previously determined sets and the minimal expression is the one previously determined. Note that only one term in the sum need be used to cover the minterms.

There are other reduction techniques that are, at times, less tedious. However, the one discussed here will always find all possible coverings.

5.9 DON'T CARE CONDITIONS WITH THE QUINE–McCLUSKEY METHOD

If there are don't care conditions, they can be used to reduce the switching function. This was done in Sec. 5.5 using Karnaugh maps. To see how such don't care conditions are handled when the Quine-McCluskey method is used, we shall first reconsider the Karnaugh map procedure. We shall use the Karnaugh map of Fig. 5-23 in this discussion. We start by assuming that *all* don't care conditions are 1's. This allows us to form the largest possible subcubes. For instance, for the Karnaugh map of Fig. 5-23 we have

$$F_1 = w\bar{x}\bar{y}\bar{z} + xz + \bar{w}\bar{x}y\bar{z} \qquad (5\text{-}66)$$

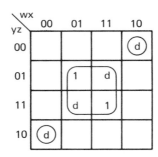

Figure 5-23 A Karnaugh map with don't care conditions

Since all ds are assumed to be 1's, it may now be possible to discard some terms in the sum that do not affect the values of the minterms that are designated to be 1. That is, we can discard the subcubes that *only cover don't care conditions*. Doing this often leads to reduced expressions.

$$F = xz \tag{5-67}$$

Note that (5-66) and (5-67) are different functions but since they only differ in the value of the don't care minterms, either one can be used. These ideas are central to the Quine-McCluskey procedure. The don't care conditions are treated as 1's when the reduction table is formed. This is equivalent to treating the don't care conditions as 1's in the Karnaugh map and it can result in additional combination of rows when the reduction table is generated. However, when the second chart, which is used to determine which prime implicants to include in the sum, is constructed, the *don't care minterms are omitted*. This is equivalent to removing the unnecessary terms from Eq. (5-66) to obtain (5-67).

As an example, minimize the function whose Karnaugh map is given in Fig. 5-24. (Note that this is the same as Fig. 5-18 except that a don't

yz \ wx	00	01	11	10
00	1		d	
01			1	
11		1	1	
10	1	1	1	1

Figure 5-24 A Karnaugh map with a don't care condition

care condition has been added.) The prime implicant determining table is

Table 5-6 Prime Implicants for Fig. 5-24

$wx\ y\ z$	*First Reduction* $wx\ y\ z$	*Second Reduction* $wx\ y\ z$
0 0 0 0 ✓	0 0 – 0	
0 0 1 0 ✓	0 – 1 0 ✓	– – 1 0
1 1 0 0 ✓	– 0 1 0 ✓	————
0 1 1 0 ✓	1 1 0 – ✓	1 1 – –
1 0 1 0 ✓	1 1 – 0 ✓	– 1 1 –
0 1 1 1 ✓	0 1 1 – ✓	
1 1 0 1 ✓	– 1 1 0 ✓	
1 1 1 0 ✓	1 – 1 0 ✓	
1 1 1 1 ✓	– 1 1 1 ✓	
	1 1 – 1 ✓	
	1 1 1 – ✓	

If we compare this with Table 5-4 (for the Karnaugh map of Fig. 5-18), we see that one of the prime implicants has become simpler. That is, 11–1 has been replaced by 11––.

The prime implicant determining chart is shown in Fig. 5-25. Note that the (don't care) 1100 minterm is *not* included. Again we have put checks next to the essential prime implicants. All minterms have been

Prime implicants \ Minterms	0000	0010	0110	1010	0111	1101	1110	1111
✓ 00–0	✓	✓						
✓ ––10		✓	✓	✓			✓	
✓ 11––						✓	✓	✓
✓ –11–			✓		✓		✓	✓

Figure 5-25 A chart used to obtain the set of optimum prime implicants from Table 5-6

covered by the essential prime implicants, which are the only prime implicants. Thus, the minimized expression is

$$f(wxyz) = \bar{w}\bar{x}\bar{z} + y\bar{z} + wx + xy \tag{5-68}$$

Thus, we have taken advantage of the don't care conditions to reduce the expression somewhat.

Note that the third row of Table 5-5 has only three checks even though the prime implicant has two dashes. This is because one of the checks covers a don't care condition which is omitted from the table.

5.10 MULTIPLE OUTPUT NETWORKS

Thus far we have considered the minimization of a single switching function. However, there are occasions where we have more than one function of the same variables. For instance, consider the full adder of Fig. 3-20. In that case the inputs were three variables a_j, b_j, and c_{j-1}. There were two outputs s_j and c_j. The switching functions for these were given by Eqs. (3-28) and (3-29) where

$$s_j(a_j,b_j,c_{j-1}) = \bar{a}_j\bar{b}_jc_{j-1} + \bar{a}_jb_j\bar{c}_{j-1} + a_j\bar{b}_j\bar{c}_{j-1} + a_jb_jc_{j-1} \tag{5-69a}$$

$$c_j(a_j,b_j,c_{j-1}) = \bar{a}_jb_jc_{j-1} + a_j\bar{b}_jc_{j-1} + a_jb_j\bar{c}_{j-1} + a_jb_jc_{j-1} \tag{5-69b}$$

The most straightforward procedure for minimizing these functions would be to consider each one individually. However, this often will not yield the simplest network. For instance, even in the unminimized form, we see that both functions have a minterm in common $(a_jb_jc_{j-1})$. Thus, one gate could realize this minterm for both functions. This saving of a gate by a sharing between two functions is illustrated in Fig. 3-20.

One procedure that can be used is to reduce each individual function separately. Then, the reduced functions are examined to see if they have any prime implicants in common. Only one gate should be used to realize each set of common prime implicants. Thus, a saving results. This is not a general procedure, as we shall subsequently discuss. However, it often yields satisfactory results. Let us illustrate it.

Consider the two functions whose Karnaugh maps are given in Fig. 5-26. If each is minimized separately, we obtain

$$f_1(w,x,y,z) = xyz + wyz \tag{5-70a}$$

and

$$f_2(w,x,y,z) = \bar{w}xz + xyz \qquad\qquad (5\text{-}70\text{b})$$

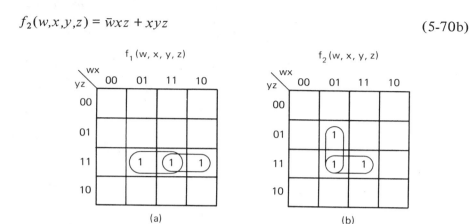

Figure 5-26 Karnaugh maps for two functions of the same variables: (a) $f_1(w, x, y, z)$; (b) $f_2(w, x, y, z)$

These can share an AND gate, the one whose output is xyz. This can be seen easily since both maps have a common subcube. The gate circuit realization is illustrated in Fig. 5-27. The drawing of these circuits has been simplified. We assume that the complement is available. Also, we have not drawn common bus bars for the independent variables but have just labeled the gate inputs. It is assumed that all the x inputs are connected together, etc. This reduces the clutter of the diagrams and makes them easier to read. As discussed, Fig. 5-27b has one less gate than does Fig. 5-27a.

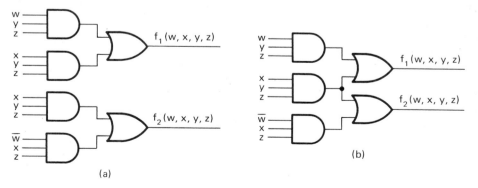

Figure 5-27 (a) A realization of Eqs. (5-70); (b) another realization that uses one less gate

Often, when we have multiple output functions, we can obtain reductions other than these simple ones. Let us illustrate this and then determine a procedure for implementing this reduction. Consider the Karnaugh maps of Fig. 5-28. The values of these functions are

$$f_1(w,x,y,z) = xyz + wyz \qquad (5\text{-}71a)$$

and

$$f_2(w,x,y,z) = \bar{w}xz + x\bar{y}z \qquad (5\text{-}71b)$$

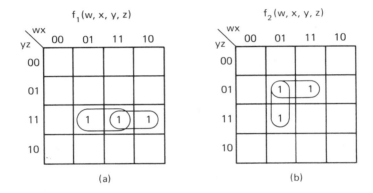

(a) (b)

Figure 5-28 Karnaugh maps of two functions of the same variables: (a) $f_1(w, x, y, z)$; (b) $f_2(w, x, y, z)$

These have no prime implicants in common. However, a reduction can still be made. We can write

$$f_1(w,x,y,z) = \bar{w}xyz + wyz \qquad (5\text{-}72a)$$

Note that, see Eq. (4-24a)

$$\bar{w}xyz + wyz = yz(\bar{w}x + w) = yz(w + x) = xyz + wyz$$

We can also write

$$f_2(w,x,y,z) = \bar{w}xyz + x\bar{y}z \qquad (5\text{-}72b)$$

Note that

$$\bar{w}xyz + x\bar{y}z = xz(\bar{w}y + \bar{y}) = xz(\bar{w} + \bar{y}) = \bar{w}xz + x\bar{y}z$$

Now, f_1 and f_2 have the minterm $\bar{w}xyz$ in common. Thus, both functions can be realized using three AND gates instead of four. Note that $\bar{w}xyz$ is an implicant of both f_1 and f_2. However, it is a prime implicant of f_1f_2.

If the switching function is complex, it may not be easy to identify modifications such as that which resulted in Eqs. (5-72). The following procedure will enable us to systematically reduce the functions.

Suppose that we want to realize several functions of the same variables: $f_1(w,x,y,z, \ldots)$, $f_2(w,x,y,z, \ldots)$, and $f_3(w,x,y,z, \ldots)$. Now consider their product $f_1f_2f_3$. If b is a prime implicant of the product, then it will also be an implicant of each of f_1, f_2, and f_3, since when $b=1$, $f_1=1$, $f_2=1$, and $f_3=1$. Thus, when we minimize, we need consider not only the prime implicants of f_1, f_2 and f_3, but we can also consider the prime implicants of f_1f_2, f_1f_3, f_2f_3, and of $f_1f_2f_3$. Note that if we can find prime implicants of the product of two or more functions, then these will be implicants of each of those functions. We can realize these with common gates which will probably result in a saving of gates. Thus, a useful minimization procedure would be to consider the functions and all combinations of their products when obtaining the prime implicants. A chart-type of simplification can then be used to obtain an optimum network.

A word of caution is in order here. When we *combine minterms*, we must keep track of the *functions from which they originated*. In general, we should keep the minterms for f_1f_2, f_1f_3, f_2f_3 and $f_1f_2f_3$ separate and obtain the prime implicants for each separately. Then, when f_1 is realized, the prime implicants from f_1, f_1f_2, f_1f_3, and $f_1f_2f_3$ can be used in its realization. *However, the others* (i.e., those from f_2, f_3, and f_2f_3) *cannot be used* in realizing f_1. A simple example will illustrate this. Suppose that $f_1=x$ and $f_2=\bar{x}$. $\bar{x}$ is a prime implicant of f_2. However, it cannot be used in realizing f_1.

Let us now consider an example. Suppose that

$$f_1(wxyz) = m_0 + m_1 + m_5 + m_9 \tag{5-73a}$$

$$f_2(wxyz) = m_0 + m_1 + m_7 + m_{10} \tag{5-73b}$$

$$f_3(wxyz) = m_0 + m_1 + m_5 + m_7 \tag{5-73c}$$

where we have used the notation of Fig. 5-8. Then we form the prime implicant-determining table.

Table 5-7 Prime Implicants for Eqs. (5-73)

f_1	0 0 0 0 √	0 0 0 - ✕
	0 0 0 1 √	0 - 0 1 ✕
	0 1 0 1 √	- 0 0 1
	1 0 0 1 √	
f_2	0 0 0 0 √	0 0 0 - ✕
	0 0 0 1 √	
	1 0 1 0	
	0 1 1 1 ✕	
f_3	0 0 0 0 √	0 0 0 - ✕
	0 0 0 1 √	0 - 0 1 ✕
	0 1 0 1 √	0 1 - 1
	0 1 1 1 √	
$f_1 f_2$	0 0 0 0 √	0 0 0 - ✕
	0 0 0 1 √	
$f_2 f_3$	0 0 0 0 √	0 0 0 - ✕
	0 0 0 1 √	
	0 1 1 1	
$f_1 f_3$	0 0 0 0 √	0 0 0 - ✕
	0 0 0 1 √	0 - 0 1
	0 1 0 1 √	
$f_1 f_2 f_3$	0 0 0 0 √	0 0 0 -
	0 0 0 1 √	

(Note that we can find the canonical sum of minterms for the product of two functions simply by determining the minterms which they have in common.)

We form the table for f_1, f_2, f_3, $f_1 f_2$, $f_2 f_3$, $f_1 f_3$, and $f_1 f_2 f_3$ as though they were independent functions. That is, we form the prime implicants of each without regard to the others. Now examine the table to see if there is any duplication of prime implicants. For instance, $f_1 f_3$ has a prime implicant 0–01. This can be used in the realizations of both f_1 and f_3. Both f_1 and f_3, individually, have 0–01 as a prime implicant.

It is desirable to use the f_1f_3 prime implicant since this results in a saving of gates. Thus, we place $\times$ marks next to the 0–01's in f_1 and f_3. This means that we shall not use them. Instead, we shall use the 0–01 from f_1f_3. Similarly, we can eliminate the 000– from each of f_1, f_2 and f_3, since 000– appears in f_1f_2, f_2f_3, and f_1f_3. Furthermore, 000– can be eliminated from these since it also appears in $f_1f_2f_3$. In general, when there are repeated prime implicants, we keep the one in the term which has the largest number of "factors."

Now we form the chart used to obtain the optimum prime implicants, see Fig. 5-29. Note that only the appropriate boxes are checked. For instance, the prime implicants of f_1 only cover f_1, the prime implicants of f_2f_3 only cover f_2 and f_3, etc. Next we check off the essential prime implicants. Then, the covered minterms are checked. Now we look for

Prime implicants	f_1				f_2				f_3			
	√ 0000	√ 0001	√ 0101	√ 1001	√ 0000	√ 0001	√ 1010	√ 0111	√ 0000	√ 0001	√ 0101	√ 0111
√ f_1 { −001		√		√								
√ f_2 { 1010							√					
× f_3 { 01−1											√	√
√ f_2f_3 { 0111								√				√
√ f_3f_1 { 0−01		√	√							√	√	
√ $f_1f_2f_3$ { 000−	√	√			√	√			√	√		

Figure 5-29 A chart used to obtain the optimum prime implicants from Table 5-7

absolutely eliminable prime implicants. In this example, 01–1 for f_3 is an absolutely eliminable prime implicant. Since all the remaining prime implicants are essential prime implicants, there is no need to draw a reduced chart. To obtain each function we add all the prime implicants used in covering that function. Hence,

$$f_1(w,x,y,z) = \bar{w}\bar{x}\bar{y} + \bar{w}\bar{y}z + \bar{x}\bar{y}z \tag{5-74a}$$

$$f_2(w,x,y,z) = \bar{w}\bar{x}\bar{y} + \bar{w}xyz + w\bar{x}y\bar{z} \tag{5-74b}$$

$$f_3(w,x,y,z) = \bar{w}\bar{x}\bar{y} + \bar{w}\bar{y}z + \bar{w}xyz \tag{5-74c}$$

A realization of this is shown in Fig. 5-30.

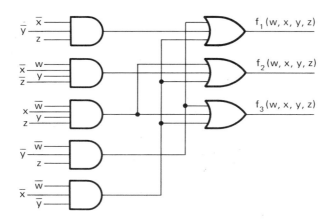

Figure 5-30   A gate circuit realization of Eqs. (5-74)

When tables such as Table 5-7 are formed, there is much duplication of minterms. Condensed forms of the tables can be used where a tag is added to the minterms indicating to which function(s) they belong. Then, minterms do not have to be repeated. Such condensed forms are given in the McCluskey reference cited in the bibliography at the end of this chapter. This method is not shown here since it is felt that the noncondensed form of table is simpler to learn. The ideas are the same in both cases.

5.11 USE OF DECIMAL NOTATION IN PRIME IMPLICANT TABLES

When we formed the tables to determine the prime implicants, a binary code was used to represent the minterms. It is more compact to use the decimal (or octal) representation of these numbers in the tables. For instance,

$$w\bar{x}yz \rightarrow 1011 \rightarrow 11_{10}$$

Let us see how to work with decimal representations in the tables. When we combine the minterms, the upper minterm is combined with the one below it which has exactly one more 1. All the 1's of the upper minterm exactly coincide with 1's of the lower minterm. Thus, on a *decimal* basis, two minterms that combine will be in adjacent groups and the lower minterm will be exactly

$$2^n \quad (n = 0, 1, 2, \ldots) \tag{5-75}$$

more than the lower one.

Let us illustrate this. As an example, we shall work with Table 5-4.

Table 5-8 A Modification of Table 5-4 Using Decimal Numbers

$wx\ y\ z$		*First Reduction*		*Second Reduction*
0 0 0 0	0 ✓	0, 2 (2)		
0 0 1 0	2 ✓	2, 6 (4)	✓	2, 6, 10, 14 (4, 8)
0 1 1 0	6 ✓	2, 10 (8)	✓	
1 0 1 0	10 ✓	6, 7 (1)	✓	6, 7, 14, 15 (1, 8)
0 1 1 1	7 ✓	6, 14 (8)	✓	
1 1 0 1	13 ✓	10, 14 (4)	✓	
1 1 1 0	14 ✓	13, 15 (2)		
1 1 1 1	15 ✓	7, 15 (8)	✓	
		14, 15 (1)	✓	

Consider the first column. We have shown both the decimal and the binary number there. In the first group, the difference between 0 and 2 is 2^1. Thus, in the next column we write 0, 2 to represent the combination of the minterms. We also write (2). The difference between the two decimal numbers indicates the column where the 0 and 1 differ in the two minterms. Thus, it supplies the same information as did the dash in Table 5-4.

We proceed similarly; for instance, 2 and 6 combine with a difference of 4. As usual, we check the combined terms. Again, lines are drawn to separate the new groups. Remember that, when a new group is formed, the number of 1's is the same as the number of 1's in the original upper group.

Now consider the second reduction. If terms in adjacent groups are to combine, then the dashes (i.e., the number in parentheses) must be the same. Hence, 0 2 (2) cannot combine with any number in the next group. However, 2, 6 (4) can possibly combine with 10, 14 (4) in the next group since both have (4). We must now check the differences between sets of corresponding numbers in each group. They must be the same and the lower group must be 2^n more than the upper group. Let us clarify this. If a term in a group is combined with a term in the group below it, the 1's of the binary representation in the upper must coincide with the 1's in the binary representation of the lower group and there will be an additional 1 in each lower group term. The dashes (numbers in parentheses) in the upper and lower groups must coincide. In each group, the dashes originally represented either 0 or 1. The smallest number in each group represents the case where the dashes originally represented 0. The larger number represents the case where the dash represented a 1. Thus, if we are to combine terms, the difference between the smallest number in the upper group and the smallest number in the lower group must exactly equal the difference between the largest number in the upper group and the largest number in the lower group. In addition, this difference bust be 2^n where n is an integer. For instance, in the case of 2, 6, (4) and 10, 14 (4), each has (4) as the number in parentheses and the pairs of differences are $10-2 = 8$ and $14-6 = 8$. These are equal and an integral power of 2. Hence, the combination of terms can be made. The resulting combined term is written in the second reduction column as

2, 6, 10, 14 (4, 8)

Note that 2, 10 (8) and 6, 14 (8) combine as 2, 6, 10, 14 (8, 4) which is the same as 2, 6, 10, 14 (4, 8). Also note that (4, 8) is equivalent to (8, 4) since they indicate dashes in the same positions. We can proceed in this way if further reductions were possible. Note that if there were a third reduction, there would have to be the same number within the parentheses and the dfference between the four pairs of numbers would have to be the same and equal to 2^n.

The unchecked terms in Table 5-8 represent the prime implicants. They are, in decimal code:

0, 2 (2)

13, 15 (2)

2, 6, 10, 14 (4, 8)

6, 7, 14, 15 (1, 8)

$$(5\text{-}76)$$

Now let us express these in terms of literals. 0, 2 (2) represnts either 0 or 2 expressed in binary with the "2-column" missing. That is, 00–0. Note that $0_{10} = 0000_2$ and $2_{10} = 0010_2$. Thus, this represents $\bar{w}\bar{x}\bar{z}$.

Similarly, 2, 6, 10, 14 (4, 8) can be obtained from the binary representation of any of 2, 6, 10, 14 with the 4 and 8 columns deleted.

$$2_{10} = 0010_2$$

$$6_{10} = 0110_2$$

$$10_{10} = 1010_2$$

$$14_{10} = 1110_2$$

Deleting the 4 and 8 columns, we have $--10 \rightarrow y\bar{z}$. Proceeding in this way, we obtain

$$f = \bar{w}\bar{x}\bar{z} + wxz + y\bar{z} + xy \tag{5-77}$$

The decimal notation is a more compact way of representating the minterm-determining tables. Note that the binary representation is more straightforward.

5.12 USE OF NAND OR NOR GATES IN THE REALIZATION OF SWITCHING FUNCTIONS

We have discussed minimization procedures that lead to switching functions which could be directly realized by an interconnection of AND, OR, and NOT gates. However, the most desirable gates to use from a circuits standpoint are often NAND or NOR gates. (We shall discuss this circuitry of gates in Appendix A.) Moreover, it is often desirable to realize the switching functions using only NAND gates or only NOR gates. In this section we shall see how this can be done using switching functions that have been minimized using the procedures discussed in this chapter.

Realization Using NAND Gates

In Sec. 3-3 we demonstrated that the NAND gate was a functionally complete set. Thus, we could realize the AND, OR and NOT functions using only NAND gates as shown in Fig. 3-18. These representations could simply replace all the AND, OR and NOT gates in any realization.

However, this would result in the use of an excessive number of gates.
We shall now discuss a procedure which will *not*, in general, result in an
increase in the number of gates used. Thus, we take advantage of the
previously discussed minimization techniques.

We shall start by obtaining an equivalent representation for the
NAND gate. From Eq. (4-25b), we have

$$\overline{xy} = \bar{x} + \bar{y} \tag{5-78}$$

The left side of this equation is the NAND operation. The right side
represents the OR operation applied to the complements of the variables.
These two equivalent forms of the NAND operation is shown in Fig. 5-
5-31. We shall call the representation of Fig. 5-31b the NOT–OR form,
and shall call Fig. 5-31a the conventional form. (Note that the circle
indicates that the complement is taken.)

Figure 5-31 Two representations of an NAND gate. (a) conventional form;
(b) NOT–OR form

We shall now show how AND, OR and NOT gates can be replaced by
NAND gates without changing the switching functions or increasing the
number of gates. A sum of products realization shall be used here.
Consider Fig. 5-32a which is the AND, OR, NOT realization of the sum
of products

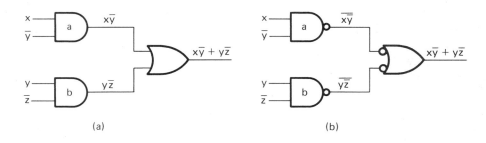

Figure 5-32 Realizations of Eq. (5-79) (a) using sum of products form (AND
and OR gates); (b) using NAND gates

$$f(x,y,z) = x\bar{y} + y\bar{z} \tag{5-79}$$

We shall demonstrate that Fig. 5-32b, where we have simply replaced all gates by NAND gates, realizes the same function. The gates which replace the AND gates have been drawn in the conventional form while the NAND gate which replaced the OR gate has been drawn in the NOT–OR form.

Consider that the NAND gates that are drawn in conventional form are actually a cascade of AND and NOT gates. (The circle represents the NOT gate.) Similarly consider that the NOT–OR form actually consists of an OR gate whose inputs are complemented by NOT gates. If two consecutive NOT operations are performed on a variable, the result is the original variable, see Eq. (4-31). Hence, the cascade of two NOT gates does not change the value of a variable. In Fig. 5-32b, each NOT gate is cascaded with another one. In effect, the NOT gates can be removed. Thus, Fig. 5-32b is equivalent to Fig. 5-32a.

Therefore, we can make the following general statement: If a network is realized using a sum of products form (using one AND gate for each product and one OR gate to sum the products) then *all* the AND gates and the single OR gate can be replaced by NAND gates, without changing the network function. Let us consider this. If the sum of products form is used, then a number of AND gates will always drive a single output OR gate. When all of these gates are replaced by NAND gates, a representation such as that of Fig. 5-32b will result. Thus, all NOT gates will appear in cascade pairs. Therefore, these "NOT gates" can be mathematically eliminated and, hence, the original switching function will be obtained.

Now let us discuss the case where the network is realized using a product of sums form. For instance, consider Fig. 5-33a. This uses many OR gates (two in this case) driving a single AND gate. Now consider the realization of Fig. 5-33b where we have replaced the AND and OR gates by NAND gates. *Note that the input variables are replaced by their complements.* This replacement cancels the effect of the NOT portions of the NOT–OR representation of the NAND gates. Thus, the outputs of the NAND gates *a* and *b* are the same as those of the OR gates *a* and *b*, respectively. Then, the output of the NAND gate *c* will be the complement of the desired output. Hence, a NOT gate is added to obtain the desired output. We have not used NAND gates exclusively here. Note, however, that if an actual NOT gate is not available, then the NOT gate can be realized using NAND gates, see Fig. 3-18a.

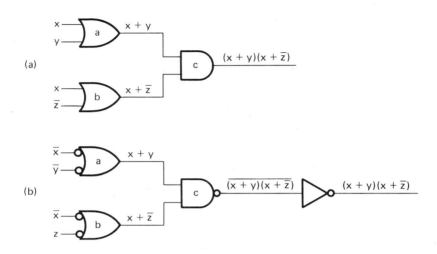

Figure 5-33 (a) A canonical product of sums realization; (b) an equivalent switching network using NAND gates

In general, if a network is realized using a product of sums form (using many OR gates and one AND gate) then all the gates can be replaced by NAND gates without changing the switching function, provided that *all the inputs and the output are complemented*. Note that taking the complement increases both the number of gates and the number of levels. Thus, it may be desirable to use a sum of products realization when NAND gates are used, unless the product of sums form uses many less gates.

Realization Using NOR Gates

The discussion of realization using NOR gates essentially parallels, on a dual basis, the previous discussion of the realization using NAND gates. Thus, we shall omit some of the details. We start by considering Eq. (4-25a).

$$\overline{x + y} = \bar{x}\bar{y} \tag{5-80}$$

The left side of this expression is just a statement of the NOR operation. Thus, Figs. 5-34a and b are equivalent representations of the NOR operation. We shall call Fig. 5-34a the conventional representation and Fig. 5-34b the NOT–AND representation.

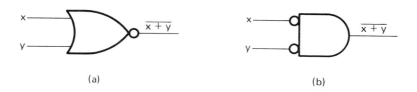

Figure 5-34 Two representations of a NOR gate; (a) Conventional form; (b) NOT–AND form

Suppose that we have a product of sums form such as

$$f(x,y,z) = (x+y)(x+\bar{z}) \tag{5-81}$$

Its realization is shown in Fig. 5-35a. Now consider Fig. 5-35b. As before, the cascading of two NOT operations yields the original function, and Figs. 5-35a and b are equivalent. We can therefore state: If a network is realized using a product of sums form (using many OR gates driving a single AND gate) then all OR and AND gates can be replaced by NOR gates without changing the network function.

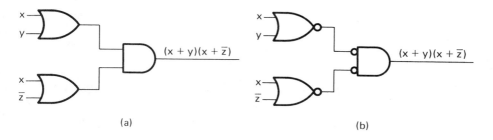

Figure 5-35 Realization of Eq. (5-81). (a) Product of sums form using OR and AND gate; (b) using NOR gates

When the sum of products form is used, we can replace all AND and OR gates by NOR gates, provided that *all* the input and the output are complemented. For instance, Figs. 5-36a and b have equivalent switching functions. Note that the sum of products form realization using OR gates will use more gates and levels than the product of sums would. Thus, if an OR gate realization is to be used, the product of sums form is more desirable, unless the sum of products form uses many less gates.

We have considered examples containing single output networks in this section. However, all the ideas carry directly over to multiple output networks.

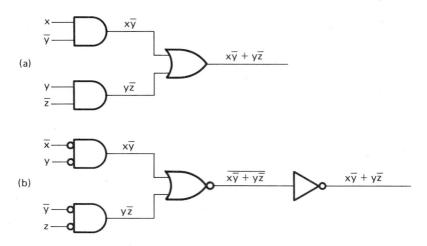

Figure 5-36 (a) A sum of products realization; (b) an equivalent switching
network using NOR gates

BIBLIOGRAPHY

Booth, T.L., *Digital Networks and Computer Systems*, 2nd. Ed., Chapter
7, John Wiley and Sons, Inc., New York 1978

Kohavi, Z., *Switching and Finite Automatic Theory*, Chapters 3 and 4,
McGraw-Hill Book Co., Inc., New York 1970

Mano, M.M., *Digital Logic and Computer Design*, Chapter 3, Prentice
Hall, Inc., Englewood Cliffs, NJ 1979

McCluskey, E.J., *Introduction to the Theory of Switching Circuits*,
McGraw-Hill Book Co., Inc., New York 1965

Rhyne, V.T., *Fundamentals of Digital System Design*, Chapters 2 and 4,
Prentice Hall, Inc., Englewood Cliffs, NJ 1973

PROBLEMS

5-1. A switching function is characterized by the following truth table:

x	y	z	f
0	0	0	0
0	0	1	1
0	1	0	1
0	1	1	0
1	0	0	1
1	0	1	0
1	1	0	1
1	1	1	1

Express the switching function as a sum of minterms.

5-2. Express the switching function of Prob. 5-1 as a sum of maxterms.

5-3. Draw a gate circuit diagram which realizes the function of Prob. 5-1.

5-4. Repeat Prob. 5-3 for the function of Prob. 5-2.

5-5. Use the designations of Table 5-1 to express the function of Prob. 5-1.

5-6. Express the following function in a canonical sum of products form.

$$f(x,y,z) = x \cdot (y+z) + x \cdot \bar{y}(z+\bar{x}) + y \cdot \bar{x}(z+\bar{x}y)$$

5-7. Repeat Prob. 5-6 for

$$f(w,x,y,z) = w \cdot (x+\bar{y}) + (z+\bar{x})(y+\bar{w}) + (x+\bar{y})(\bar{w}+z)$$

5-8. Express the function of Prob. 5-6 in a canonical product of sums form.

5-9. Repeat Prob. 5-8 for the function of Prob. 5-7.

5-10. A logic circuit is to be built for an automatic change machine for a rapid transit system. The fare is 55 cents. The money deposited can consist of nickels, dimes and quarters. If there is too much deposited, nickels or dimes are to be returned. The outputs are to consist of a signal for money paid, and signals to the nickel or the dime change releases. Obtain a switching function for the system. Hint: Assume that binary variables express the number of nickels, dimes and quarters. For instance, N_1N_2 represents the number of nickels.

5-11. Obtain a Karnaugh map for the switching function

$$f(x,y) = x + x\bar{y}$$

5-12. Repeat Prob. 5-11 for

$$f(x,y) = xy + \bar{x}\bar{y}$$

5-13. Obtain a switching function in a canonical sum of products form for the Karnaugh map of Fig. 5-37.

Figure 5-37

5-14. Obtain a Karnaugh map for the switching function

$$f(x,y,z) = \bar{x}y + \bar{x}\bar{y}z + \bar{x}yz + xyz$$

5-15. Obtain a function in a canonical sum of products form for the Karnaugh map of Fig. 5-38.

z\\xy	00	01	11	10
0	1			1
1	1		1	1

Figure 5-38

5-16. Simplify the switching function of Prob. 5-12 using the Karnaugh map.

5-17. Obtain a Karnaugh map for the switching function

$$f(w,x,y,z) = wxyz + wx\bar{y}z + w\bar{x}yz + w\bar{x}y\bar{z}$$

5-18. Obtain a switching function in canonical sum of products form for the Karnaugh map of Fig. 5-39.

yz\\wx	00	01	11	10
00	1			1
01		1	1	1
11		1	1	1
10	1			1

Figure 5-39

5-19. Obtain a Karnaugh map for the switching function

$$f(w,x,y,z) = m_0 + m_1 + m_2 + m_4 + m_{15} + m_{14}$$

5-20. Repeat Prob. 5-19 for

$$f(w,x,y,z) = m_0 + m_4 + m_1 + m_5 + m_3 + m_7 + m_{15} + m_{11} + m_2 + m_6$$

5-21. Use the Karnaugh map to minimize the switching function of Prob. 5-17. Draw a gate circuit that realizes this network.

5-22. Repeat Prob. 5-21 for the switching function of Prob. 5-19.

5-23. Repeat Prob. 5-21 for the switching function of Prob. 5-20.

5-24. Obtain the truth table for the switching function of Prob. 5-20.

5-25. Modify Table 5-2 for a 6-variable Karnaugh map.

5-26. Use the Karnaugh map to minimize the switching function

$$f(w,x,y,z) = m_0 + m_8 + m_2 + m_{10} + m_5 + m_{13} + m_7 + m_{15} + m_{11} + m_1$$

5-27. Repeat Prob. 5-26 for the switching function

$$f(u,v,w,x,y,z) = m_0 + m_4 + m_{32} + m_{36} + m_{17} + m_{49} + m_{27} + m_{59}$$

5-28. Repeat Prob. 5-26 for the switching function

$$f(u,v,w,x,y,z) = m_1 + m_9 + m_{41} + m_{33} + m_5 + m_{13} + m_{45} + m_{37} + m_{18}$$
$$+ m_{50} + m_{22} + m_{54} + m_{31} + m_{63} + m_{59} + m_{21} + m_{24} + m_{25}$$
$$+ m_{27} + m_{26} + m_{30} + m_{31} + m_{29} + m_{28}$$

5-29. Minimize the switching function whose Karnaugh map is given in Fig. 5-40.

yz \ wx	00	01	11	10
00	d			1
01		1	1	d
11		1	1	
10	d	d		

Figure 5-40

5-30. Use the Karnaugh map to minimize the following switching function

$$f(w,x,y,z) = m_0 + m_5 + m_{13}$$

where the following positions are "don't care" positions: m_7, m_{15}, m_2, m_8, and m_{12}.

5-31. Use the Karnaugh map to minimize the following switching function.

$$f(u,v,w,x,y,z) = m_0 + m_{36} + m_{18} + m_{54} + m_9 + m_{34}$$

where the following are "don't care" positions: m_4, m_5, m_{22}, m_2, m_{10}, m_{50}, and m_{32}.

5-32. Repeat Prob. 5-31 but now omit the "don't care" terms.

5-33. Use the Karnaugh map to minimize the switching function of Prob. 5-17. Use maxterms and 0's in this reduction.

5-34. Repeat Prob. 5-33 for the map of Fig. 5-39.

5-35. Repeat Prob. 5-33 for the switching function of Prob. 5-19.

5-36. Repeat Prob. 5-33 for the switching function of Prob. 5-20.

5-37. Repeat Prob. 5-33 for the switching function of Prob. 5-27.

5-38. Repeat Prob. 5-33 for the switching function of Prob. 5-30.

5-39. Determine an implicant of

$$f(x,y,z) = x + yz + xyz$$

5-40. If $f_1, f_2, \ldots, f_n$ are prime implicants of a function g, and if $g = 0$, what are the values of $f_1, f_2, \ldots f_n$?

5-41. Discuss the meaning of essential prime implicants and absolutely eliminable prime implicants.

5-42. Use the Quine-McCluskey procedure to minimize the switching function of Prob. 5-17. Draw a gate circuit that realizes this function.

5-43. Repeat Prob. 5-42 for the switching function of Fig. 5-40.

5-44. Repeat Prob. 5-42 for the switching function of Prob. 5-19.

5-45. Repeat Prob. 5-42 for the switching function of Prob. 5-20.

5-46. Repeat Prob. 5-42 for the switching function of Prob. 5-26.

5-47. Repeat Prob. 5-42 for the switching function of Prob. 5-27.

5-48. Repeat Prob. 5-42 for the switching function of Prob. 5-28.

5-49. Repeat Prob. 5-42 for the switching function of Fig. 5-40.

5-50. Repeat Prob. 5-42 for the switching function of Prob. 5-30.

5-51. Repeat Prob. 5-42 for the switching function of Prob. 5-31.

5-52. Minimize the number of gates used to realize the following switching functions:

$$f_1(x,y,z) = m_0 + m_1 + m_3 + m_5$$

$$f_2(x,y,z) = m_0 + m_1 + m_7$$

5-53. Repeat Prob. 5-52 for the switching functions

$$f_1(w,x,y,z) = m_0 + m_1 + m_5 + m_9 + m_{15}$$

$$f_2(w,x,y,z) = m_0 + m_1 + m_9 + m_{12} + m_{14}$$

$$f_3(w,x,y,z) = m_0 + m_1 + m_5 + m_{12} + m_{14} + m_{15}$$

5-54. Repeat Prob. 5-52 but now assume that the following are "don't care" positions: in f_1, m_{14}; in f_2, m_5.

5-55. Repeat Prob. 5-42 but now use decimal numbers in the prime implicant-determining table.

5-56. Repeat Prob. 5-55 for the switching function of Prob. 5-19.

5-57. Repeat Prob. 5-55 for the switching function of Prob. 5-20.

5-58. Repeat Prob. 5-55 for the switching function of Prob. 5-28.

5-59. Repeat Prob. 5-55 for the switching function of Prob. 5-30.

5-60. Repeat Prob. 5-21 but now use NAND gates to realize the network.

5-61. Repeat Prob. 5-60 but now use NOR gates to realize the network.

5-62. Repeat Prob. 5-22 but now use NAND gates to realize the network.

5-63. Repeat Prob. 5-62 but now use NOR gates to realize the network.

5-64. Repeat Prob. 5-60 for the network of Prob. 5-52.

5-65. Repeat Prob. 5-61 for the network of Prob. 5-52.

6

Sequential Circuits

Thus far, we have only considered combinational logic circuits where the output signals, at any one time, depended only upon the input signals, at that time. Of course, there is a small time delay between the application of the input signal and the establishment of the output signal, but those output signals were *not* functions of any signals which occurred at past times. In other words, a combinational circuit does not have *memory*, so past signals do not affect it, and it usually does not feed back signals from the output to the input. A combinational circuit is illustrated in block diagram form in Fig. 6-1a.

We shall now consider circuits where these restrictions are removed. That is, they have memory, allowing output signals to be functions of past values of signals. This implies that output signals can be fed back so that, at any one time, the output signals may be functions of their past values. The block diagram for such a circuit, called a *sequential circuit*, is shown in Fig. 6-1b. This name is chosen because we can perform operations in sequence. A very simple example of a sequential circuit is the telephone dialing circuit, where there is one dial, or set of push-buttons. The first number is dialed and stored, then the same dial is used for the second number, etc. Note that this results in a great saving. If there were no memory, then a separate dial would have to be used for each digit of the telephone number.

Another example of a sequential circuit is a sequential adder. In Sec. 3-4 we discussed adder circuits. If combinational circuits are used, then one adder must be used for each digit. When sequential circuits are used, only one adder need be constructed. It is used to add each digit in turn with the results being stored in the register (memory). We shall discuss this circuit in Chapter 10. Thus, sequential circuits can result in a great saving of components. More important, the calculations per-

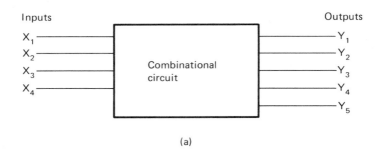

(a)

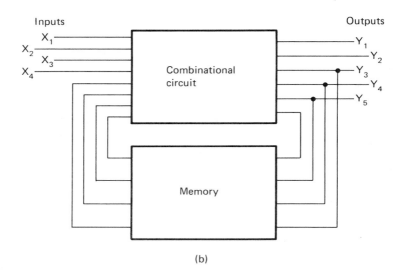

(b)

Figure 6-1 Block diagram representation of logic circuits. (a) combinational;
(b) sequential

formed by the modern digital computer would be impractical or
impossible to perform without using memory.

Sequential circuits are often classified as being either *synchronous*
or *asynchronous*. In a synchronous sequential circuit, the system
responds to signals only at discrete intervals of time. A device called a
master clock generates pulses and the circuits can respond to signals
only when allowed to by these clock pulses. Different circuits can
respond at different rates. The clock pulses should be slow enough so
that every circuit has time to respond to its own input signals. Of
course, the slowest circuit now limits the speed of all the circuits. (We
shall modify this statement subsequently.)

In asynchronous circuits, each circuit responds at its own rate. Thus, they are faster than synchronous circuits but, as we shall see, there are considerable design difficulties introduced in asynchronous sequential circuits. Therefore, at the present time, many digital compuers utilize synchronous sequential circuits. However, there are those that utilize asynchronous ones or combinations of both synchronous and asynchronous ones. In this chapter we shall consider the various types of sequential circuits and the techniques used in their analysis and design.

6.1 FLIP-FLOPS

A basic sequential circuit is one called the *flip-flop* or *bistable multivibrator*. This is a unit whose output remains either a 0 or a 1 until it is switched to the other by one or more input signals. This accounts for the name flip-flop. The output of the flip-flop (either a 0 or a 1) is called the *state* of the flip-flop. A block diagram representation of a flip-flop is shown in Fig. 6-2. The output Q represents the state of the flip-flop. Usually, there is another output $\overline{Q}$ which is the complement of Q. A single flip-flop can be in only one of the two states. A memory, or register, can be made up of many flip-flops. The contents of the register or memory is determined by the states of all of its flip-flops. Thus, if a memory has n flip-flops, it can have 2^n possible states. The variable Q is called the *state variable* of the flip-flop.

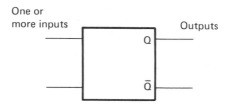

Figure 6-2 A general representation of a flip-flop

Now we shall consider some specific flip-flops. We shall do this because they are important in their own right and also because it will lay the groundwork for a general discussion of sequential circuits.

The R–S Flip-Flop

A fundamental form of flip-flop is called the R–S flip-flop. Its block diagram is shown in Fig. 6-3a. The input leads are called R and S, which stand for reset and set, respectively. Note that the names of the signals

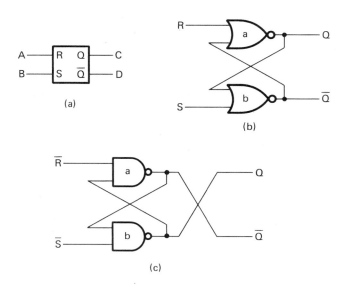

Figure 6-3 R–S flip-flops. (a) block diagram; (b) NOR gate realization; (c) NAND gate realization. The function of the 0's and 1's is interchanged in gates (b) and (c).

need not, necessarily, be R and S. For instance, in Fig. 6-3a, we have called the R input A and the S input B. However, the interior of the R–S flip-flop is always labeled with an R and an S, as shown. The value of Q is called the state of the flip-flop. We assume that the inputs are in the form of signals which are at levels corresponding to either 0 or 1. If $Q=0$, and a 1 input pulse arrives at S, then the flip-flop will change its state so that $Q=1$. If $Q=1$, then a 1 pulse at S will cause no change in Q. We assume that $R=0$ in both of the cases. Note that the S input sets the state of the flip-flop to 1.

The R input is the reset. That is, it resets the state of the flip-flop to 0. In all of the following we assume that $S=0$. If $R=0$ and $Q=1$, then the state does not change. If $R=1$ and $Q=1$ then the state resets to $Q=0$. If $R=1$ and $Q=0$, then the state remains at 0, and if $R=0$ and $Q=0$, then the state remains at 0.

In the previous discussion, we did not consider the case $R=1$, $S=1$. This would attempt to simultaneously set and reset the flip-flop. Thus, this is not allowed and the circuitry which supplies the input signals should be such that it does not occur.

All of the preceding information can be summarized in a table called a *transition table*, see Table 6-1, which indicates the changes in the state in response to all possible (or allowed) inputs. The first column indicates the state prior to the application of the input signal. The second column indicates the input signals and the last column indicates the state after response to the input signals. Note that we assume that it takes τ seconds for the flip-flop to change state.

Table 6-1 Transition Table for R–S Flip-Flop

Present State $Y(t)$	R	S	Next State $Y(t+\tau)$
0	0	0	0
0	0	1	1
0	1	0	0
0	1	1	Indeterminate (not allowed)
1	0	0	1
1	0	1	1
1	1	0	0
1	1	1	Indeterminate (not allowed)

Another table, called the *excitation table*, provides the same information in a different format. It gives the signal required to change the state in a specified way.

Table 6-2 Excitation Table for R–S Flip-Flop

State Change from	to	Required Input R	S
0	0	d	0
0	1	0	1
1	0	1	0
1	1	0	d

where the d's represent "don't cares." For instance, $R=1$, $S=0$ and $R=0$, $S=0$ will result in no change in a state of 0.

We have characterized the R–S flip-flop. Now let us discuss a circuit which implements it. The circuit of Fig. 6-3b uses two NOR circuits in conjunction with feedback. Suppose that $Q=0$, $\bar{Q}=1$. Now if we apply the signal which results in $S=0, R=0$, then Q, the output of NOR a, will remain 0 since $\overline{Q}=1$. Then, both inputs to NOR b will be 0 and $\overline{Q}$ remains a 1.

Now consider that the S input shifts to a 1 while R remains a 0. One input of NOR b is now a 1. Thus, its output changes to a 0. Note that this does not take place instantaneously, but requires T_1 seconds. Hence, after T_1 seconds, $\overline{Q}=0$ and both inputs to NOR a become 0's. Therefore, after another T_1 seconds, $Q=1$. Note that it requires $2T_1$ seconds for the circuit to change its state. We shall call

$$\tau = 2T_1 \tag{6-1}$$

the required time for the change of state.

To clarify these ideas, let us plot the various signals as functions of time. To simplify the diagrams we shall omit the exponential rise of the pulses and just assume that the gates delay the signals. Here we assume that $Q=0$ and $\overline{Q}=1$. At $t=t_a$ we apply $S=1$, keeping $R=0$. The resulting waveforms, which are explained in the previous paragraph, are shown in Fig. 6-4.

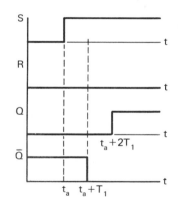

Figure 6-4 Some idealized time responses in an R-S flip-flop which illustrate delays

Now assume that the input shifts to the case $S=0, R=0$ (the output prior to application of the signals is $Q=1, \overline{Q}=0$). Both inputs to NOR a will be 0. Thus, Q remains at 1. This maintains one input of NOR b at 1. Hence, $\overline{Q}$ remains at 0. Therefore, an $S=1, R=0$ input followed by an $S=0, R=0$ input changes the state only once (following the application of $S=1, R=0$). Note that the $S=1, R=0$ input must be maintained for at least τ seconds in order to allow the state to change.

Now assume that we apply the input $S=0, R=1$ with the state $Q=1$, $\bar{Q}=0$. Since one input to NOR a is a 1, it will switch Q to 0 after T_1 seconds. Now both inputs to NOR b are 0. Hence, after another T_1 seconds, $\bar{Q}=1$.

If we now apply $R=1, S=0$, the input to NOR a will have two inputs of 1. Thus, Q remains at 0 and both inputs to NOR b remain at 0 and $\bar{Q}$ remains at 1.

Suppose that we apply the forbidden inputs $S=1, R=1$ (with $Q=0$, $\bar{Q}=1$). Then, one input to each NOR circuit becomes 1, and the output is $Q=0, \bar{Q}=0$. This violates our condition that $\bar{Q}$ be the complement of Q, which is a requirement in many logic circuits. Thus, such an input is not allowed. Different realizations of the R–S flip-flop may produce other outputs. There is no "standard" behavior for the R–S flip-flop with the forbidden input.

An R–S flip-flop using NAND gates is shown in Fig. 6-3c. In this flip-flop, 0's activate the circuit. That is, a 0 on the set input (i.e. the one labeled $\bar{S}$) causes Q to become a 1, while a 0 on the reset ($\bar{R}$) input causes Q to become a 0. For this reason, the inputs are labeled as complements, i.e., as $\bar{S}$ and $\bar{R}$. (Note that the actual inputs are $\bar{S}$ and $\bar{R}$.) For instance, when there is a 1 on the lower input of NAND b, then $\bar{S}=1$. We can say that $S=0$ since this will conform with the R–S flip-flop of Fig. 6-3b. However, the actual input is a 1.

If $Q=0, \bar{Q}=1$, and we apply $\bar{S}=0, \bar{R}=1$ (equivalent to $S=1, R=0$) then one input to NAND b is a 0. Then, Q becomes a 1. Now all inputs to NAND a are 1's and $\bar{Q}$ becomes a 0.

Next let us apply $\bar{R}=0, \bar{S}=1$ with $Q=1, \bar{Q}=0$. Hence, one input to NAND a is a 0 so that $\bar{Q}$ becomes a 1. Now both inputs to NAND b are 1's so that Q becomes 0. We have considered the response of the flip-flop for a pair of inputs. Proceeding similarly, we can determine the state for all combinations of inputs and states. Note that, in this case, the disallowed input is $\bar{R}=0, \bar{S}=0$.

Let us now illustrate the use of the R–S flip-flop of Fig. 6-3b with a simple example. Suppose that we have a decimal number between 0 and 9, which is represented in binary form. Consider that this is the last digit of a number and we want to use this to control roundoff. That is, we want the circuit which uses an R–S flip-flop to store a 1 if the digit is 5 through 9 and to store a 0 otherwise. Thus, we want the input to the R–S flip-flop to be $R=0, S=1$ if the digit is 5 through 9, and $R=1$, $S=0$ if the digit is 0 through 4. Let us assume that the number is given by the binary digits $X_3 X_2 X_1 X_0$. Then, the truth table for the input to the flip-flop is

Decimal Number	X_3	X_2	X_1	X_0	S	R
0	0	0	0	0	0	1
1	0	0	0	1	0	1
2	0	0	1	0	0	1
3	0	0	1	1	0	1
4	0	1	0	0	0	1
5	0	1	0	1	1	0
6	0	1	1	0	1	0
7	0	1	1	1	1	0
8	1	0	0	0	1	0
9	1	0	0	1	1	0

(6-2)

The remaining six rows of the truth table are don't care conditions. A Karnaugh map for this table is shown in Fig. 6-5a. The functions that realize R and S are

$$S = X_0 X_2 + X_1 X_2 + \bar{X}_1 X_3 \tag{6-3a}$$

$$R = \bar{S} \tag{6-3b}$$

A combinational logic gate circuit is used to realize the R and S signals which drive the R–S flip-flop. The complete circuit is shown in Fig. 6-5b.

Clocked R–S Flip-Flop

The flip-flop that we considered responded to changes in its inputs no matter when they arrived. We now want to obtain an R–S flip-flop which can be controlled by clock pulses. At the start, let us assume that we want a flip-flop that will only respond to the input signals when pulses from a master clock are present. A simple form of clocked R–S flip-flop is shown in Fig. 6-6. The basic part of the circuit is an R–S flip-flop. However, the R and S inputs are applied through separate AND gates. One of the AND gate inputs is the clock pulse (i.e., that input is a 1 when $CP=1$). Consider AND c. If the clock pulse is present, then its output is R. That is, if $R=1$, then the output is 1; when $R=0$, the output is 0. When the clock pulse is 0, the output of gate c is 0. AND gate d serves the same function with the S signal. Thus, the input to the internal unclocked R–S flip-flop is 0 unless the clock pulse is present. Then, the flip-flop can only change its state when the clock pulse is present. Remember that an input $R=0$, $S=0$ leaves the state of the flip-

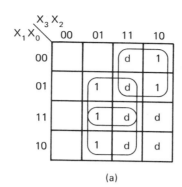

(a)

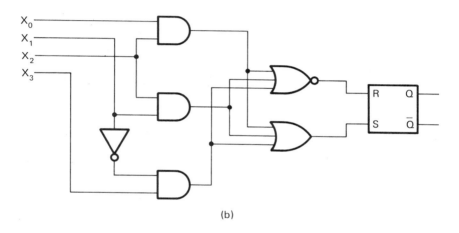

(b)

Figure 6-5 (a) The Karnaugh map for the roundoff control memory; (b) the circuit

flop unchanged. The time of the clock pulse τ_p must be long enough to allow the state of the flip-flop to change. In the next section we shall consider this in much greater detail. A block diagram representation of the R–S flip-flop is shown in Fig. 6-6c. This type of clock control is simple and results in simple circuits. However, we shall see that problems usually arise and more complex clocking circuits must be used. These will be discussed in the next section.

Now let us consider a clocked R–S flip-flop which uses NAND gates, see Fig. 6-6d. Remember that the output of a NAND gate is 1 unless *all* its inputs are 1's. Thus, when the clock pulses are absent, the output of NANDs c and d are 1's, i.e., $\bar{R}_1 = 1$, $\bar{S}_1 = 1$. This input causes no change in the values of Q and $\bar{Q}$. (Note that, in this case, $\bar{R}_1 = 0$, $\bar{S}_1 = 0$ is the disallowed case.)

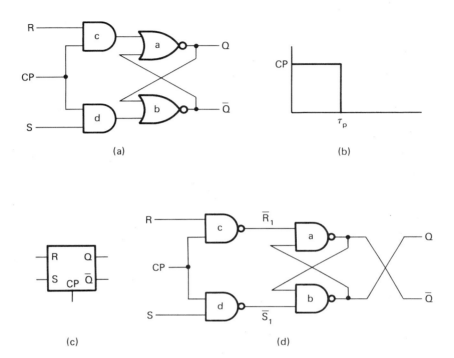

Figure 6-6 (a) A clocked R–S flip-flop using NOR gates; (b) a single clock
pulse; (c) the block diagram for a clocked R–S flip-flop; (d) a
clocked R–S flip-flop using NAND gates

When a clock pulse is present, then $\bar{R}_1 = 0$ if $R = 1$. On the other hand,
$\bar{R}_1 = 1$ if $R_1 = 0$. Thus, during a clock pulse, the output of NAND c is the
complement of R, i.e., $\bar{R}$. Similarly, when a clock pulse is present, the
output of NAND d is the complement of S, i.e., $\bar{S}$. Thus, during a clock
pulse, the input to the nonclocked portion of this R–S flip-flop is the
complement of the actual input. Hence, the clocked NAND gate R–S
flip-flop responds to 1's. (The unclocked portion of the flip-flop
responds to 0's.) An alternative way of stating this is to say that the
flip-flops of Fig. 6-6a and d perform identically as far as the polarity
of the input pulses is concerned.

The R–S flip-flop is not the only type of flip-flop. We shall now
consider some other forms of flip-flops that are commonly used.

The D Flip-Flop

We can modify the R–S flip-flop so that it has only one input which is called D. A block diagram for such a flip-flop is shown in Fig. 6-7a. If $D=1$, then $Q=1$ and $\bar{Q}=0$; if $D=0$, then $Q=0$ and $\bar{Q}=1$. The transition table is

Table 6-3 Transition Table for the D Flip-Flop

Present State Q(t)	D	Next State Q(t+τ)
0	0	0
0	1	1
1	0	0
1	1	1

The D flip-flop can be simply implemented using an R–S flip-flop and a NOT gate, see Fig. 6-7b. If $D=0$, then the input circuit including the NOT gate sets $R=1$ and $S=0$. This will always reset the R–S flip-flop so that $Q=0$ and $\bar{Q}=1$. If $D=1$ then $R=0$, $S=1$. This will always set the R–S flip-flop so that $Q=1$, $\bar{Q}=0$.

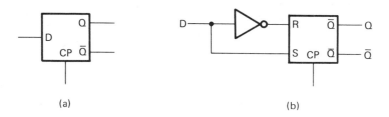

(a) (b)

Figure 6-7 The D flip-flop. (a) block diagram; (b) an implementation

The J–K Flip-Flop

At times, it is desirable to have a flip-flop which responds to its two inputs in the same way as the R–S flip-flop does, but which allows the $R=1$, $S=1$ input. A flip-flop called the J–K flip-flop does this. The K (clear) input replaces the R input and the J input replaces the S input. The J–K flip-flop functions as does the R–S flip-flop except that now, if the $J=1$, $K=1$ input occurs, the output state changes. A block diagram for the J–K flip-flop is shown in Fig. 6-8a. We assume that the J–K flip-flop is always clocked. The transistion and excitation tables for the J–K flip-flop are

Table 6-4 Transition Table for the J–K Flip-Flop

Present State $Q(t)$	J	K	New State $Q(t+\tau)$
0	0	0	0
0	0	1	0
0	1	0	1
0	1	1	1
1	0	0	1
1	0	1	0
1	1	0	1
1	1	1	0

where τ is the time between clock pulses.

Table 6-5 Excitation Table for the J–K Flip-Flop

State Change from	to	Required Input J	K
0	0	0	d
0	1	1	d
1	0	d	1
1	1	d	0

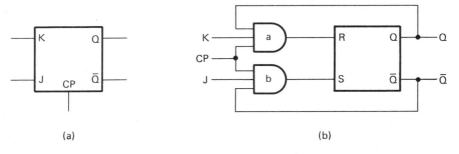

(a)   (b)

Figure 6-8 The clocked J–K flip-flop. (a) block diagram; (b) a circuit implementation

Note that there are many don't cares in this table. This often simplifies the logic design.

An implementation of the J–K flip-flop using AND gates and an R–S flip-flop is shown in Fig. 6-8b. Note that when the clock pulse is absent, the outputs of the AND gates are 0 and no input is applied to the R–S flip-flop.

Now suppose that $Q=0$ and $K=0$ and $J=0$ and a clock pulse is applied. Then, $R=0$ and $S=0$ and the outputs are unchanged. Suppose now that $J=0$ and $K=1$ and that a clock pulse is applied. Since $Q=0$, then $R=0$, and $S=0$ since $J=0$; again the outputs are unchanged.

Now suppose that with $Q=0$, we have $J=1$ and $K=0$ during a clock pulse. All inputs of AND b will be 1. Thus, $S=1$. Also, $R=0$ since the K and Q inputs of AND a are 0. Then, the R–S flip-flop changes its state to $Q=1$ and $\bar{Q}=0$. Note that this changes one input of AND b to 0. Then, $S=0$. The output of AND a is 0 since $K=0$. Thus, the output remains $Q=1$, $\bar{Q}=0$.

With the input $Q=1$, suppose that, during a clock pulse, $J=1$, $K=0$. Then, $S=0$ and $R=0$, and the output does not change. Now with $Q=1$, assume that, during a clock pulse, we apply $J=0$, $K=1$. Then, the output of AND a is 1 while that of AND b is 0. Thus, $R=1$ and $S=0$. The R–S flip-flop switches back to $Q=0$ and $\bar{Q}=1$. Again, both AND gates have at least one input 0 and the output remains at $Q=0$, $\bar{Q}=1$.

Now let us consider the input $K=1$, $J=1$. Suppose that $Q=0$ and $\bar{Q}=1$ and that this input is applied during a clock pulse. In this case, the output of AND a is 0 while that of AND b is 1. Thus, the feedback and the input AND gates cause $R=0$ and $S=1$, and the R–S flip-flop changes its state to $Q=1$ and $\bar{Q}=0$. Now the output of AND a is a 1 and that of AND b is a 0, therefore, $R=1$, $S=0$. Thus, the R–S flip-flop would reset itself to $Q=0$, $\bar{Q}=1$. So we can see that if $J=1$, $K=1$ is applied, the flip-flop will *continuously* change its state as long as a clock pulse is applied. This is an *unstable operation* or *oscillation*. One way of preventing it is to make τ, the length of the clock pulse, exactly equal to the switching time of the R–S flip-flop. However, this requires precise adjustment. In addition, each output may not change at the same time, which can cause further problems. In the next section we shall consider practical procedures for eliminating such unstable operation. For the time being, assume that the clock pulses are such that instability does not occur.

The T Flip-Flop

We shall consider one more basic form of flip-flop called the T or toggle flip-flop. This has only one input and is such that the application of an input 1 signal causes it to change its state. That is, a 0 input leaves the output state unchanged while a 1 input causes it to change. The block diagram for this flip-flop is given in Fig. 6-9a. The transition table is as follows:

Table 6-6 Transition Table for the T Flip-Flop

Present State Y(t)	T	Next State Y(t + τ)
0	0	0
0	1	1
1	0	1
1	1	0

A circuit implementation using a J-K flip-flop is shown in Fig. 6-9b. Here we connect the two inputs of the J-K flip-flop together. If $T=0$, then $J=0$ and $K=0$ and the output is unchanged. If $T=1$ we have the case $J=1$, $K=1$ which causes the output to change. Of course, changes can only occur when the clock pulses are present. The instability problem discussed in conjunction with the J-K flip-flop is also a problem here. In the next section we shall see how the problem is resolved.

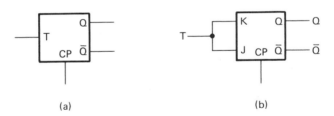

(a)   (b)

Figure 6-9 The clocked T flip-flop. (a) block diagram; (b) circuit implementation using a J-K flip-flop

6.2 SYNCHRONIZING AND TRIGGERING OF FLIP-FLOPS — MASTER-SLAVE FLIP-FLOPS; EDGE-TRIGGERED FLIP-FLOPS

In the last section we considered the basic ideas of some clocked flip-flop circuits. Now we shall consider the timing in greater detail and discuss some additional circuits. Also, techniques for eliminating some problems that were discussed in Sec. 6.1 shall be covered.

T Q

0 1

1 0

In Sec. 6.1 we assumed that the clock pulses were on "long enough" for a change in state to take place. Let us see what is involved here. We shall work with the clocked R–S flip-flop of Fig. 6-6a. Normally, the clock pulse is a periodic signal whose duration is short in comparison to the period, see Fig. 6-10a. The period of the clock pulse is τ_p and its duration is τ_c where τ_p is considerably greater than τ_c.

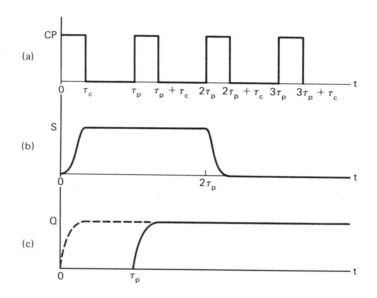

Figure 6-10 A timing diagram for an R–S flip-flop which is triggered on the leading edge of the clock pulse. (a) clock pulse; (b) S-input; (c) X–output, note the ambiguity in X.

Now consider that an S input is applied at the same time as the clock pulse (with $X=0$). This is shown in Fig. 6-10b. In general, the signals will be the outputs of other flip-flops or of other electronic circuits and will not build up instantaneously. The AND gates of Fig. 6-6a allow signals to be applied between 0 and τ_c. However, S is just building up during that time and it may not be large enough to cause the flip-flop to change its state. Thus, action may not take place until time τ_p when the next clock pulse allows the inputs to activate the R–S flip-flop. Now, Q changes as shown by the solid curve of Fig. 6-10c. However, it is possible that the circuit might have responded in the first timing pulse interval. Then, Q would be of the dashed form. Thus, Q is potentially ambiguous. Such ambiguities occur if the input signal is changing when the clock pulse is on. Figure 6-10 illustrates what is called *leading-edge triggering*, since the leading edge of the pulse (i.e., the edge that occurs first in time) determines when the flip-flops start to output their signals. This, in turn, determines when input signals are applied to the flip-flop and when switching starts.

One way of avoiding the ambiguity of simple leading-edge triggering is to delay the output signals from the flip-flops so that they do not change until the clock pulse has passed, see Fig. 6-11. We shall consider the implementation of these circuits subsequently. For the time being, let us just consider that the delay is available. Since the output signal is delayed by τ_c seconds, then the input signal must also be delayed by at least τ_c seconds. That is, the intput signal cannot change until the *trailing edge* of the clock pulse has passed. The possible ambiguity has been removed, since the S input is 0 until after the first clock pulse. At $t=\tau_p$, the second clock pulse arrives. Then, $S=1$ and the flip-flop changes state so that $Q=1$.

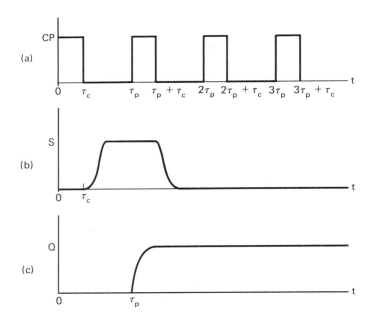

Figure 6-11 A timing diagram for an R–S flip-flop which introduces delay to avoid ambiguous response. (a) clock pulses; (b) S–input delayed to trailing edge of clock pulse; (c) X–output

One way of implementing the delayed output is to put a delay at the output of each flip-flop or other logic element. This is illustrated in Fig. 6-12. The delay network could consist of a length of transmission line. This is awkward since, if the delay is relatively long, very long lengths of transmission line are needed. The delay could be established by driving a cascade of NOT circuits. Since each one takes a finite time to respond, the output would be delayed. In either of these cases, any shift in the timing of the clock pulses might cause the delay to be too short, reinstating ambiguity. Thus, flip-flops with built-in delay could only be used with specific clock pulses.

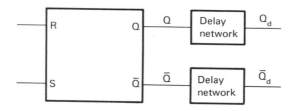

Figure 6-12 An R–S flip-flop with delay networks

Master-Slave Flip-Flops

Usually, the best type of delay is derived from the clock pulses. Suppose that we had a flip-flop which would start to respond with the leading edge of the clock pulse but whose *output did not change until the trailing edge of the clock pulse had arrived.* Assume that the circuit is made up of flip-flops of this type. Then, the inputs to any gates (which come from the outputs of other gates) would not change until after the clock pulse. This is the desired situation. It is relatively easy to implement. Consider the circuit of Fig. 6-13, which consists of one R–S

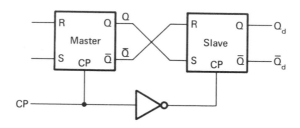

Figure 6-13 An R–S master-slave flip-flop

flip-flop driven by another one. For the time being, let us ignore the clock pulse. The Q output of the first flip-flop is connected to the S input of the second. Similarly, the $\bar{Q}$ output of the first is connected to the R input of the second.

Now, if the first flip-flop has an output $Q=1$, $\bar{Q}=0$, then the input to the second is $S=1$, $R=0$. The output will be $Q_d=1$, $\bar{Q}_d=0$. That is, the second flip-flop will be set by its $S=1$, $R=0$ input. If the first flip-flop has an output $Q=0$, $\bar{Q}=1$, then the second has an input $R=1$, $S=0$. This will reset the second flip-flop so that $Q_d=0$, $\bar{Q}_d=1$. Thus, the output of the second flip-flop will always be the *same* as that of the first. For this reason, the second flip-flop is called the *slave* while the first is called the *master*.

Now let us consider the effect of clock pulses on the flip-flop. The clock pulses are directly applied to the clock terminal of the master but they are passed through a NOT circuit before they are applied to the slave. Thus, there will be a 1 applied to the CP terminal of the slave only when a 0 is applied to the CP terminal of the master. Assume that an input is applied to the master. When the clock pulse arrives, the state of the master changes. During the clock pulse, Q and $\bar{Q}$ of the master achieve their final values. However, the slave *cannot* change because there is a 0 input at its clock pulse terminal. After the clock pulse is over, a 1 will appear at the slave's clock pulse terminal. Thus, it can now change its state. Note that the slave does not change its state until after the clock pulse is past. Then the master switches on the leading edge of the clock pulse while the slave switches on the trailing edge. This is the operation we desired. That is, if all the flip-flops are master-slave types, then no flip-flop output can change during the clock pulse. If these outputs are inputs to other flip-flops, then these inputs cannot change during the clock pulse.

A timing diagram is shown in Fig. 6-14. Assume that the S input is from another master-slave flip-flop, so that it does not start to change until the trailing edge of the clock pulse. The master flip-flop does not start to change its state until time $t=2\tau_p$. (Note that at $t=\tau_p$, the S input was 0.) The output of the slave, Q_d, does not start to change until $t=2\tau_p + \tau_c$, at which point Q has become established. Thus, the master-slave flip-flop eliminates the problems associated with flip-flops' being triggered by changing signals. The time delay of the slave is exactly fixed at τ_c. Thus, if there is some shift in the timing signals, then the slave delay also shifts.

We have illustrated master-slave flip-flops using the R-S flip-flop. Actually, any flip-flop can be put into a master-slave configuration. Let us consider the J–K master-slave flip-flop, see Fig. 6-15. The basic

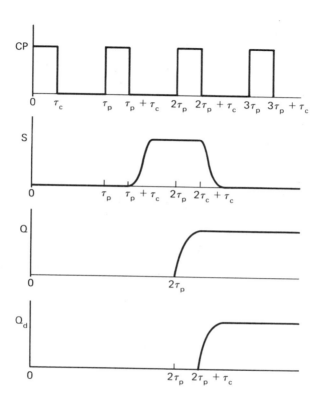

Figure 6-14 A timing diagram for a master-slave R–S flip-flop. (a) clock pulse; (b) S-input; (c) X-response; (d) X_d-output

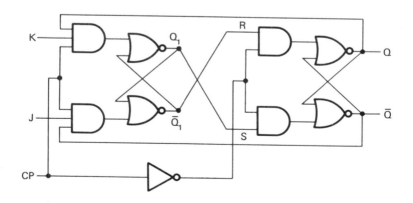

Figure 6-15 A J–K master-slave flip-flop

circuit of Fig. 6-8b is used, except that the R–S flip-flop of that circuit is replaced by a master-slave R–S flip-flop, and the internal AND gates are combined with the input AND gates of the clocked R–S flip-flop.

This circuit functions in a manner similar to that of the ordinary J–K flip-flop except that the slave does not change state until after the clock pulse is passed and the master can only change its state during a clock pulse.

Now consider the operation. Suppose that $Q=0$, $\bar{Q}=1$, and we apply $J=1$, $K=0$. Then, during the clock pulse, the output of the master becomes $Q_1=1$, $\bar{Q}_1=0$. Once the clock pulse is over, the slave can change its state. Since the R input is 0 and the S input is 1, the output becomes $Q=1$, $\bar{Q}=0$. The operation follows that for the ordinary J–K flip-flop with the exception of the time delay between master and slave. Thus, all the other possible inputs will not be considered.

One case we will consider is $J=1$, $K=1$. Suppose that $Q=0$, $\bar{Q}=1$. Then, during the clock pulse, Q_1 becomes a 1 and $\bar{Q}_1$ a 0. However, the slave does not change its state. After the clock pulse, the slave is able to change its state and Q becomes a 1 and $\bar{Q}$ becomes a 0. In this case, instability does not occur because now the master cannot change its state until the next clock pulse occurs. Thus, the output remains constant. By the next clock pulse, the inputs will be changed since it is assumed that they have also come from a master-slave flip-flop. Of course, if the logic maintains the $K=1$, $J=1$ input, then the output state will again change after the next clock pulse.

A J–K master-slave flip-flop using NAND gates is shown in Fig. 6-16. This uses the R–S flip-flop of Fig. 6-6d as both the master and slave portions of the R–S flip-flop. Otherwise, the basic ideas are essentially the same as those of Fig. 6-15. Since the input of the master R–S flip-flop is a NAND gate, we do not have to use a separate AND gate to combine the output feedback signal with the input signal.

We have illustrated master-slave J–K and R–S flip-flops. Actually, any of the flip-flops discussed could be implemented in a master-slave configuration. For instance, in Figs. 6-15 or 6-16, if we connect the J and K inputs together, a T master-slave flip-flop is obtained.

Edge-Triggered Flip-Flops

The master-slave is one form of flip-flop whose output changes after the trailing edge of the clock pulse has occurred. However, there are other flip-flops where output is triggered by a transition or change of the clock pulse. That is, a change in the clock signal, rather than merely its presence or absence, produces an effect. Such flip-flops are called

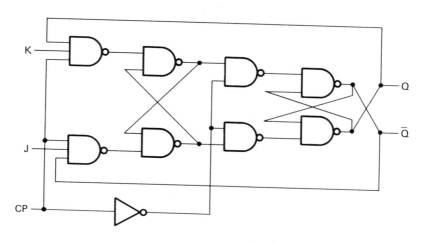

Figure 6-16 A J–K master-slave flip-flop using NAND gates

edge-triggered. Usually, the magnitude and sign of the derivative of the clock pulse is the determining factor. For instance, some edge-triggered flip-flops change state when the clock voltage changes in a positive direction, while other edge-triggered flip-flops change state when the clock voltage changes in a negative direction. By choosing the proper clock voltage polarity and the proper edge-triggered flip-flop, we can obtain trailing-edge triggering.

One form of edge-triggered flip-flop makes use of the differentiating ability of an RC circuit, see Fig. 6-17a. Let us assume that the input signal v_i is of the form shown in Fig. 6-17b. This is an idealization since actual signals do not have straight-line slopes as shown. However, such idealizations are widely used to characterize digital circuits.

The output signal will be of the form shown in Fig. 6-17c. (It is assumed that the reader is familiar with simple transient analysis.) Thus, the circuit of Fig. 6-17a only produces an output in response to changes in the input signal.

Now suppose that we modify the circuit as shown in Fig. 6-18a. The incorporation of the direct voltage power supply V_1 shifts the output voltage by an amount equal to V_1. A typical input voltage and the resulting output voltage is shown in Figs. 6-18b and c, respectively. Consider the levels marked 0 and 1 in Fig. 6-18c. Suppose that any v_0 equal to or greater than the 1-level represents a 1 while any voltage equal to or less than the 0-level represents a 0. Then, the output of the circuit of Fig. 6-18a will be a 1 unless the input signal decreases. Then the output becomes a 0. (This assumes that the decrease in v_i is large enough to produce the desired change in v_0.)

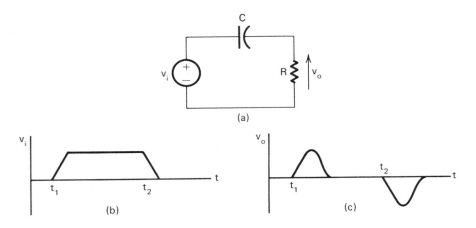

Figure 6-17 (a) A simple RC circuit; (b) an input waveform; (c) the output waveform

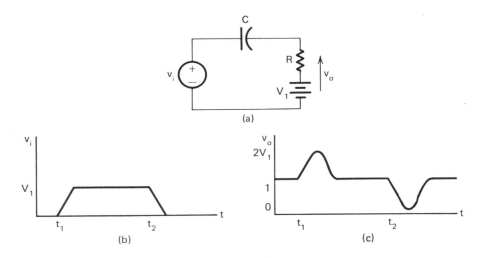

Figure 6-18 (a) A modification of Fig. 6-17 that produces an output of 1 unless v_i decreases; (b) an input waveform; (c) the resulting output waveform

In Fig. 6-19 we use the RC circuit of Fig. 6-18a to implement an edge-triggered flip-flop. The flip-flop is the same as that shown in Fig. 6-3b. Remember that this flip-flop is activated by 0's and not by 1's. The RC circuit and the 1-level voltage are analogous to Fig. 6-18a. Now consider the output of AND gate c. If $R = 1$ and remains unchanged, and if $CP = 1$, then the voltage v_c will be 1 unless the clock signal

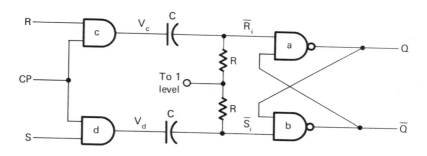

Figure 6-19 An edge-triggered flip-flop

changes from 1 to 0, in which case v_c will change to 0. Now v_c plays the same role as v_i did in Fig. 6-18a as far as the RC circuit is concerned. Hence, if $R=1$, then $\bar{R}_i$ will become 0 when the clock signal decreases in voltage. On the other hand, if $R=0$, then $\bar{R}_i$ will be a 1 for all values of the clock signal. Note that $\bar{R}_i$ only becomes 0 in response to negative transitions of the clock when $R=1$. A similar set of statements can be made about S and $\bar{S}_i$. Then, the output of the flip-flop can only change after the clock pulses have made a negative-going transition. If the clock signal is of the form shown in Fig. 6-11a, then the proper operation will result.

We have discussed one type of edge-triggered flip-flop. There are others that utilize capacitors in other types of circuits, while still others utilize the delay of gate circuits to achieve the desired results. Such circuits are discussed in detail in the electronic circuit text cited at the end of the chapter.

Timing Diagrams for Master-Slave and Edge-Triggered Flip-Flops

Master-slave and edge-triggered flip-flops are fabricated in integrated circuit form. Manufacturers supply data concerning them. Some data indicates the proper operation conditions such as voltage levels. If these conditions are not met, the circuit will not function properly. Of particular importance are the *timing diagrams*, since these indicate to the circuit designer those relations that must exist among the various signals if the circuit is to function properly.

A typical set of timing diagrams is shown in Fig. 6-20. The waveforms are usually drawn in idealized form with straight, sloping sides. The clock waveform is shown in Fig. 6-20a. The time that it should take for the clock signal to rise to its maximum value is specified by the manufacturer. Note that, in actual circuits, the rise and fall of signals

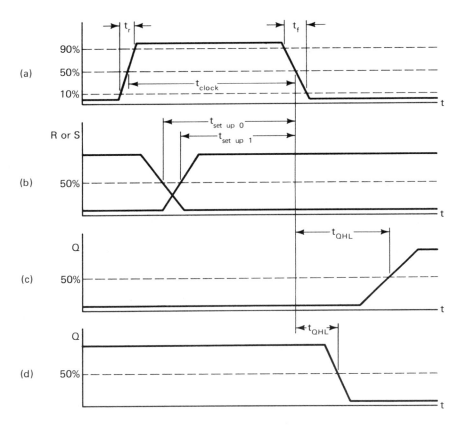

Figure 6-20 Timing diagrams for flip-flops. (a) clock signal; (b) input signal; (c) and (d) output signal

are often exponential in nature and thus, they never reach their peak value. Because of this *rise times* are specified as the time it takes for the signal to rise from 10 to 90 percent of its peak value. In a similar way, *fall time* is the time that it takes for a signal to fall from 90 to 10 percent of its peak value. Manufacturers specify maximum and minimum allowed values of rise and fall time. If these specifications are not met, the circuit will not function properly. The length of the clock pulse is nominally measured between 50 percent points as shown. An allowable range of clock pulse lengths is also specified.

If an edge-triggered flip-flop is to function properly, then the input signals must stabilize at their final values at a time sufficiently long before the occurrance of the trailing edge of the clock pulse. For instance, consider the circuit of Fig. 6-19. The input signals to the flip-flop must stabilize before CP becomes 0 if the values of $\bar{R}_i$ and $\bar{S}_i$ are to be proper. This time is called the *set-up time*, and it is illustrated in Fig. 6-20b. In general, the minimum set-up times are specified by the manufacturer. Note that the set-up time for a 0 is, in general, different from the set-up time for a 1. If the set-up times in an actual circuit are too short, the circuit may not function properly.

In addition to specifying requirements on the input signals, manufacturers also describe the response of the circuit. For instance, the time that it takes for the flip-flop to change its state is given in the data sheets. This is shown in Figs. 6-20c and 6-20d. The response time is measured from the 50 percent points of the clock and output signals as shown. Often, digital circuits can switch from a high to a low value at a faster rate than they can switch from a low to a high value. Thus, two different switching specifications are given. Note that $t_{QHL} < t_{QLH}$.

Flip-Flop Symbols

The symbol shown in Fig. 6-21 is a block diagram for an R–S master-slave or edge-triggered flip-flop. The clock terminal is indicated by the V-shaped symbol for the clock input. If the output changes on the positive-going edge of the clock pulse, then the symbol is as shown in Fig. 6-21a. If the output changes at the negative-going edge of the clock pulse, then the symbol is that shown in Fig. 6-21b (the circle is added). Note that different polarities of clock signals can be used so that

(a) (b)

Figure 6-21 Symbols for master-slave or edge-triggered R–S flip-flops. (a) output changes on positive-going edge of clock pulse; (b) output changes on negative-going edge of clock pulse

different circuits can be constructed. Also, we have assumed that a high voltage represents a 1 and that a low voltage represents a 0. However, the converse may be true. This will be discussed in greater detail subsequently.

Preset and Clear Leads

We have assumed that the changes of state of a flip-flop can only occur in accordance with the clock pulses. This is usually the case, but integrated-circuit flip-flops are provided with leads that can be used to set the flip-flop ($Q=1$, $\bar{Q}=0$), or clear it ($Q=0$, $\bar{Q}=1$), independent of the presence or absence of the clock pulse. Such leads are called *PRESET* and *CLEAR*. In Fig. 6-22 we illustrate the block diagram for an R–S flip-flop with such leads. In the circuit of Fig. 6-22a, the PRESET and CLEAR leads would normally be connected to a signal represented by a 0-voltage level. If we want to change the state, for instance, to set $Q=1$, $\bar{Q}=0$, then the PRESET lead would momentarily be set equal to a voltage equal to the level of a 1. Similarly, if we want to set the state $Q=0$, $\bar{Q}=1$, then the CLEAR lead would momentarily be set to the level of a 1. In some flip-flops a converse condition exists, where the clear levels are normally kept at a 1-level, and are momentarily brought to the 0-level to actuate them. For in-

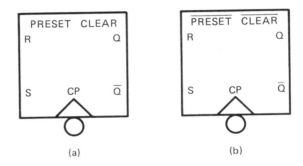

Figure 6-22   R–S flip-flops with preset and clear leads. (a) high signals preset and clear; (b) low signals preset and clear

stance, if we want to set the state $Q=0$, $\bar{Q}=1$, then the $\overline{CLEAR}$ lead would be momentarily brought to the level of a 0. The bar over the words $\overline{PRESET}$ or $\overline{CLEAR}$ in the block diagram indicates this situation.

In the remainder of the book, unless we state otherwise, it will be assumed that master-slave, or other trailing edge-triggered flip-flops are used when we work with synchronous sequential circuits.

6.3 SOME BASIC ANALYSIS TOOLS FOR SYNCHRONOUS SEQUENTIAL CIRCUITS

In general, sequential circuits consist of a combination of a great many flip-flops and logic elements. In this section we shall consider an efficient procedure that can be used to analyze synchronous sequential circuits.

Transition Tables

We can characterize the operation of a sequential circuit by a table. called a *transition table*. Suppose that we have the clocked sequential circuit of Fig. 6-23. Note that, for simplification, we often omit the clock connection. However, all clock terminals are connected to the clock circuit. The boxes represent master-slave, or trailing-edge-triggered flip-flops. We use D and T flip-flops here. Remember that the state of the T flip-flop changes whenever a 1 is applied and the state of the D flip-flop is a 1 if the input is a 1 and is 0 if the input is 0.

There are three classes of quantities with which we are concerned: the input variables, in this case just one, $x(t)$; the output variables, in this case just one, $z(t)$; and the internal states of the flip-flops which, in this case, are $y_1(t)$ and $y_2(t)$.

An applied input will, in general, cause the state to change. In characterizing the network, we list the present values of the input and of the current states. Then, the next state is listed. The *current* value of the output is also listed. Note that the output does not change until after the next clock pulse. In Table 6-7 we characterize the circuit of Fig. 6-23.

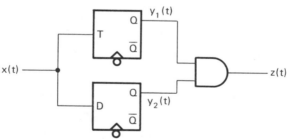

Figure 6-23 A simple sequential network

Table 6-7 A Tabular Characterization of the Sequential Network of Fig. 6-23

Present Input $x(t)$	Present State $y_1(t)$ $y_2(t)$		Next State $y_1(\tau+\tau_p)$ $y_2(\tau+\tau_p)$		Present Output $z(t)$
0	0	0	0	0	0
0	0	1	0	0	0
0	1	0	1	0	0
0	1	1	1	0	1
1	0	0	1	1	0
1	0	1	1	1	0
1	1	0	0	1	0
1	1	1	0	1	1

This table provides the pertinent information. However, since all possible combinations of inputs and states are listed, a table of this type will be very long when complex circuits are encountered. Thus, a shorthand notation is used to condense the table. The next state and present output are combined into a single term, as illustrated in Fig. 6-24. Note

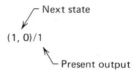

Figure 6-24   An illustration of the notation used in the transition table

that the slash separates the *next* state from the *present* output. Table 6-8 is such a representation; it is called a *transition table*.

Table 6-8   Transition Table for the Network of Fig. 6-23

Current State y_1y_2	Current Input x	
	0	*1*
(0,0)	(0,0)/0	(1,1)/0
(0,1)	(0,0)/0	(1,1)/0
(1,0)	(1,0)/0	(0,1)/0
(1,1)	(1,0)/1	(0,1)/1

Next state and current output

Table 6-8 provides all the information of Table 6-7 in a much more condensed form. At times, the current state is represented by a single subscripted variable. For instance, (0,0) could be written as q_0, (0,1) as q_1, (1,0) as q_2, and (1,1) as q_3. This is especially helpful in design procedures when we may not have a specific binary representation of the state. In this case the table is called a *state table*. For instance, we can rewrite Table 6-8 as

Table 6-9   State Table for the Network of Fig. 6-23

Current State	Current Input x	
	0	*1*
q_0	$q_0/0$	$q_3/1$
q_1	$q_0/0$	$q_1/1$
q_2	$q_2/0$	$q_1/0$
q_3	$q_2/1$	$q_1/1$

It often clarifies the state (or transition) table if we represent it by a drawing. Examples of such drawings, called *state transition diagrams*, are shown in Fig. 6-25. The circles represent the states; the lines repre-

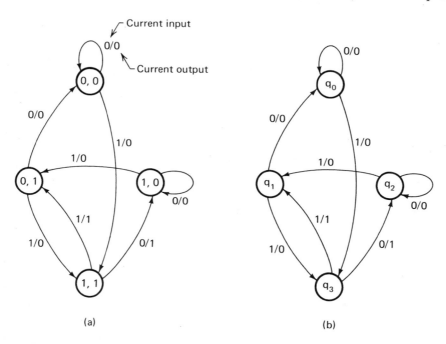

(a)

(b)

Figure 6-25 State transistion diagrams for the network of Fig. 6-23. (a) from the transition table; (b) from the state table

sent transition from one state to another. The arrowheads indicate the direction of the transition. The numbers separated by a slash represent the current input and currrent output, in the following manner:

current input/current output

Note that this is a different notation from the representation in the tables. Now suppose that we start in state (0,1) and we apply the input signals in the following sequence (each input coincides with a clock pulse) 1, 1, 0, 1, 0, 1, that is

$$x(0) = 1$$

$$x(\tau_p) = 1$$

$$x(2\tau_p) = 0 \tag{6-4}$$

$$x(3\tau_p) = 1$$
$$x(4\tau_p) = 0$$
$$x(5\tau_p) = 1$$

Now we want to determine the output signal $z(0)$, $z(\tau_p)$, $z(2\tau_p)$, . . . The first transition is to state $(1,1)$ and the current output is 0. The next transition is to state $(0,1)$ and the "current" output then is 1. Proceeding in this way, we obtain

$$z(0) = 0$$
$$z(\tau_p) = 1$$
$$z(2\tau_p) = 0$$
$$z(3\tau_p) = 0 \qquad\qquad (6\text{-}5)$$
$$z(4\tau_p) = 1$$
$$z(5\tau_p) = 0$$

The state sequence is

$$q_1 \rightarrow q_3 \rightarrow q_1 \rightarrow q_0 \rightarrow q_3 \rightarrow q_2 \rightarrow q_1 \qquad\qquad (6\text{-}6)$$

Note that the sequence depends upon the assumed starting state. That is, if we started with state q_3, a different sequence of states and outputs would be obtained.

We have considered a network with a single input and a single output. This need not be the case. For instance, a more complex case is shown in Fig. 6-26. In order to analyze this network, we must be able to express the behavior of flip-flops in terms of switching functions. Let us write general logical expressions for the output of the commonly-used flip-flops. Then we can obtain the logical expressions for the network.

D Flip-Flop

The state of a D flip-flop is 1 if the input is 1 and the state is 0 otherwise. Thus, we can write

$$Q(t+\tau_p) = D(t) \qquad\qquad (6\text{-}7)$$

where Q and D are identified in Fig. 6-7a. Thus, if $D(t) = 1$, then $Q(t+1) = 1$, etc.

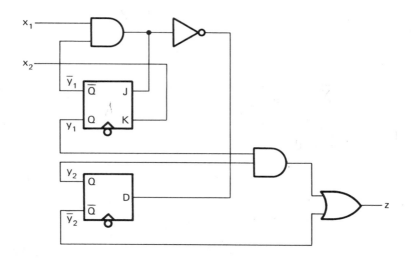

Figure 6-26 A more complex sequential network. The clock connections are not shown.

J-K Flip-Flop

The logic equation is

$$Q(t+\tau_p) = \bar{Q}(t)J(t) + Q(t)\bar{K}(t) \tag{6-8}$$

see Fig. 6-8a. Note that if $Q=0$ and we apply $J=1$, $K=0$, then $\bar{Q}J=1$. Thus, $Q(t+\tau_p) = 1$ as it should. If we check all possible inputs and outputs, the validity of Eq. (6-8) will be demonstrated. We leave this as an exercise for the reader.

R-S Flip-Flop

In this case we have

$$Q(t+\tau_p) = S(t) + Q(t)\bar{R}(t) \qquad R(t)S(t) = 1 \text{ not allowed} \tag{6-9}$$

Again, this can be verified by checking all possible combinations.

T Flip-Flop

In this case, the state changes if $T=1$. Thus, a logical expression that represents this is

$$Q(t+\tau_p) = \bar{Q}(t)T(t) + Q(t)\bar{T}(t) \tag{6-10}$$

Note that this can be obtained from Eq. (6-8) by setting

$J = K = T$

Now let us analyze Fig. 6-26. Using the circuit, we obtain

$$J(t) = x_1(t)\bar{y}_1(t) \tag{6-11a}$$

$$K(t) = x_2(t) \tag{6-11b}$$

$$D(t) = \bar{J}(t) \tag{6-11c}$$

input

Then, substitution in Eqs. (6-7) and (6-8) yields

$$y_1(t+\tau_p) = \bar{y}_1(t)x_1(t)\bar{y}_1(t) + y_1(t)\bar{x}_2(t) \tag{6-12a}$$

$$y_2(t+\tau_p) = \bar{x}_1(t) + y_1(t) \tag{6-12b}$$

From Fig. 6-26, we have

$$z(t) = y_1(t)y_2(t) + \bar{y}_2(t) \tag{6-12c}$$

Simplifying Eqs. (6-12), we obtain

$$y_1(t+\tau_p) = x_1(t)\bar{y}_1(t) + y_1(t)\bar{x}_2(t) \tag{6-13a}$$

$$y_2(t+\tau_p) = \bar{x}_1(t) + y_1(t) \tag{6-13b}$$

$$z(t) = y_1(t) + \bar{y}_2(t) \tag{6-13c}$$

output

Then, substituting in Eqs. (6-13) we can determine the transition table.

Table 6-10 Transition Table for the Network of Fig. 6-26

Current State (y_1, y_2)	Current Input (x_1, x_2)			
	(0,0)	(0,1)	(1,0)	(1,1)
(0,0)	(0,1)/1	(0,1)/1	(1,0)/1	(1,0)/1
(0,1)	(0,1)/0	(0,1)/0	(1,0)/0	(1,0)/0
(1,0)	(1,1)/1	(0,1)/1	(1,1)/1	(0,1)/1
(1,1)	(1,1)/1	(0,1)/1	(1,1)/1	(0,1)/1

We can use the table to plot the state transition diagram for this network, see Fig. 6-27. The notation used is

$$x_1 x_2/z \qquad (6\text{-}14)$$

If there is more than one set of inputs that will cause the same transition between the same two states, then they are enclosed in parentheses and only one arrow is drawn. For instance, both $x_1=0$, $x_2=1$ and $x_1=0$, $x_2=0$ will cause the transition from $y_1=0$, $y_2=0$ to $y_1=0$, $y_2=1$. The output sequence for a sequence of inputs is determined for this diagram just as it was for Fig. 6-26.

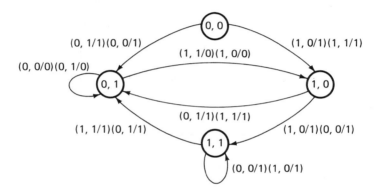

Figure 6-27 A state transition diagram for the network of Fig. 6-26

Equations (6-12) were relatively short. Thus, Eqs. (6-13) could be obtained by inspection. In general, this is not the case, and all the techniques of Chapter 5 can be used to simplify the expressions. Note that the right side of the equation, which gives the state or output, just consists of simple combinational logical expressions. Thus, these simplification techniques can be applied to them. (We shall illustrate this subsequently.)

6.4 DESIGN OF SYNCHRONOUS SEQUENTIAL CIRCUITS I — DETERMINATION OF THE NETWORK FROM THE TRANSITION TABLE

In this and the next sections we shall discuss a procedure used to design synchronous sequential circuits. There are several distinct steps and each shall be considered in turn. First we shall obtain the network from a specified transition table or diagram. Then, we shall consider how the transition table is obtained from general specifications. The simplification of the table will also be discussed.

We start by considering that the transition table is available and that the network is to be determined. Let us illustrate this with Table 6-8. (That table was obtained from the network of Fig. 6-23. However, we shall assume that this is not known and shall obtain the network from the table.) Repeating Table 6-8,

Table 6-11 A Transition Table

Current State (y_1,y_2)	Current Input x 0	1
(0,0)	(0,0)/0	(1,1)/0
(0,1)	(0,0)/0	(1,1)/0
(1,0)	(1,0)/0	(0,1)/0
(1,1)	(1,0)/1	(0,1)/1

In order to synthesize the network, we shall use a truth-table-type of formulation. This should include *all inputs to the flip-flops*. In general, the designer can choose any type of flip-flop to realize the network. However, the choice is not arbitrary in that the proper choice can result in a much simpler network. Unfortunately, it is not usually obvious which type of flip-flop is to be used. The design can be repeated using different flip-flops and then all the networks can be studied to see which is optimum. This is a tedious procedure. Nevertheless, it is often necessary. This ambiguity is common to most design procedures, not only to those involved with digital circuits. One of the designers most important tasks is to choose the best design when many designs which will do the job are available. (Determination of what is meant by "best design" is often a difficult job in itself.)

Let us assume that y_1 is the state of a T flip-flop and that y_2 is the state of a D flip-flop. This is an arbitrary choice and is done, for illustrative purposes, to conform with the original network. Now let us obtain the truth table. This will contain the inputs, present states, next states, present inputs to the flip-flops, and the present outputs. That table is

Table 6-12 Truth Table for the Network Whose Transition Table is given in Table 6-11

Input $x(t)$	Current State $y_1(t)$ $y_2(t)$		Next State $y_1(t+\tau_p)$ $y_2(t+\tau_p)$		Flip-Flop Input $T_1(t)$ $D_2(t)$		Current Output $z(t)$
0	0	0	0	0	0	0	0
0	0	1	0	0	0	0	0
0	1	0	1	0	0	0	0
0	1	1	1	0	0	0	1
1	0	0	1	1	1	1	0
1	0	1	1	1	1	1	0
1	1	0	0	1	1	1	0
1	1	1	0	1	1	1	1

The columns for $y_1(t+\tau_p)$, $y_2(t+\tau_p)$ and $z(t)$ are obtained directly from the transition table. The columns for $T_1(t)$ and $D_2(t)$ are obtained from a knowledge of the excitation tables for the T and D flip-flops. For instance, a T flip-flop changes its state when $T=1$ and the output of a D flip-flop is equal to D. Now we use the truth table to obtain logical expressions for all the flip-flop inputs and for the output. The procedures of Chapter 5 are used here. Then, considering the 1 terms in the T_1 column, we have

$$T_1(t) = x(t)\bar{y}_1(t)\bar{y}_2(t) + x(t)\bar{y}_1(t)y_2(t)$$

$$+ x(t)y_1(t)\bar{y}_2(t) + x(t)y_1(t)y_2(t) \qquad (6\text{-}15)$$

Since these all occur simultaneously, it is convenient to omit time specifications. Thus, we write

$$T_1 = x\,\bar{y}_1\bar{y}_2 + x\,\bar{y}_1 y_2 + x\,y_1\bar{y}_2 + x\,y_1 y_2 \qquad (6\text{-}16a)$$

Similarly,

$$D_2 = x\,\bar{y}_1\bar{y}_2 + x\,\bar{y}_1 y_2 + xy_1\bar{y}_2 + x\,y_1 y_2 \qquad (6\text{-}16b)$$

$$z = \bar{x}\,y_1 y_2 + x\,y_1 y_2 \qquad (6\text{-}16c)$$

The first step is to simplify these functions. Since they represent a multiple-output network, the procedures discussed in Sec. 5.10 can be used. Actually, the network is so simple that we can consider the functions separately. Let us use Karnaugh maps for the simplification. Then, from Eq. (6-16a) we obtain the Karnaugh map of Fig. 6-28a.

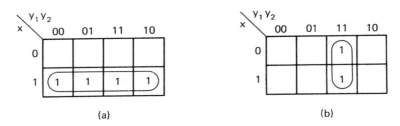

(a)   (b)

Figure 6-28 Karnaugh maps. (a) used to minimize Eqs. (6-16a) and (6-16b); (b) used to minimize Eq. (6-16c). Note that the labeling is not conventional

(Note that, to avoid confusion, we placed the x on the left side and y_1 and y_2 across the top. The conventional notation would have x and y_1 across the top and y_2 on the left side.) From these Karnaugh maps, we obtain

$$T_1 = x \qquad\qquad (6\text{-}17a)$$

and

$$D_2 = x \qquad\qquad (6\text{-}17b)$$

The Karnaugh map of Fig. 6-28b is used to minimize Eq. (6-16c). This yields

$$z = y_1 y_2 \qquad\qquad (6\text{-}17c)$$

Equations (6-17) allow us to construct the network. The resulting network is shown in Fig. 6-23.

The designer can arbitrarily specify the type of flip-flops. To illustrate this, realize the network using all D flip-flops. Table 6-12 must now be modified by changing T_1 to D_1. In this case, we have

Table 6-13 Truth Table Using D Flip-Flops for the Network Whose Transition Table is Table 6-11

Input $t=t$	Current State $t=t$		Next State $t=t+\tau$		Flip-Flop Input $t=t$		Present Output $t=t$
x	y_1	y_2	y_1	y_2	D_1	D_2	z
0	0	0	0	0	0	0	0
0	0	1	0	0	0	0	0
0	1	0	1	0	1	0	0
0	1	1	1	0	1	0	1
1	0	0	1	1	1	1	0
1	0	1	1	1	1	1	0
1	1	0	0	1	0	1	0
1	1	1	0	1	0	1	1

All the columns are the same as before except the one for D_1. Now we have

$$D_1 = \bar{x}\,y_1\bar{y}_2 + \bar{x}\,y_1y_2 + x\,\bar{y}_1\bar{y}_2 + x\,\bar{y}_1y_2 \qquad (6\text{-}18)$$

Minimizing this, we obtain

$$D_1 = \bar{x}y_1 + x\bar{y}_1 \qquad (6\text{-}19)$$

(This is the XOR operation.) From the previous results, we have

$$D_2 = x$$

$$z = y_1y_2$$

Thus, the network has the form shown in Fig. 6-29. Comparing this with Fig. 6-23, we see that the complexity of the network can vary greatly with the type of flip-flop used.

As a final example illustrating the possible choice of gate type, let us realize this network using J–K flip-flops. Now the truth table becomes

Table 6-14 Truth Table Using J–K Flip-Flops for the Network Whose Transition Table is Table 6-11

Input	Current State		Next State		Flip-Flop Input		Current Output
x	y_1	y_2	y_1	y_2	J_1K_1	J_2K_2	z
0	0	0	0	0	0 d	0 d	0
0	0	1	0	0	0 d	d 1	0
0	1	0	1	0	d 0	0 d	0
0	1	1	1	0	d 0	d 1	1
1	0	0	1	1	1 d	1 d	0
1	0	1	1	1	1 d	d 0	0
1	1	0	0	1	d 1	1 d	0
1	1	1	0	1	d 1	d 0	1

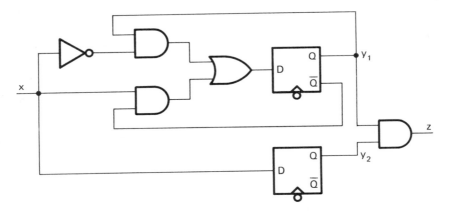

Figure 6-29 A realization, using D flip-flops, of the network characterized by transition Table 6-11. The clock connections are not shown.

We can now write the logical expressions. However, these steps can be skipped and the Karnaugh maps obtained directly from the truth table. For instance, if we are obtaining the map for J_1 then, corresponding to each $J_1 = 1$, we mark a 1 directly on the Karnaugh map. The maps for the various functions are shown in Fig. 6-30. Thus, we have

$$J_1 = x \qquad\qquad\qquad (6\text{-}20a)$$

$$K_1 = x \qquad\qquad\qquad (6\text{-}20b)$$

$$J_2 = x \tag{6-20c}$$

$$K_2 = \bar{x} \tag{6-20d}$$

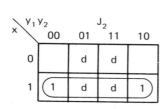
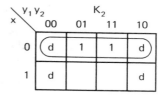

Figure 6-30 Karnaugh map used to minimize the J and K functions in Table 6-14. Note that the labeling is not conventional.

The expression for z is the same as before,

$$z = y_1 y_2$$

The realization for the network is given in Fig. 6-31.

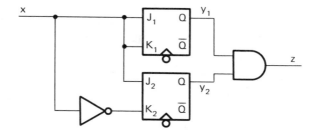

Figure 6-31 A realization, using J-K flip-flops, of the network whose transition table is given in Table 6-11. The clock connections are not shown.

6.5 DESIGN OF SYNCHRONOUS SEQUENTIAL CIRCUITS II – ASSIGNMENT OF BINARY VALUES TO STATES

The networks that we have discussed all reflect the transition diagram of either Fig. 6-25a or 6-25b. In Fig. 6-25a, the states do not have binary values assigned to them. Note that we do not have to know the binary values assigned to the states in order to determine the output sequence for a given input sequence. That is, the sequence is independent of the specific binary values assigned to the states. Thus, we can use different states of binary variables and still obtain the same input and output responses. Let us illustrate this. Using Fig. 6-25b we can obtain the following transition table.

Table 6-15 A Transition Table

Current State $(y_1 y_2)$	Current Input x	
	0	1
q_0	$q_0/0$	$q_3/0$
q_1	$q_0/0$	$q_3/0$
q_2	$q_2/0$	$q_1/0$
q_3	$q_2/1$	$q_1/1$

Although the assigned binary values of the states do not affect the input and output sequence, they will affect the complexity of the network. However, there is no way of telling beforehand which assignment to use. Just as in the case of the choice of flip-flops, we can try all possible assignments of binary values to the states. However, in a complex network, there will be very many choices and this will become an extremely tedious procedure.

If there are n states, then we need at least K bits, where K is the smallest integer satisfying

$$K \geqslant \log_2 n$$

Here we have four states, so two bits will be required. If there are many states, there can be many bits and, thus, very many choices. (Sometimes more than the minimum number of bits is used since such realizations are occasionally superior to the ones using the minimum number of bits. However, there is no algorithm for generating these realizations.)

Let us now consider an example where we assign states which are different than those used before. In particular, let

$$q_0 = 1,1 \qquad (6\text{-}21\text{a})$$

$$q_1 = 0,1 \qquad (6\text{-}21\text{b})$$

$$q_2 = 0,0 \qquad (6\text{-}21\text{c})$$

$$q_3 = 1,0 \qquad (6\text{-}21\text{d})$$

Then, the transition table becomes

Table 6-16

Current State $(y_1 y_2)$	Current Input x 0	1
(1,1)	(1,1)/0	(1,0)/0
(0,1)	(1,1)/0	(1,0)/0
(0,0)	(0,0)/0	(0,1)/0
(1,0)	(0,0)/1	(0,1)/1

Let us realize this using J–K flip-flops. The truth table is

Table 6-17 Truth Table for Table 6-16 Using J–K Flip-Flops

Input	Current State		Next State		Current Flip-Flop Input		Current Output
x	y_1	y_2	y_1	y_2	$J_1 K_1$	$J_2 K_2$	z
0	0	0	0	0	0 d	0 d	0
0	0	1	1	1	1 d	d 0	0
0	1	0	0	0	d 1	0 d	1
0	1	1	1	1	d 0	d 0	0
1	0	0	0	1	0 d	1 d	0
1	0	1	1	0	1 d	d 1	0
1	1	0	0	1	d 1	1 d	1
1	1	1	1	0	d 0	d 1	0

In Fig. 6-32 we have the various Karnaugh maps which are used to minimize the switching functions for J_1, K_1, J_2, K_2, and z. Thus we have

$$J_1 = y_2 \tag{6-22a}$$

$$K_1 = \bar{y}_2 \tag{6-22b}$$

$$J_2 = x \tag{6-22c}$$

$$K_2 = x \tag{6-22d}$$

$$z = y_1 \bar{y}_2 \tag{6-22e}$$

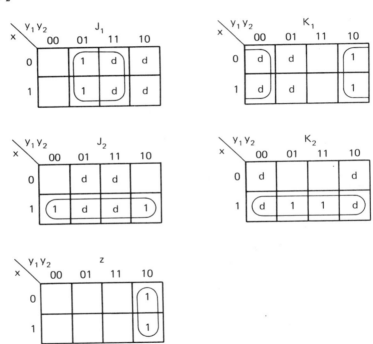

Figure 6-32 Karnaugh maps used to minimize the J, K and z functions of Table 6-17. Note that the labeling is not conventional.

The realization of the network is shown in Fig. 6-33. It is somewhat simpler than that of Fig. 6-31. In large, complex networks, it is possible that differences in assignments of states can result in substantial differences in the complexity of the final network.

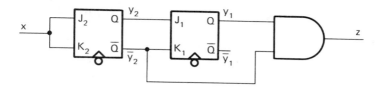

Figure 6-33 A realization, using J–K flip-flops, of the network whose transition table is given in Table 6-15. The assignments of Eqs. (6-21) are used. The clock connections are not shown.

6.6 DESIGN OF SYNCHRONOUS SEQUENTIAL CIRCUITS III – REDUCTION OF THE TRANSITION TABLE AND DIAGRAM

Sometimes, there are two identical rows in the transition table, such as the rows for q_0 and q_1 in Table 6-15. Note that the same inputs produce *both* the *same transitions* and the *same outputs*. When this happens, we can eliminate one state. For instance, if we are in state q_1 and the current input is a 0, then the next state will be q_0 with a current output of 0. This same statement could be made for state q_0. An equivalent set of statements could be made for $x = 1$. Thus, there would be no change in the results if we replaced q_1 with q_0 wherever it occurred and eliminated the row for q_1. (In general, we try to eliminate more than one row by repeating this procedure, if possible.) Note that, in this case, this procedure does not reduce the number of flip-flops required because although there are now only three states, two flip-flops are still required. If we could reduce the number of states by one more, then one less flip-flop would be necessary. In a practical design, it is unlikely that this will happen since the amount of storage required can usually be determined by the statement of the problem. We shall discuss this in the next section. Note that, in Table 6-15, we cannot combine q_2 and q_3 since the output terms are different.

Let us illustrate this procedure with an example. Suppose that, in Table 6-15, we eliminated state q_1 since the q_1 and q_0 rows are identical. Then the q_1 row of the table is closed out and all references to q_1 in the remainder of the table are replaced by references to q_0. The reduced table then becomes:

Table 6-18 A Reduced Form of Table 6-15

Current State (y_1, y_2)	Current Input x	
	0	1
q_0	$q_0/0$	$q_3/0$
q_2	$q_2/0$	$q_0/0$
q_3	$q_2/1$	$q_0/1$

We can use this table to draw a new state transition diagram. It is shown in Fig. 6-34. Note that this can be obtained from Fig. 6-25b in the following way: Remove all branches that point away from the q_1 circle.

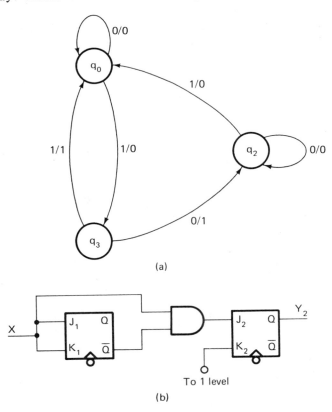

(a)

(b)

Figure 6-34 (a) A reduced form of Fig. 6-25b; (b) a realization using J–K flip-flops

This is equivalent to removing the "q_1 row" from the transition table. Now move the ends of any branch that point *toward* the q_1 circle so that they now point to the q_0 circle. This is equivalent to changing the references in the body of the table from q_1 to q_0. Let us use the state assignments:

$q_0 = 0,0$

$q_2 = 1,0$

$q_3 = 1,1$

The reduced transition table becomes:

Table 6-19.

	Current State (y_1,y_2)	Current Input x	
		0	1
	(0,0)	(0,0)/0	(1,1)/0
	(1,0)	(1,0)/0	(0,0)/0
	(1,1)	(1,0)/1	(0,0)/1

The truth table then becomes:

Input	Current State		Next State		Current Flip-Flop Input		Current Output
x	y_1	y_2	y_1	y_2	$J_1 K_1$	$J_2 K_2$	
0	0	0	0	0	0 d	0 d	0
0	0	1	d	d	d d	d d	d
0	1	0	1	0	d 0	0 d	0
0	1	1	1	0	d 0	d 1	1
1	0	0	1	1	1 d	1 d	0
1	0	1	d	d	d d	d d	d
1	1	0	0	0	d 1	0 d	0
1	1	1	0	0	d 1	d 1	1

There are new additional don't care conditions (because we have eliminated a state) that may simplify some of the functions. A study of the Karnaugh maps yields the following functions:

$J_1 = x$

$K_1 = x$

$J_2 = x\bar{y}_1$

$K_2 = 1$

$z = y_2$

The realization is shown in Fig. 6-34b. In this case we have not saved any gates by reducing the transition table, although it is possible that a different state assignment could eliminate the need for the AND gate. However, in many cases, combining rows in the transition table will reduce the number of gates required.

6.7 DESIGN OF SYNCHRONOUS SEQUENTIAL CIRCUITS IV — DETERMINATION OF STATE TRANSITION TABLES AND DIAGRAMS

We shall now consider some more complex examples where we shall determine the state transition tables from some design specifications. That is, given the task the circuit is to perform, we shall determine the state transition table and, finally, the network.

Before we consider some specific examples, let us consider some definitions. Certain sequential circuits' outputs are functions only of the present states and not of the present input. For instance, in the examples just considered, $z = y_1 y_2$ is a function of the state alone, and not of the input. Of course, the states *are* functions of the input. Such a device is called a *Moore machine* or *Moore network*. If the output is a direct function of the input and the states (e.g., $z = xy_1 y_2$) then the device is called a *Mealy machine* or *Mealy network*. G. H. Mealy and E. F. Moore are the names of the early investigators of these devices.

If a Moore circuit has no input, then it is called an *autonomous network*. (Note that a clock pulse is not considered to be an input.) It might seem as though such a network would have no use. However, they are actually very important. We shall consider several of them here. A circuit such as this could be used as a counter. Suppose that we have a synchronous sequential network which changes its state with each clock pulse. By determining the state of the network, we can determine how many clock pulses have elapsed. Such a circuit can have many uses. It can be used to keep track of the number of steps performed by a computer. Or, suppose that the clock pulses come from impulses received from an electric eye on a production line. The counter can then determine the number of items produced. Similarly, the number of dimes passing a point on a change sorter could be counted. Another use of a counter is to determine the frequency of an oscillator very accurately. The output of the oscillator is first shaped to be a pulse and then applied to the clock pulse terminal through an AND gate. The output input to the AND gate is a single pulse whose time duration is known very accurately. Thus, the oscillator is effectively connected to the clock pulse terminal for a standard length of time. The counter is then used to determine the number of cycles of the oscillator and the frequency can, thus, be accurately determined.

Let us now design a counter that will count from 0 to 7 and then reset itself. This is a *modulo 8* counter. Actual counters may count to much higher values. However, this would just involve a duplication of circuitry. We will keep the number small to avoid obscuring the design with unnecessary details. There must be eight different states and the circuit must change state with each clock pulse. No state can repeat until all states have been used. Then the circuit must return to the original state and the process is repeated. In this way each state will indicate a different count.

The state transition diagram for a modulo 8 counter is shown in Fig. 6-35. No input values are given since there are no inputs. The output values are the same as the past state. There are eight states and we shall use the following state assignments.

Table 6-21 Transition Table for a Modulo 8 Counter

Current State (y_1,y_2,y_3)	Current Input None
(0,0,0)	(0,0,1)/(0,0,0)
(0,0,1)	(0,1,0)/(0,0,1)
(0,1,0)	(0,1,1)/(0,1,0)
(0,1,1)	(1,0,0)/(0,1,1)
(1,0,0)	(1,0,1)/(1,0,0)
(1,0,1)	(1,1,0)/(1,0,1)
(1,1,0)	(1,1,1)/(1,1,0)
(1,1,1)	(0,0,0)/(1,1,1)

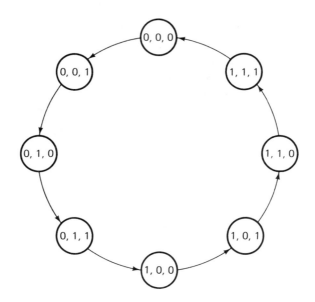

Figure 6-35   State transition diagram for a modulo 8 counter

Note that three states are needed. In general, if there are n flip-flops, then

$$2^n$$

values can be stored. Since eight values are to be stored, we must use $n=3$.

Now let us obtain the truth table for this device. We shall omit the output since it is the same as the current state. Thus, the y_1, y_2, and y_3 leads can be used as the output leads. We shall build the counter using J–K flip-flops.

Table 6-22 Truth Table for the Modulo 8 Counter

| *Current State* | | | *Next State* | | | *Current Flip-Flop Inputs* | | |
y_1	y_2	y_3	y_1	y_2	y_3	$J_1 K_1$	$J_2 K_2$	$J_3 K_3$
0	0	0	0	0	1	0 d	0 d	1 d
0	0	1	0	1	0	0 d	1 d	d 1
0	1	0	0	1	1	0 d	d 0	1 d
0	1	1	1	0	0	1 d	d 1	d 1
1	0	0	1	0	1	d 0	0 d	1 d
1	0	1	1	1	0	d 0	1 d	d 1
1	1	0	1	1	1	d 0	d 0	1 d
1	1	1	0	0	0	d 1	d 1	d 1

The Karnaugh maps corresponding to this truth table are given in Fig. 6-36. Thus,

$$J_1 = y_2 y_3 \tag{6-23a}$$

$$K_1 = y_2 y_3 \tag{6-23b}$$

$$J_2 = y_3 \tag{6-23c}$$

$$K_2 = y_3 \tag{6-23d}$$

$$J_3 = 1 \tag{6-23e}$$

$$K_3 = 1 \tag{6-23f}$$

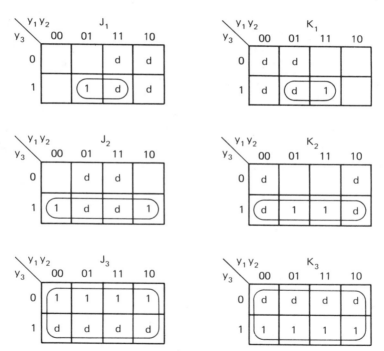

Figure 6-36 Karnaugh maps used in the design of the modulo 8 counter

The network that realizes this is shown in Fig. 6-37. Note that the J–K flip-flops are actually all connected as T flip-flops.

The inputs J_3 and K_3 are always 1's. They are connected to the clock pulse terminal so that whenever a clock pulse occurs, $J_3 = 1$ and $K_3 = 1$.

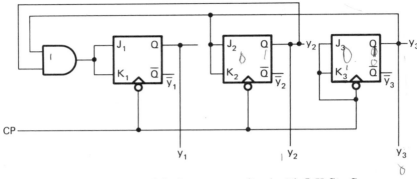

Figure 6-37   A modulo 8 counter realized with J-K flip-flops

We designed the counter so that the binary numbers corresponding to 0–7 were generated in sequence. Actually, we need not do this. For instance, the sequence could be 000, 101, 001, 110, 010, 111, 011, and 100 or any other sequence. This is equivalent to changing the assignment of the binary values of the states as discussed earlier in the section. It is not necessary that the counter have eight states. For instance, using three flip-flops, we could count the sequence 000, 001, 010, 011, 100, 101 and then repeat. The details of generating such networks follow those already discussed.

We have considered circuits that generate a sequence of numbers. Now let us design a circuit that will identify when a specific sequence occurs. For instance, suppose that we want the output of a network to be a 1 each time that the sequence 010 appears at the input. For instance, the sequence

01101010

would produce two 1's.

Now let us see how the state transition diagram can be obtained. If a 1 follows a 1, then the desired sequence cannot be initiated. Let us call this state the starting state q_0. A first segment of the state transition diagram is shown in Fig. 6-38a. If there are any successive 1 inputs, the circuit should remain in state q_0 with no output, see Fig. 6-38b. Any 0 input is a possible first term in the desired 010 sequence. This should initiate a new state, q_1. Of course, the output should remain a 0 at this time. This next step is shown in Fig. 6-38c. A 0 input at this time should leave the state unchanged. This is also shown in Fig. 6-38c. (Note that *any* sequence of 0's followed by a 10 produces the desired 010 sequence.)

If a 1 input *follows* a 0 input, then a *possible* 010 sequence may occur. Thus, this should lead to a new state q_2, see Fig. 6-38d. Finally, if the state is q_2, and the next input is a 0, then the sequence 010 has been received. Thus, a 1 should now be output. This should also result in the next state's being q_1, since a 0 was input, see Fig. 6-38e. On the other hand, if we are in state q_2, and the next input is a 1, then we must return to state q_0, with 0 output, since the 010 sequence cannot be initiated until the next 0 is input. This transition is also shown in Fig. 6-38e. Figure 6-38e is the complete state transition diagram since all possible input sequences have been considered. The state transition table is

Table 6-23 Transition Table for Fig. 6-38

Current State	Current Input x	
	0	1
q_0	$q_1/0$	$q_0/0$
q_1	$q_1/0$	$q_2/0$
q_2	$q_1/1$	$q_0/0$

Let us use the following binary assignments (for purposes of example, we do not make the assignments "agree" with the subscripts).

$q_0 = 1,0$ $\qquad\qquad q_1 = 0,0 \qquad\qquad q_2 = 0,1$

The truth table is then

Table 6-24 Truth Table Corresponding to Table 6-23

Current Input	Current State		Next Input		Current Flip-Flop Input		Current Output
x	y_1	y_2	y_1	y_2	D_1	D_2	z
0	0	0	0	0	0	0	0
0	0	1	0	0	0	0	1
0	1	0	0	0	0	0	0
0	1	1	-------		d	d	d
1	0	0	0	1	0	1	0
1	0	1	1	0	1	0	0
1	1	0	1	0	1	0	0
1	1	1	-------		d	d	d

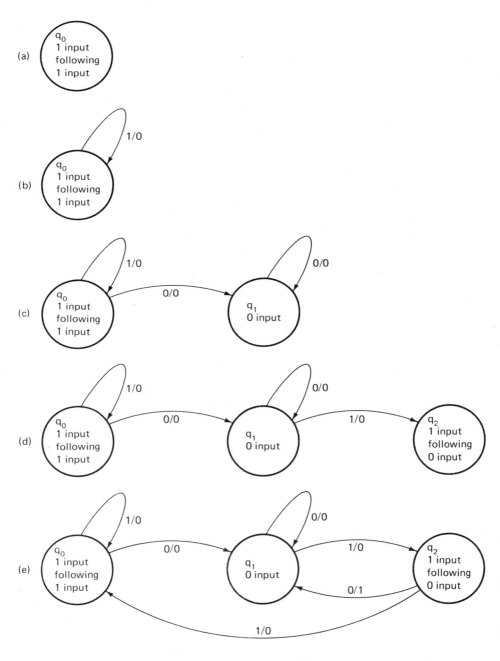

Figure 6-38 State transition diagram for the 010 sequence detector. (a) first step; (b) second step; (c) third step; (d) fourth step; (e) final diagram

Note that the state 1,1 will never occur. Thus, we "don't care" about the flip-flop inputs or the z outputs in these cases. These "don't cares" in the truth table result in a simplification of the logic functions. Proceeding as before, we obtain

$$D_1 = x\,y_2 + x\,y_1 = x\,(y_1 + y_2) \tag{6-24a}$$

$$D_2 = x\,\bar{y}_1\bar{y}_2 \tag{6-24b}$$

$$z = \bar{x}\,y_2$$

The network which realizes the 010 sequence detector is shown in Fig. 6-39. Some typical voltages are shown in Fig. 6-40.

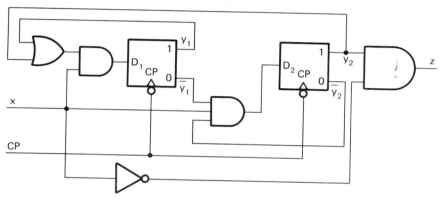

Figure 6-39 A realization of the 010 sequence detector using D flip-flops

6.8 ASYNCHRONOUS SEQUENTIAL CIRCUITS — RACES

Now let us consider unclocked or asynchronous sequential circuits. Actually, we have already discussed these. For instance, a flip-flop is an asynchronous circuit. Even if it is a clocked flip-flop its internal behavior is asynchronous. For instance, consider Fig. 6-6a. Suppose that $Q=1$ and $\bar{Q}=0$, and we apply the input $S=0$, $R=1$. During a clock pulse, Q becomes 0 and after that $\bar{Q}$ becomes 1. These changes are asynchronous. That is, although they take place during a clock pulse, the rate at which Q and then $\bar{Q}$ changes is not controlled by the clock pulse. Of course, the rate of change has to be fast enough if proper operation is to result. Portions of sequential circuits operate asynchronously. For instance, in Fig. 6-39, the logic gates respond to their inputs without regard to the clock pulses whereas the flip-flops are synchronized. Of course, there are many sequential circuits which operate independently of any clock pulse. We shall discuss these circuits here.

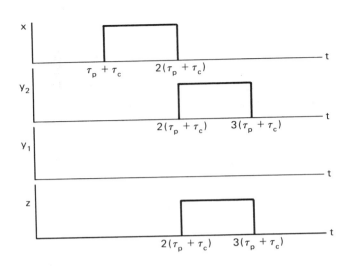

Figure 6-40 Some typical idealized waveforms in the circuit of Fig. 6-39

Let us illustrate the analysis of an asynchronous circuit by considering one which has the following switching expressions.

$$y_1 = \bar{x}_1 \bar{y}_2 + \bar{x}_1 x_2 y_1 \tag{6-25a}$$

$$y_2 = \bar{x}_1 \bar{x}_2 \bar{y}_1 + x_1 x_2 \bar{y}_1 + \bar{x}_1 x_2 y_1 y_2 \tag{6-25b}$$

Let us consider that y_1 and y_2 are the states of the system and that x_1 and x_2 are the independent inputs. Now let us construct a transition table for this system. In it we list the present state and present input and the succeeding (next) state, or *successor* state, is given for every possible combination of inputs. The values in the table are found by substitution in Eqs. (6-25). For instance, if the present state is $y_1=0$, $y_2=0$, and the present input is $x_1=0$, $x_2=0$, then the next state is $y_1=1, y_2=1$.

Table 6-25 Transition Table for Eqs. (6-25)

Present State $y_1 y_2$	Present Input $x_1 x_2$			
	0 0	*0 1*	*1 1*	*1 0*
(0,0)	(1,1)	(1,0)	(0,1)	(0,0)
(0,1)	(0,1)	(0,0)	(0,1)	(0,0)
(1,1)	(0,0)	(1,1)	(0,0)	(0,0)
(1,0)	(1,0)	(1,0)	(0,0)	(0,0)

successor states

Note that, in the table, the ordering of the present state y_1y_2 and the present input x_1x_2 is such that there is only a *single* change of variable between adjacent terms.

The details of the operation are considerably different from the synchronous case. For instance, suppose that the present state is $(0,1)$, i.e., $y_1=0$, $y_2=1$, and the present input is $x_1=1$, $x_2=1$. Now suppose that the input switches to $x_1=0$, $x_2=1$ and then does not change. Then the next state will be

$$y_1 = 0 \qquad\qquad\qquad\qquad\qquad (6\text{-}26a)$$

$$y_2 = 0 \qquad\qquad\qquad\qquad\qquad (6\text{-}26b)$$

The network will not remain in this state. If Eqs. (6-26) are substituted into Eqs. (6-25) with $x_1=0$, $x_2=1$, then the state becomes $y_1=1$, $y_2=0$ $(1,0)$. Note that we do not have to substitute in Eqs. (6-25) since this can be determined from Table 6-25. Now, if this is the state, then the next state will be the same, $(1,0)$. Thus, the system remains in this state. Therefore, an asynchronous network may go through a sequence of states before it comes to its final state. This is called a *cycle*. Let us illustrate this on the transition table by drawing the transitions that take place. To reduce the clutter we shall omit the parentheses.

Table 6-26 Excitation Table Illustrating Races and Cycles

Present State y_1y_2	Present Input x_1x_2			
	0 0	*0 1*	*1 1*	*1 0*
0 0	1 1	1 0	0 1	0 0
0 1	0 1	0 0	0 1	0 0
1 1	0 0	1 1	0 0	0 0
1 0	1 0	1 0	0 0	0 0

We have marked the transition that takes place with the arrows labeled *a*. The $(1,0)$ is encircled in the table to indicate that it is a final state. That is, this state will not change as long as the input does not change.

Now suppose that we are in state $(0,0)$ and the input switches from $x_1 = 1$, $x_2 = 0$ to $x_1 = 0$, $x_2 = 0$. We start from the state $(0,0)$. The state will change to $(1,1)$. It will then change to $(0,0)$ which, in turn, will change to $(1,1)$. This transition is marked with a b in Table 6-26. This will repeat continuously. Thus, an *oscillation* will be set up. This *unstable condition* is indicated by the dashed lines in the first column of Table 6-26. Such operation should always be avoided.

Another situation can exist. Assume that the state is $(0,0)$ and that the inputs are $x_1 = 1$, $x_2 = 0$. Now the inputs are switched as before so that the new inputs are $x_1 = 0$, $x_2 = 0$. The next state should be $(1,1)$. However, y_1 and y_2 *might not change simultaneously*. Suppose that y_1 changes before y_2. The state will shift to $(1,0)$. Now, if the state is $(1,0)$ and the input is $x_1 = 0$, $x_2 = 0$, then the next state should be $(1,0)$. That is, the state remains at $(1,0)$ which is stable operation. This is indicated by the arrow marked c in Table 6-26. On the other hand, suppose that y_2 changes faster than y_1. Then, the next state will be $(0,1)$. If the state is $(0,1)$ and the input is $x_1 = 0$, $x_2 = 0$, then the next state will be $(0,1)$. This, then, is a stable state and the circuit remains in this state. This is indicated on Table 6-26 by the arrow marked d. We see that the transition from $(0,0)$ can take one of three paths depending upon the speed of the various transitions. This is called a *race*. Note that races result when *more* than one of the binary variables must change simultaneously, since these variables may not all change at the same time. This type of race is called a *critical race* because the final state can have more than one value and, thus, erroneous results can occur.

To avoid clutter, the table is often rewritten omitting the numerical values for the successor states, with the various transitions marked. Table 6-27 is such a table and all the stable states are encircled as before.

Table 6-27 Transition Table for Eqs. (6-25) with Transitions of Table 6-26 Shown

Present State y_1y_2	Present Input x_1x_2			
	0 0	0 1	1 1	1 0
0 0			·	⊙
0 1	⊙	⊙ ← ⊙		·
1 1		⊙	·	·
1 0	⊙	⊙	·	·

The intermediate states are indicated by dots. The stable final states are encircled. To avoid cluttering the representation, not every possible transition is shown.

The circuit which realizes Eqs. (6-25) is shown in Fig. 6-41. Let us see how the race condition can be eliminated. Suppose that we switch from an input $x_1 = 1$, $x_2 = 0$ to $x_1 = 0$, $x_2 = 0$. In this case, the three race

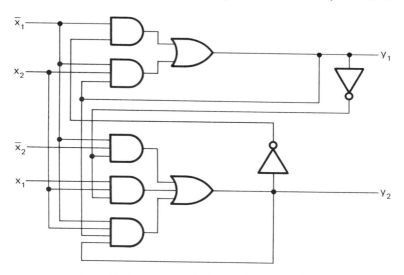

Figure 6-41 A network that realizes Eqs. (6-25)

conditions indicated in the $x_1 = 0$, $x_2 = 0$ column result. If we add delay appropriately to y_1 or y_2, then the "winner of the race" will be determined and the ambiguity will be removed. For instance, suppose that we add an additional state y_3 such that

$$y_3 = \bar{y}_1$$

and then modify Eqs. (6-25) to read

$$y_1 = \bar{x}_1 \bar{y}_2 + \bar{x}_1 x_2 y_3 \tag{6-27a}$$

$$y_2 = \bar{x}_1 \bar{x}_2 \bar{y}_3 + x_1 x_2 \bar{y}_3 + \bar{x}_1 x_2 y_3 y_2 \tag{6-27b}$$

$$y_3 = \bar{\bar{y}}_1 \tag{6-27c}$$

These are equivalent to Eqs. (6-25) in that, upon substitution of y_1 for y_3, we obtain the same y_1 and y_2 as was obtained in Eqs. (6-26). However, now the realization is as in Fig. 6-42. Suppose we are in the state $(0,0)$ and the input changes from $x_1 = 1$, $x_2 = 0$ to $x_1 = 0$, $x_2 = 0$; then the delay introduced by the two added NOT gates will cause y_2 to change before y_1 and the output will be $y_1=0$, $y_2=1$. Thus, the race has been eliminated. So we see that the addition of additional states can eliminate races. That is equivalent to saying that the addition of delay can eliminate races. The delay added by the additional gates must be sufficiently long. For instance, in Fig. 6-42, we assume that the two added NOT gates add enough delay so that y_2 changes before the upper input to the upper AND gate changes.

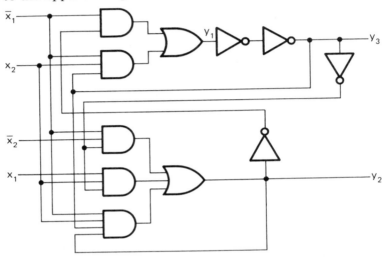

Figure 6-42 A network that realizes Eqs. (6-25) without the race condition

We have eliminated the race by the addition of delay. At times, the reace is eliminated by the assignment of different binary variables to the states. This will be discussed subsequently.

We have thus far considered that races must be avoided. Actually, there is one type of race that can be tolerated. Consider the simple excitation table, Table 6-28.

Table 6-28 A Simple Transition Table Illustrating A Noncritical Race

Present State $y_1 y_2$	Present Input x	
	0	1
0 0	1 1	0 0
0 1	0 1	1 1
1 1	0 1	1 1
1 0	0 1	1 0

Now suppose that we have an initial state (0,0) with an input $x=1$. Then the input switches to $x=0$. Then, the next state is (1,1) and then (0,1), which is a stable state. The transition from (0,0) to (1,1) involves a change of more than one variable. Thus, a race exists. Now suppose that the transition from (0,0) is to (0,1). This is the same stable state and it does not change. If the transition from (0,0) is to (1,0), then the next state will be (0,1), again the same stable state. Thus, the result of all the races is the same stable state. This is called a *noncritical race*. If the intermediate states are not important, then noncritical races can be tolerated and nothing need be done to eliminate them.

6.9 HAZARDS

Another problem can arise in asynchronous circuits or in the asynchronous portion of a synchronous one or in combinational circuits. Consider the circuit of Fig. 6-43. An expression for the output y is

$$y = x a + \bar{x} b \tag{6-28}$$

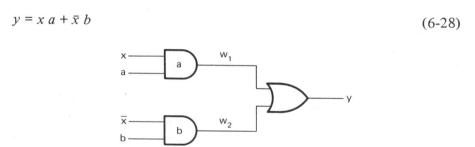

Figure 6-43 A simple gate network

Now suppose that x switches from a 1 to a 0 while $a=b=1$. The output of AND a becomes a 0 while that of AND b becomes a 1. Thus, y should remain constant at a 1. However, it is possible that the switching of the AND gates may not occur at exactly the same time. This is illustrated in the timing diagram of Fig. 6-44. Suppose that, as illustrated, w_1 switches at T_1 while w_2 switches at T_2 where $T_2 > T_1$. Then, there will be a short period of time $T_2 - T_1$ when both inputs to the OR gate will be 0. In this interval y will be 0 (or at least tend to decrease). Thus, we have the following situation. A change in x should produce no change in y (with $a=1$, $b=1$). However, because of unequal time delays, y becomes 0 for a short time. Such short changes in the variable can produce spurious responses and usually should be avoided. If the change in a *single* variable causes a momentary change in other

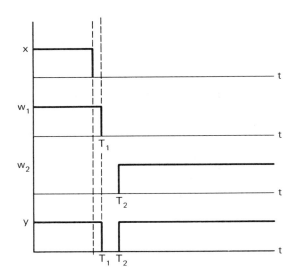

Figure 6-44 A timing diagram for the network of Fig. 6-43

variables, which should not occur, then a *hazard* is said to exist. This is actually called a *static hazard*.

Now let us see if we can eliminate the hazard. First we shall review some terminology. Two products of variables are said to be *adjacent* if they differ by just one variable which is complemented in one and is uncomplemented in the other. For instance, xab and $\bar{x}ab$ are adjacent. Now we can determine if a hazard exists by studying the Karnaugh map. If there are two adjacent products of input variables which do not lie in the same subcube, then a hazard exists. When the network is realized, each subcube results in an AND gate. Thus, if two adjacent input products are in different subcubes, then a situation such as that in Fig. 6-43 will exist and a hazard will result. That is, the change in a single input variable can cause the output of one AND gate to switch from 0 to 1 while the output of the other gate switches from a 1 to a 0 and a hazard results. Let us consider this. If we write Eq. (6-28) in the unminimized canonical sum of products form, it becomes

$$y = xa(b+\bar{b}) + \bar{x}b(a+\bar{a}) = xab + xa\bar{b} + \bar{x}ab + \bar{x}\bar{a}b \tag{6-29}$$

The Karnaugh map which corresponds to this is shown in Fig. 6-45. In Fig. 6-45a we have chosen subcubes which yield the minimal realization of Eq. (6-28) as shown in Fig. 6-43. This contains a hazard. Note that the arrow shows the two adjacent product terms which lie in different subcubes.

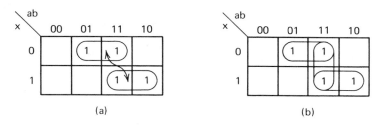

Figure 6-45 Karnaugh maps for Eq. (6-28). (a) minimial realization;
(b) hazard-free realization

To eliminate the hazard, we can add a third subcube as shown in Fig. 6-45b. Now the adjacent product terms lie in the same subcube. Thus we have

$$y = xa + \bar{x}b + ab \tag{6-30}$$

This realization is shown in Fig. 6-46. If $a=1$ and $b=1$, then the output of AND c will constantly be a 1 independent of x. Thus, we have eliminated the hazard.

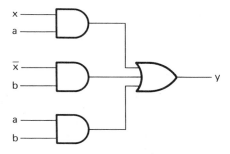

Figure 6-46 A hazard-free realization of Eq. (6-28)

Note that Eq. (6-30) is a redundant expression. That is, it can be reduced. However, it is not reduced and, to eliminate the hazard, the circuit is realized exactly according to Eq. (6-30). This increases the network complexity. In general, the price paid for the elimination of hazards is increased network complexity. In other words, the elimination of hazards requires that a compromise be made. We have considered hazards based on only one input variable since hazards are usually eliminated only for single input variables.

Another form of hazard can exist. Suppose that, upon switching the input variable, the output variable is to switch from 0 to 1. There may be a brief period when it shifts back and forth, e.g., the output sequence is 0101. The intermediate 0 and 1 persist only for a short time. This is called a *dynamic hazard*. Circuits which have been made free of static hazards are also free of dynamic hazards.

A third type of hazard occurs when a change in input variable affects one part of a circuit before another. Thus, the circuit can take on one stable state rather than another. This is analogous to a race condition, and is called an *essential hazard*. Such hazards can always be eliminated by incorporation of the appropriate delay.

6.10 DESIGN OF ASYNCHRONOUS SEQUENTIAL CIRCUITS

We shall now consider the design of asynchronous sequential circuits. Although there are some similarities between this and the design of synchronous circuits, there are important differences. We shall start by formalizing some definitions. In a synchronous circuit, there are pulses applied to the inputs. These, in conjunction with the clock pulses, cause changes in state. In an asynchronous circuit, we can speak of the inputs as *levels* which are either 0 or 1. These levels can persist for, or change at, arbitrary times. To make the design tractable, and to eliminate race-type operation, the operation is usually restricted to one of several modes.

Fundamental Mode Operation

A basic sequential circuit was shown in Fig. 6-1b. We shall redraw it here is a somewhat more specific way, see Fig. 6-47. Here we have a combinational circuit with feedback through delay elements. In the last section we discussed that such a circuit could cycle through a sequence of states before reaching a stable state. When a stable state has been reached, the system stops changing. Thus, the inputs and outputs of each delay have become equal.

$$y_{1d} = y_1 \qquad\qquad\qquad\qquad (6\text{-}31a)$$

$$y_{2d} = y_2 \qquad\qquad\qquad\qquad (6\text{-}31b)$$

$$y_{3d} = y_3 \qquad\qquad\qquad\qquad (6\text{-}31c)$$

$$\vdots$$

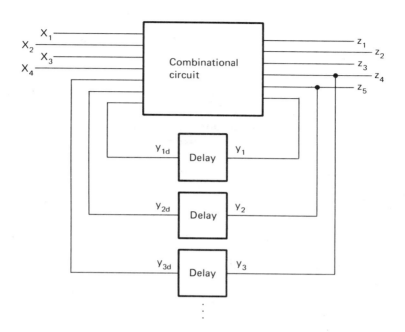

Figure 6-47 The representation of a simple asynchronous sequential circuit

Then, when all the inputs and output of the delays are equal, the system is said to be in its stable state. The set of variables $x_1, x_2, \ldots$ is called the *input state* while the $y_1, y_2, \ldots$ make up the *internal state*.

If a system is operating in a stable state, then it can only change its state if the input state changes. We shall restrict our operation in the following ways: (1) only one input variable can change at a time; (2) no other change in any input variable can occur until the system has settled down to a stable state; (3) each state will be stable for at least one set of inputs. When the operation satisfies conditions (1), (2), and (3), it is said to be exhibiting *fundamental mode operation*.

Now let us consider the design of a fundamental mode, sequential circuit. The specifications for sequential circuits are often stated by specifying the output signal for a given sequence of input signals. For instance, suppose that we have two input signals x_1 and x_2 and two output signals y_1 and y_2, and that we want a logic circuit to perform in the following way:

$y_1 = 1$ if and only if $x_1 = 1$ and $x_2 = 1$, following $x_1 = 0$, $x_2 = 1$ (6-32a)

$y_2 = 1$ if and only if $x_1 = 1$ and $x_2 = 1$, following $x_1 = 1$, $x_2 = 0$ (6-32b)

There are five possible states of input variable combinations: $(x_1 = 0,$ $x_2 = 0)$, $(x_1 = 1,$ $x_2 = 0)$, $(x_1 = 0,$ $x_2 = 1)$, $(x_1 = 1,$ $x_2 = 1$, following $x_1 = 0,$ $x_2 = 1)$ and $(x_1 = 1,$ $x_2 = 1$, following $x_1 = 1,$ $x_2 = 0)$. Note that we only consider different states for different input *sequences* for the states immediately preceding the input state $x_1 = 1$, $x_2 = 1$, since no other sequence has any effect on the output. A timing diagram for a typical sequence is shown in Fig. 6-48. We assume that the initial state is $x_1 = 0$, $x_2 = 0$. The five states are designated by number as shown in Fig. 6-48. We shall list them here.

1	$x_1 = 0 \; x_2 = 0$	
2	$x_1 = 0 \; x_2 = 1$	
3	$x_1 = 1 \; x_2 = 1$	following $x_1 = 0, x_2 = 1$
4	$x_1 = 1 \; x_2 = 0$	
5	$x_1 = 1 \; x_2 = 1$	following $x_1 = 1, x_2 = 0$

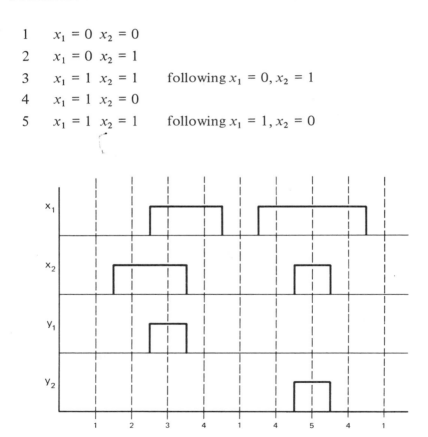

Figure 6-48   A typical timing diagram showing the assigned states for the network specified in (6-32)

The next step in the design is to set up some tables, called *primitive flow tables*. These show the transition between the internal states as a function of input as well as the stable states and the output.

Table 6-29 Partial Primitive Flow Table

"Present State"		*Input $x_1 x_2$*		
	0 0	*0 1*	*1 1*	*1 0*
1	①/0,0 → 2		—	
2		②/0,0 → 3		
3			③/1,0	

successor state/output $y_1 y_2$

In this table we show the "present state," then the "next state," and the resulting state. The output is also shown. However, we only list the output for stable states. We shall discuss this subsequently.

Now let us consider Table 6-29 in more detail. Let us start with a present input state of 1 (i.e., $x_1 = x_2 = 0$). Now suppose that the input is switched to $x_1 = 0$, $x_2 = 1$. This will lead to state 2. We note this by drawing a horizontal move into the $x_1 = 0$, $x_2 = 1$ column. This is labeled 2. However, it is not a stable state since this corresponds to a present state of 1. Thus, the next move is a vertical one to the "present state" 2-line. Now we encircle the state since the present state and the successor state are the same. Thus, this is a stable state. The output is also shown since this is a stable state. Next we assume that the input changes to $x_1 = 1$, $x_2 = 1$. Again, a similar set of transitions is shown.

Note the dash — in the 1 1 portion of the first row. We cannot make the change from $x_1 = 0$, $x_2 = 0$ to $x_1 = 1$, $x_2 = 1$ since only one input is allowed to change at a time. Thus, we put a dash in that position. Note that this is a type of "don't care" since this input condition can never occur. We can put down any state we wish there.

Now let us form the complete primitive flow table. We do not show arrows in this table to avoid cluttering it.

Table 6-30 Primitive Flow Table

"Present Input State"	Input x_1x_2			
	0 0	0 1	1 1	1 0
①	①/0,0	2	–	4
②	1	②/0,0	3	–
③	–	2	③/1,0	4
④	1	–	5	④/0,0
⑤	–	2	⑤/0,1	4

This flow table can be used in the synthesis of the network. However, before this is done, we shall reduce the table since this will result in a simplified network.

Reduction of the Primitive Flow Table

Suppose that two rows of the flow table are identical. This would indicate that the same new inputs would result in the same successor states, independent of the original states. Thus, it does not matter which state we were in originally. In this case, we can combine rows. Then there will be fewer states and the table will be simpler. This is similar to the discussion in Sec. 6-6. There we indicated that the merging of rows could only take place if the rows, including the output, were identical. Essentially the same holds true here. However, except for stable states, the output is not specified. Note that there will be one, and only one, stable state in each row. We will not be able to merge rows which have different stable states in the same column. However, for the unstable state, we do not have to consider the output. Actually, there are times when we want to specify the outputs during the transition times. We shall consider this subsequently.

Since Table 6-30 is short, the merging of rows can be done by inspection. However, it is desirable to systematize the procedure, since this will be helpful when large tables are considered.

We draw a diagram called a *merger diagram*. In this, we put down all the states. Then, each row is compared with all the others and any combinable pair of states is connected by a straight line, see Fig. 6-49.

Any two states can be combined if they are connected by a straight line in the merger diagram. Similarly, if there are three states, *each* one of which is connected to *both* of the other two, then these three rows can be merged. In addition, if there are four states, *each* of which is

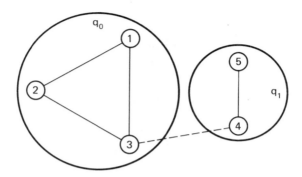

Figure 6-49 Merger diagram for the primitive flow table, (6-30). The dashed line should be ignored.

connected to *all* of the other three states, then these can be merged. In general, if there are n states, *each* of which is connected to *all other* $n{-}1$ states, then these states can be merged.

The merger indicated in Fig. 6-49 is the obvious one to use. However, suppose that the extra line shown dashed were added to the diagram. Then, we could divide into two states q_0 and q_1 as shown, or divide into three states, 1 and 2, 3 and 4, and 5. Probably, the first division (indicated on the figure) is preferable since only two states are needed. However, in complex cases, several trial designs may have to be made. We shall merge as shown in Fig. 6-49. Now we merge the rows as indicated. This yields

Table 6-31 Merged Flow Table

Present State	Present Input $x_1 x_2$				
	0 0	*0 1*	*1 1*	*1 0*	
q_0	$\boxed{q_0}/0,0$	$\boxed{q_0}/0,0$	$\boxed{q_0}/1,0$	q_1	successor states
q_1	q_0	q_0	$\boxed{q_1}/0,1$	$\boxed{q_1}/0,0$	and outputs

Remember that these tables give the input states plus the output. Now let us consider q_0 and q_1 which represent the internal state of the system. That is, each row in the merged flow table represents a different stable mode of response of the system. Hence, it must represent a different internal state. A transition diagram representing this is shown in Fig. 6-50. Just as in the case of synchronous sequential circuits, we must now assign binary variables to the states. To avoid having races, we must attempt to do this in such a way that no two internal state

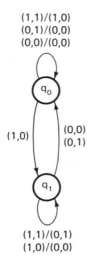

$$(1,1)/(1,0)$$
$$(0,1)/(0,0)$$
$$(0,0)/(0,0)$$

$$q_0$$

$$(1,0) \qquad (0,0)$$
$$(0,1)$$

$$q_1$$

$$(1,1)/(0,1)$$
$$(1,0)/(0,0)$$

Figure 6-50 State transition diagram for the states of Table 6-31

variables can change simultaneously. Of course, since we have only two internal state variables, we cannot have a race. Let us assign

$$q_0 = 0 \tag{6-33a}$$

$$q_1 = 1 \tag{6-33b}$$

We call the internal state y. To synthesize the network, we shall use Table 6-31 to set up the following truth table.

Table 6-32 Truth Table Used to Realize the Network Whose Merged Flow Table is Given in Table 6-31

Input		Current State	Next State
x_1	x_2	y	y
0	0	0	0
0	1	0	0
1	0	0	1
1	1	0	0
0	0	1	0
0	1	1	0
1	0	1	1
1	1	1	1

The Karnaugh map for this table is given in Fig. 6-51. Thus,

$$y = x_1 \bar{x}_2 + x_1 y \tag{6-34}$$

Note that, when the Karnaugh map minimization procedure is used, it should be done in such a way that hazards are avoided, see Sec. 6-9.

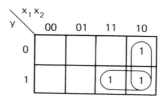

Figure 6-51 The Karnaugh map for Table 6-32

Now let us consider the output. From Table 6-31 we have: The z_1 output will only be 1 when $y=0$, $x_1=1$, $x_2=1$; similarly, z_2 will be 1 only when $y=1$ and $x_1=1$, $x_2=1$. In this simple case we do not have to minimize the functions. Thus,

$$z_1 = x_1 x_2 \bar{y} \tag{6-35a}$$

$$z_2 = x_1 x_2 y \tag{6-35b}$$

The network realization then takes the form shown in Fig. 6-52.
To provide a more complex example, let us not combine the last two rows of Table 6-30. Thus, we have

Table 6-33 Partially Merged Flow Table

Present State	Present Inputs $x_1 x_2$			
	0 0	0 1	1 1	1 0
q_0	$\widehat{q_0}$/0,0	$\widehat{q_0}$/0,0	$\widehat{q_0}$/1,0	q_1
q_1	q_0	—	q_2	$\widehat{q_1}$/0,0
q_2	—	q_0	$\widehat{q_2}$/0,1	q_1

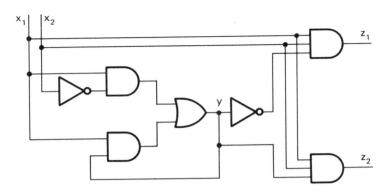

Figure 6-52 A network that realizes the requirements of Eqs. (6-32)

Then, the state transition diagram is shown in Fig. 6-53. There are three states. Thus, we need two binary variables. We did not assign the state

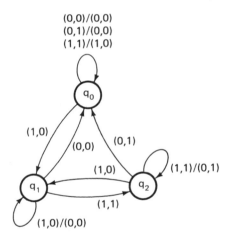

Figure 6-53 A state transition diagram for the states of Table 6-33

$(1,1)$ to avoid having a race condition. In general, the state assignment should be such that there are no races. If this cannot be done, then extra delay can be incorporated, see Sec. 6-8. The states are

$$q_0 = (0,0) \tag{6-36a}$$

$$q_1 = (0,1) \tag{6-36b}$$

$$q_2 = (1,0) \tag{6-36c}$$

Then, the truth table is

Table 6-34 Truth Table Used to Realize the Network Whose Merged Flow Table is Given in Table 6-33

Input		Present State		Next State	
x_1	x_2	y_1	y_2	y_1	y_2
0	0	0	0	0	0
0	1	0	0	0	0
1	0	0	0	0	1
1	1	0	0	0	0
0	0	0	1	0	0
0	1	0	1	—	—
1	0	0	1	0	1
1	1	0	1	1	0
0	0	1	0	—	—
0	1	1	0	0	0
1	0	1	0	0	1
1	1	1	0	1	0
0	0	1	1	—	—
0	1	1	1	—	—
1	0	1	1	—	—
1	1	1	1	—	—

The Karnaugh maps for y_1 and y_2 (which are obtained from the lines where $y_1 = 1$ and $y_2 = 1$, respectively) are given in Fig. 6-54. Thus, we have

$$y_1 = x_2 y_2 \qquad (6\text{-}37a)$$

$$y_2 = x_1 \bar{x}_2 \qquad (6\text{-}37b)$$

Then, from Table 6-33, we obtain

$$z_1 = \bar{y}_1 \bar{y}_2 x_1 x_2 \qquad (6\text{-}38a)$$

$$z_2 = y_1 \bar{y}_2 \qquad (6\text{-}38b)$$

The gate circuit realization is shown in Fig. 6-55. Note that this realization uses the *same* number of gates as does Fig. 6-52. In general, the reduction of the primitive flow table may not result in the reduction

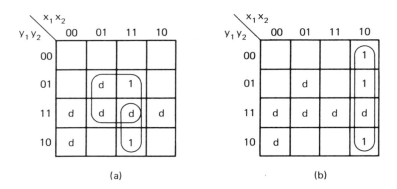

Figure 6-54   Karnaugh maps for Table 6-34. (a) for y_1; (b) for y_2

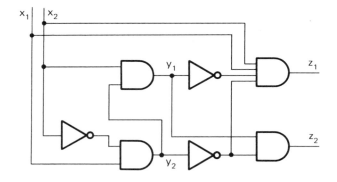

Figure 6-55   A network that realizes the requirements of Eqs. (6-32)

of the number of gates. All that *can* be stated is that reduction of the primitive flow table reduces the number of internal states.

At times, we want to specify the output during the intermediate unstable states. This can eliminate considerable output transients during the transition between states. In this case, separate states must be defined for these intermediate values. In general, fewer rows of the primitive flow table can be merged in this case.

Pulse Mode Operation

We have discussed fundamental mode operation. There is another mode of operation that is used. In this case, the input signals consist of short pulses which must satisfy two requirements: (1) they must be long enough so that the states can change; (2) they must be short enough so that all inputs are 0 after the memory elements have changed their states. This is called *pulse mode operation*. In this case, a pulse is applied at the input. The network changes its state. By the time the circuit has settled down, the pulse must be removed. The network's state then remains unchanged until the next input. A counter is an example of a pulse mode circuit.

6.11 INTEGRATED CIRCUIT IMPLEMENTATION OF SEQUENTIAL CIRCUITS

In this chapter we discussed the fundamental ideas behind the analysis and design of sequential circuits. Many complete sequential circuits can be purchased from manufacturers in the form of integrated circuits. In this section we will discuss some such commercially available, integrated-circuit sequential circuits. Since they are in integrated-circuit form, the complete circuit is fabricated on a single chip. There are many such circuits available. As an example, we shall consider a flip-flop.

In Secs. 6.1 and 6.2 we discussed flip-flops. Figure 6-56 illustrates a commercially available type CD4027A COSMOS, J-K flip-flop chip. Actually, there are two flip-flops on a single chip. Figure 6-56a shows the basic pin connections. That is, each integrated circuit is mounted and leads are brought out. This diagram specifies the lead locations. In Fig. 6-56b, a logic circuit for one of the flip-flops is shown, and in Fig. 6-56c, the actual circuit is shown. Note that the dual number, e.g., 11/5, at the terminals of Figs. 6-56b and c refers to terminal connections (i.e., terminal assignments). (There are two sets of numbers since there are two flip-flops on the chip.)

The boxes marked TG on Fig. 6-56b are toggles. That is, they function as T flip-flops. This J-K flip-flop also incorporates set and reset leads which can perform the set and reset functions, independent of the clock pulses. Of course, the J and K inputs initiate action in accordance with the clock pulses.

Manufacturers specify much of the data associated with the operating conditions of integrated circuits. For instance, power supply voltages are given. These specifications are discussed in great detail in the electronics circuits text cited in the bibliography at the end of this chapter.

We have illustrated an integrated circuit in this section. Of course, a great many other integrated circuits are available. For instance, there are counters, parallel and series input and output shift registers, etc. The data given by manufacturers may differ in detail. However, similar methods are used to present the information. We shall discuss much more complicated integrated circuitry subsequently.

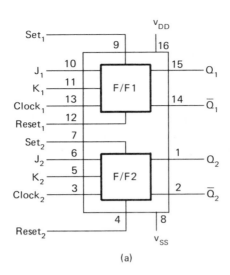

(a)

Figure 6-56   A COSMOS dual J–K flip-flop Type CD4027A. (a) pin diagram;
(b) logic circuit; (c) schematic diagram (Courtesy RCA, Inc.)

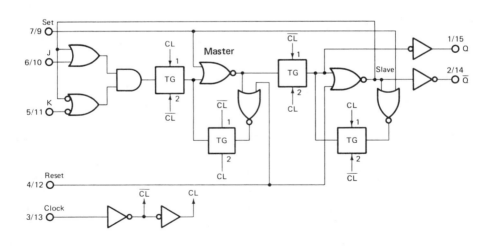

Figure 6-56   (Continued) part (b)

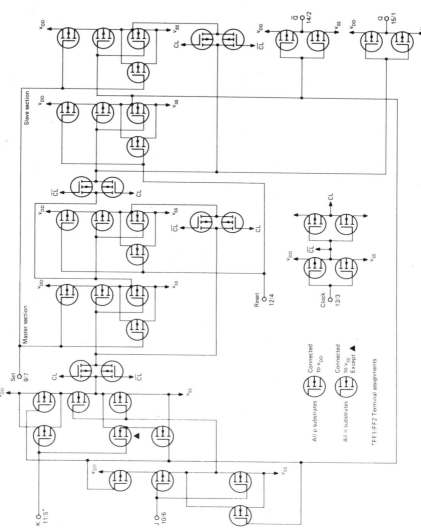

Figure 6-56 (Continued) part (c)

BIBLIOGRAPHY

Booth, T.L., *Digital Networks and Computer Systems*, Second Ed., Chaps. 8 and 9, John Wiley and Sons, Inc., New York 1978

Chirlian, P.M., *Analysis and Design of Integrated Electronic Circuits*, Chaps. 8, 9, and 10, Harper and Row, Inc., New York 1981

Givone, D.D., *Introduction to Switching Circuit Theory*, Chaps. 8-10, and 12, McGraw-Hill Book Co., Inc., New York 1970

Klir, G.J., *Introduction to the Methodology of Switching Circuits*, Chaps. 7-9, D. VanNostrand Co., Inc., New York 1972

Lee, S.C., *Modern Switching Theory and Digital Design*, Chap. 9, Prentice Hall, Inc., Englewood Cliffs, N.J. 1978

Mano, M.M., *Digital Logic and Computer Design*, Chaps. 6 and 7, Prentice Hall, Inc., Englewood Cliffs, N.J. 1979

Rhyne, V.T., *Fundamentals of Digital System Design*, Chaps. 5 and 6, Prentice Hall, Inc., Englewood Cliffs, N.J. 1973

PROBLEMS

6-1. Discuss the difference between combinational and sequential circuits.

6-2. Discuss the difference between synchronous and asynchronous sequential circuits.

6-3. Determine the behavior of the R–S flip-flop for all possible inputs and all possible states.

6-4. Determine a logic circuit which will produce an output of 1 if a four-digit binary number is equal to or greater than 7_{10}. Use an R–S flip-flop. (Actually, this could be realized by a combinational circuit but it is done in this way as an exercise.)

6-5. Repeat Prob. 6-4 but now obtain an output of 1 if the number is more than 6 or less than 3.

6-6. Repeat Prob. 6-4 using a D flip-flop.

6-7. Repeat Prob. 6-5 using a D flip-flop.

6-8. Repeat Prob. 6-4 using a J–K flip-flop.

6-9. Repeat Prob. 6-5 using a J–K flip-flop.

6-10. Repeat Prob. 6-4 using a T flip-flop.

6-11. Repeat Prob. 6-5 using a T flip-flop.

6-12. Design a logic circuit which functions as a J–K flip-flop. That is, it has two inputs J and K, etc., but uses a D flip-flop in its implementation.

6-13. Repeat Prob. 6-12 now using a T flip-flop.

6-14. Discuss the advantage of master-slave-type flip-flops.

6-15. Draw a timing diagram for all inputs and states of the R–S master-slave flip-flop.

6-16. Repeat Prob. 6-15 for the J–K master-slave flip-flop of Fig. 6-15.

6-17. Repeat Prob. 6-16 for the J–K master-slave flip-flop of Fig. 6-16.

6-18. Repeat Prob. 6-16 for the R–S edge-triggered flip-flop of Fig. 6-19.

6-19. Discuss the difference between master-slave and edge-triggered flip-flops.

6-20. Discuss the different timing requirements of edge-triggered and master-slave flip-flops.

6-21. Draw a circuit for a T master-slave flip-flop.

6-22. Repeat Prob. 6-21 for a T edge-triggered flip-flop.

6-23. Discuss the use of preset and clear leads.

6-24. Obtain the transition table and transition diagram for the network of Fig. 6-57.

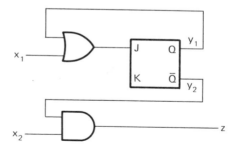

Figure 6-57

6-25. Repeat Prob. 6-24 for the network of Fig. 6-58.

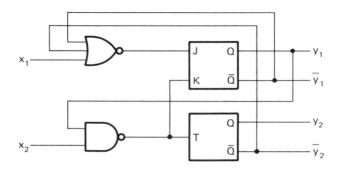

Figure 6-58

6-26. Realize a network using a D flip-flop that realizes the transition table of Prob. 6-24.

6-27. Repeat Prob. 6-26 for the transition table of Prob. 6-25.

6-28. Repeat Prob. 6-27 but now use T flip-flops.

6-29. Design a network using D flip-flops, which realizes the transition diagram of Fig. 6-59. Use the following assignments: $q_0=0,0$; $q_1=0,1$; $q_2=1,0$; and $q_3=1,1$.

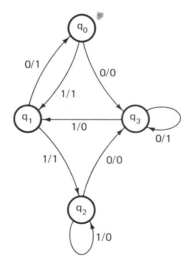

Figure 6-59

6-30. Repeat Prob. 6-29 but now use J–K flip-flops.

6-31. Repeat Prob. 6-29 but now use T flip-flops.

6-32. Repeat Prob. 6-29 but now use the state assignments $q_0 = 1,1$; $q_1 = 0,1$; $q_2 = 0,0$; and $q_3 = 1,0$.

6-33. Repeat Prob. 6-30 but now use the state assigments of Prob. 6-32.

6-34. Repeat Prob. 6-31 but now use the state assigments of Prob. 6-32.

6-35. Design a network using J–K flip-flops, which realizes the transition diagram of Fig. 6-60. Use the following state assignments: $q_0 = 0,0,0$; $q_1 = 0,0,1$; $q_2 = 0,1,0$; $q_3 = 0,1,1$; $q_4 = 1,0,0$; $q_5 = 1,0,1$; $q_6 = 1,1,0$; $q_7 = 1,1,1$

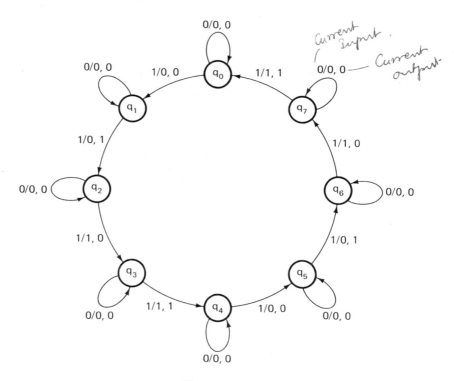

Figure 6-60

6-36. Repeat Prob. 6-35 but now use D flip-flops.

6-37. Repeat Prob. 6-35 but now use T flip-flops.

6-38. Repeat Prob. 6-35 but now use the following assignment of states: $q_0 = 0,0,1$; $q_1 = 0,0,0$; $q_2 = 0,1,0$; $q_3 = 0,1,1$; $q_4 = 1,0,1$; $q_5 = 1,0,0$; $q_6 = 1,1,1$; $q_7 = 1,1,0$

6-39. Repeat Prob. 6-36 but now use the state assignments of Prob. 6-38.

6-40. Repeat Prob. 6-37 but now use the state assignments of Prob. 6-38.

6-41. Design a counter which produces the following sequence:

   001,   000,   010,   011,   100,   111,   110,   101

and then repeats. Use J–K flip-flops.

6-42. Repeat Prob. 6-41 but now use D flip-flops.

6-43. Repeat Prob. 6-41 for the sequence

   001,   010,   111,   110,   011

6-44. Repeat Prob. 6-43 but now use T flip-flops.

6-45. Design a circuit which will produce an output of 1 whenever the sequence 0110 occurs at its input. Use J–K flip-flops.

6-46. Repeat Prob. 6-45 for the sequence 0111.

6-47. Repeat Prob. 6-45 but now use D flip-flops.

6-48. Repeat Prob. 6-46 but now use T flip-flops.

6-49. Design a circuit which receives an input of five pulses, one at each clock pulse, and produces an output of 1 if a majority of them are 1's. Assume that the input stops after five pulses.

6-50. Repeat Prob. 6-49 but have the circuit reset itself after each five clock pulses. The circuit then produces a second output after the next five input pulses, etc.

6-51. Determine an excitation table for a network characterized by the switching functions

$$y_1 = x_1 + x_2 \bar{y}_2$$

$$y_2 = \bar{y}_2 + x_2$$

Examine it for races and oscillations.

6-52. Repeat Prob. 6-51 for

$$y_1 = x_1 + \bar{x}_2 + \bar{y}_2 + \bar{y}_1$$

$$y_2 = x_1 + x_2 \bar{y}_1 + \bar{y}_2 y_1$$

6-53. The Karnaugh map of Fig. 6-61 is to be used to realize a function y. Obtain a minimal gate realization. Then obtain one which is hazard-free.

Figure 6-61

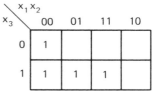

x_3 \ $x_1 x_2$	00	01	11	10
0	1			
1	1	1	1	

6-54. Repeat Prob. 6-53 for the Karnaugh map of Fig. 6-62.

x_3x_4 \ x_1x_2	00	01	11	10
00		1	1	
01			1	
11	1	1	1	
10				

Figure 6-62

6-55. Discuss the compromises made in obtaining a hazard-free network.

6-56. Design a fundamental-mode asynchronous sequential circuit with two inputs x_1 and x_2 and two outputs z_1 and z_2. The z_1 output is to be a 1 if, and only if, $x_1=1$ and $x_2=0$ following $x_1=1$, $x_2=1$, and the z_2 output is to be a 1 if, and only if, $x_1=1$, $x_2=0$ following $x_1=0$, $x_2=1$.

6-57. Repeat the design of Prob. 6-56 with the following specifications on the output. z_1 is to be a 1 if, and only if, $x_1=1$, $x_2=1$ and this state follows the sequence $x_1=0$, $x_2=1$, $x_1=0$, $x_2=0$. $z_2=1$ if, and only if, $x_1=1$, $x_2=1$ following $x_1=1$, $x_2=0$.

7

Registers and Memories

An important part of any digital computer system is the memory which stores both data and programs. In this chapter we shall discuss circuits that can be used to store information. The flip-flop is a simple form of memory that can store a single bit (i.e. 1 or 0).

A somewhat more complex form of memory is called a *register*. Typical registers can store 8, 16, 32, or 64 bits. Such a collection of bits is usually called a *word*. The terminology is such that the term "memory" is usually associated with a circuit that can store many words. Small memories can store only a few words (e.g. 64). However, large memories can store many thousands or millions of words.

Memories are made up of elements, each of which can store a single bit. The flip-flop is such an element. Special magnetic elements are also used. The semiconductor flip-flop memory has the advantage that it is fast and, since it can be fabricated using integrated circuit techniques, it is also small. Magnetic memories, however, will not lose their information in the event of a power failure. For this reason, these memories are said to be *nonvolatile*, while the semiconductor memories are called *volatile* memories. The advent of LSI (large scale integration) has caused a large increase in the use of semiconductor memories, and a corresponding drop in the use of magnetic core memories. We shall discuss these semiconductor memories in this chapter.

The memories that we have discussed make up what is called the *main memory* or *inner memory*. This stores the data and instructions used during the actual running of a computer program. There are other memories, called *auxiliary memories*, that are used to store information such as data to be used at a future time. When the computer program using this information is to be executed, the information is put into the main memory. Auxiliary memories ususally consist of storage on magnetic tape or magnetic disks. Other storage techniques such as those using magnetic bubbles and charge-coupled devices are also used for this purpose.

We shall start with a discussion of registers.

7.1 REGISTERS

Registers are used to store information in computers. In this section we shall discuss such registers, not only to consider them, but also to discuss their design.

Information is stored in a register as a sequence of 1's and 0's. This sequence can be supplied to the register in one of two ways, either serially or in parallel. In serial operation, one binary digit is supplied at each clock pulse. For instance, suppose that the register stores the output of a sequential adder, see the introduction to Chapter 6. In this case, the adder circuit supplies only one digit at a time and these are supplied to the register at successive clock pulses. Thus, this type of operation is called *series* or *serial* operation of the register. That is, data is input to the register in *series*. Similarly, data can be output from the register in series, that is one digit for each clock pulse. Again, this type of operation would be useful if the information were to be supplied to a sequential adder. We have ignored the details of addition in this simple discussion. We shall consider them again in Chapter 10.

There is another type of register operation; this is *parallel* operation. In this case, all data is input or output during a single clock pulse period. This is much faster than series operation. However, many more gates and flip-flops are required for such operation.

Shift Register

A shift register is one where the information is transmitted down the register with each clock pulse. Often, the data is entered one binary digit per clock pulse. This is illustrated in Fig. 7-1. At the start, the register has the contents 0000. On the next clock pulse, a 1 is entered. On the next pulse, a 0 is entered. The 1 that is already in the register shifts one space to the right, see Fig. 7-1c. On the next pulse a 0 is again entered and the 1 is shifted to the right again as shown in Fig. 7-1d. On the next clock pulse a 1 is entered. Thus, it appears in the leftmost position and the original 1 is again shifted to the fourth (rightmost) position. On the next clock pulse a 0 is again entered and the contents

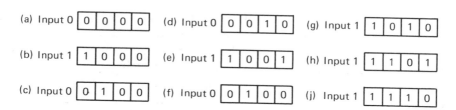

(a) Input 0 | 0 | 0 | 0 | 0 | (d) Input 0 | 0 | 0 | 1 | 0 | (g) Input 1 | 1 | 0 | 1 | 0 |

(b) Input 1 | 1 | 0 | 0 | 0 | (e) Input 1 | 1 | 0 | 0 | 1 | (h) Input 1 | 1 | 1 | 0 | 1 |

(c) Input 0 | 0 | 1 | 0 | 0 | (f) Input 0 | 0 | 1 | 0 | 0 | (j) Input 1 | 1 | 1 | 1 | 0 |

Figure 7-1 Illustration of the information stored in a shift register on successive clock pulses

of the register are shifted to the right again. Note that the previous rightmost term is now lost, see Fig. 7-1f. Of course, it could have been used by some other logic circuit before this happened.

Now let us see how we can implement a simple shift register. In Fig. 7-2a we have a three-stage shift register using J–K flip-flops. Remember that these are master-slave flip-flops. (Note that we have drawn the J on top for convenience.) Suppose that all the states are 0, i.e. $y_1=y_2=y_3=0$. If the input x is 0, then the states will remain 0. Now suppose that $x=1$. Then, $J_1=1$, $K_1=0$. Then, *after* the next clock pulse, $y_1=1$. However, the state of the second flip-flop cannot change until the next clock pulse occurs. Thus, it remains $y_2=0$. Now suppose that the input is 0. Then, $J_1=0$, $K_1=1$. Thus, *after* the (next) clock pulse, $y_1=0$. However, *during* the clock pulse, y_1 has not as yet changed. Then, $J_2=1$, $K_2=0$, during that pulse. This actuates the second flip-flop. Hence,

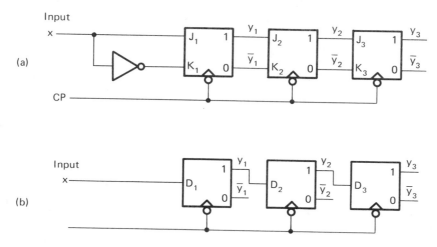

Figure 7-2 Shift registers. (a) using J–K flip-flops; (b) using D flip-flops. Note that R–S flip-flops can replace the J–K flip-flops of Fig. 7-2a.

after the clock pulse, $y_2 = 1$. Thus, the 1 has shifted to the right, and the shift register will operate as desired. Since the input $J = 1$, $K = 1$ never occurs for any of the flip-flops, we can replace the J–K flip-flops by R–R–S flip-flops here, without changing the operation. In Fig. 7-2b we have implemented a shift register using D flip-flops. The basic ideas are the same as those used in Fig. 7-2a. A state transition diagram for these flip-flops is shown in Fig. 7-3. This diagram should be compared with Fig. 7-2 to gain familiarity with the operation. (Note that y_1, y_2, and y_3 are the outputs.) We have constructed a shift register which can store three binary digits. To store more digits, we need only add more flip-flops to the right in Fig. 7-2.

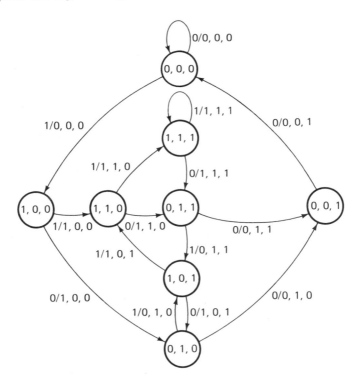

Figure 7-3 State transition diagram for the shift registers of Fig. 7-2

In an actual computer, the shift registers are more versatile than those of Fig. 7-2. For instance, at times, we do not want to shift when a clock pulse occurs. Sometimes we want to enter data on the left side and shift to the right. Other times, we may want to enter data on the right and shift to the left. Finally, there are times when we want to clear the register, that is, to set all its states to 0. Control signals are present which indicate which type of operation is desired. Let us now design such a registers which can be controlled by a set of signals.

Let us assume that there are three control leads, c_0, c_1, and c_2. The operation is to be such that if $c_0 = 1$, $c_1 = 0$ and $c_2 = 0$, then the input is at the left and the register shifts right; if $c_0 = 0$, $c_1 = 1$, and $c_2 = 0$, then the input is at the right and the register shifts left; if $c_0 = 0$, $c_1 = 0$ and $c_2 = 1$, then the register clears, i.e. all states are set to 0; and if $c_0 = 0$, $c_1 = 0$ and $c_2 = 0$, there is neither input nor shift.

Let us realize the shift register using D flip-flops. In order to obtain the logical expression needed to design the circuit, we need a truth-table type of formulation. In this case, we can obtain the table directly. With D flip-flops, the output state is the same as the input. Thus, the following table is that part of the truth table for which $D = 1$. (Remember that we need only those terms for which $D = 1$ to obtain the logical expression.) Thus, for the first flip-flop, we have

c_0	c_1	c_2	x_L	y_1	y_2	D_1
1	d	d	1	d	d	1
d	1	d	d	d	1	1
0	0	0	d	1	d	1

The don't cares in the table result because we assume that no two c's will ever be 1 at the same time. Note that D_1 will be 1 if $c_0 = 1$, and the left input $x_L = 1$. That is, if $c_0 = 1$, the input is from the left. Also, $D_1 = 1$ if $y_2 = 1$ when $c_1 = 1$. Flip-flop 2 is to the right of flip-flop 1. Thus, if we are shifting to the left, the input to flip-flop 1 is y_2, the state of flip-flop 2. The only other time that the input to the flip-flop is a 1 is if its present output is a 1 and if $c_1 = c_2 = c_3 = 0$, that is, when there are to be no shifts. Using the table we can write the logical expression

$$D_1 = c_0 x_L + c_1 y_2 + \bar{c}_0 \bar{c}_1 \bar{c}_2 y_1 \tag{7-1a}$$

Note that if the don't cares in the table were not present, then considerably more complex expressions would result.

The equation for the other two flip-flops can be obtained from Eq. (7-1a) if we replace x_L by the left input to the flip-flop, y_2 by the right input, and y_1 by the state of the flip-flop in question. Thus,

$$D_2 = c_0 y_1 + c_1 y_3 + \bar{c}_0 \bar{c}_1 \bar{c}_2 y_2 \qquad\qquad (7\text{-}1b)$$

$$D_3 = c_0 y_2 + c_1 x_R + \bar{c}_0 \bar{c}_1 \bar{c}_2 y_3 \qquad\qquad (7\text{-}1c)$$

where x_R is the input signal when the register is to be shifted to the left, i.e. the right input.

The complete circuit is shown in Fig. 7-4. Note that each flip-flop has three AND gates and an OR associated with it. These correspond to each of the terms in Eqs. (7-1). The NOR gate is used to realize $\bar{c}_0 \bar{c}_1 \bar{c}_2$.

The last term in each of Eqs. (7-1) is used when shifting does not occur. An alternative way of doing this, which reduces the number of gates required, is to prevent the clock pulses from reaching the flip-flop when we do not require shifting. If we add the logic which will do this, then the last terms can be omitted from Eqs. (7-1). A circuit which accomplishes this is shown in Fig. 7-5. Instead of redrawing the entire circuit, only one stage is drawn. AND a will apply the left input signal if $c_0 = 1$. Similarly, AND b will apply the right input signal if $c_1 = 1$. If $c_2 = 1$, then c_0 and $c_1 = 0$ and the input to the flip-flop will be 0. This will set $y = 0$. This clears the register. If c_0, c_1, and c_2 are all 0's, then the output of AND c will be 0. In this case, clock pulses will not be applied to the (any) flip-flop and states will not change at the clock pulse time.

In integrated-circuit, shift-register packages, provision is often made for parallel transfer of information into and out of the register. The parallel transfer out is made possible by having access to the y_1, y_2 and y_3 leads. To obtain parallel as well as series input, the circuit of Fig. 7-6 can be used. Again, to avoid duplication, we only show one stage. An extra control signal c_3 has been added. c_3 is 0 unless parallel input is desired. If $c_3 = 0$, then no signals pass through AND p and the operation is essentially as in the circuit of Fig. 7-5. If $c_3 = 1$, with all other control signals 0, then

$$D_j = x_j$$

where there is a separate x_j input for each flip-flop (i.e. x_1 is applied to D_1, x_2 to D_2, etc.). Thus, parallel input is obtained. If $c_3 = 0$, then series input is used and the circuit operates as an ordinary shift register.

Note that this circuit can also be used as both a *serial-to-parallel converter* and as a *parallel-to-serial converter*. For instance, data which is input serially can then be used in a parallel fashion.

Now let us consider an integrated circuit shift register. In Fig. 7-7 we illustrate a type SN54165 8-bit shift register. Note that this shift register

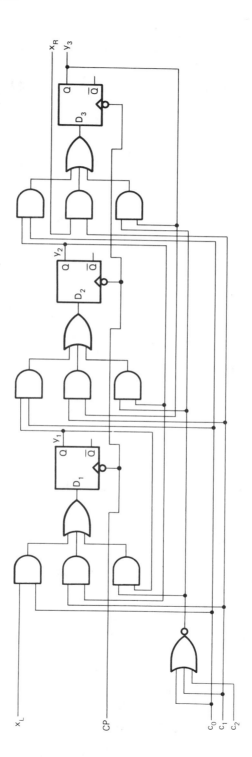

Figure 7-4 The controlled shift register

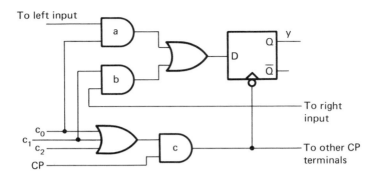

Figure 7-5   One stage of a modified shift register that uses fewer gates than Fig. 7-4

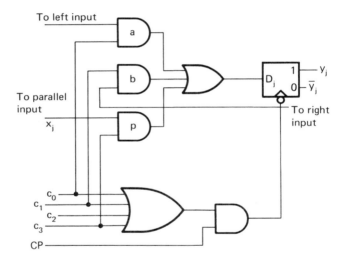

Figure 7-6   One stage of a shift register that allows parallel as well as series input

can receive its data either serially or in parallel. However, provision is only made for serial output. (Parallel output would only require access to the points marked Q_A, Q_B, . . . Q_G and their complements.) in Figs. 7-7b and c, typical input and output circuits are given. In addition, other pertinent data is provided so that the interaction of this shift register with the other elements in the circuit can be determined.

An abbreviated truth table is also given in the specifications. This is shown in Fig. 7-7d. The column entitled "internal outputs" lists the conditions of the internal flip-flop's outputs. However, for this particular integrated circuit, these (parallel) outputs are not available. In the

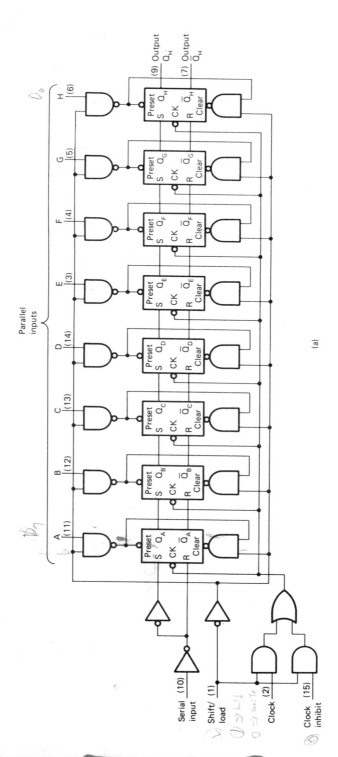

Figure 7-7 A TTL eight-bit shift register Type SN54165. (a) the logic diagram; (b) typical input circuits; (c) typical output circuit; (d) function table. (Courtesy Texas Instruments, Inc.)

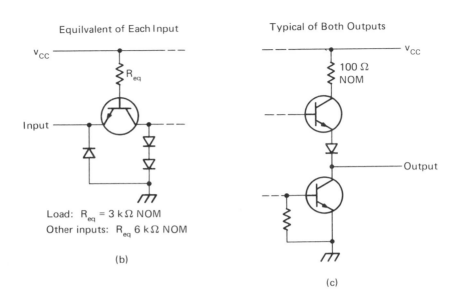

Load: R_{eq} = 3 kΩ NOM
Other inputs: R_{eq} 6 kΩ NOM

(b)

(c)

Function Table

Inputs					Internal Outputs		Output
Shift/ load	Clock inhibit	Clock	Serial	Parallel A . . . H	Q_A	Q_B	Q_H
L	X	X	X	a . . . h	a	b	h
H	L	L	X	X	Q_{A0}	Q_{B0}	Q_{H0}
H	L	↑	H	X	H	Q_{An}	Q_{Gn}
H	L	↑	L	X	L	Q_{An}	Q_{Gn}
H	H	↑	X	X	Q_{A0}	Q_{B0}	Q_{H0}

(d)

Figure 7-7 (Continued) part (b), (c) and (d)

function table L stands for low and H for high. If positive logic is used, the H corresponds to a 1 and L to a 0. With negative logic, the reverse is true. The X's in the table are "don't cares." The symbols a . . . h represent the steady-state inputs at leads A . . . H, respectively. The symbols Q_{An} . . . Q_{Gn} are equal to the levels of Q_A . . . Q_G before the most recent positive-going transition of the clock pulse. The symbols Q_{A0} . . . Q_{H0} are equal to the values of Q_A . . . Q_H *before* the indicated steady-state conditions were established. The upward-pointing arrow indicates a transition from a low to a high level. Note that this only occurs in the clock column and indicates that the transition is initiated by the leading edge of the clock pulse as it switches from low to high.

7.2 SEMICONDUCTOR MEMORIES

Memories used with microprocessors are constructed using only semiconductor devices. Many large computers use magnetic memories. However, semiconductor memories are considerably faster than magnetic-type memories (e.g. they have access times in the order of nanoseconds in comparison to $1\mu sec$ for magnetic memories). Modern semiconductors are constructed using LSI (large scale integration) integrated circuit techniques. Thus, a complete memory can be fabricated on either a single silicon chip or on several such chips. Such chips are commercially available. Thus, semiconductor integrated circuit memories are very attractive.

In the large memories, each bit is stored in an element which can be written on or read from. Figure 7-8a illustrates a *semiconductor memory element*. It uses an R–S flip-flop to store the information. Data can be *stored in* the memory. In this case we say that we *write* on the memory. Similarly, the data can be *read* from the memory. There are leads called the read lead and the write lead which determine if the memory is being read from or written on. They are often combined into a single read/write control lead. In Fig. 7-8, for simplicity, we show separate leads. If the write signal is 0, then the input to the flip-flop is $R=0, S=0$. Hence, the state y can only change when the write signal is a 1. When the write signal is a 1, and the input signal is a 0, then $R=1$, $S=0$. In this case the state of the flip-flop will be reset to $y=0$. If, during a write pulse, the input signal is a 1, then $R=0, S=1$, and the state will be set to $y=1$. Then, the state of the flip-flop will always be equal to the input signal that is present when the write control lead has a 1 on it. We say that the state of the memory element is equal to the state of its flip-flop.

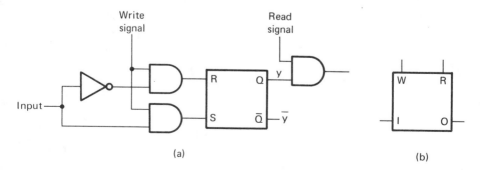

Figure 7-8 (a) A semiconductor memory element; (b) its block diagram. To simplify the diagram, all clock connections are omitted.

The output of the memory element will be 0 unless a 1 is applied along the read signal lead. In this case, the output is the state y. Thus, we can read and write on the semiconductor memory element. A block diagram representation of this memory element is shown in Fig. 7-8b.

The gates and flip-flop of the semiconductor memory can be fabricated using either transistors or MOSFETs. If transistors are used, the operation will be faster. However, the integrated circuit fabrication is simpler, and more elements can be placed on a single chip if we use MOS devices. A compromise is often made in that the memory elements themselves are MOS devices while the associated circuitry used to drive the memory and extract information from it uses (integrated) transistors which are faster. In the usual memory, there are very many memory elements and relatively few elements in the associated circuitry (e.g. there is a single memory buffer register). However, the associated circuitry may be used several times, each time that a single memory element is written on or read from. Thus, a compromise such as the one above becomes a reasonable one, since the fast circuit is used several times for each use of the slow one.

In Appendix A we shall discuss dynamic MOS gates. Here the information is retained by charge stored in the gate-to-channel capacitance of the MOSFET. This charge will slowly "leak off" and, to prevent the loss of the information, it is periodically refreshed. Such MOS gates are also used in memory devices.

Now let us consider a complete small memory unit. We shall use a 16-bit memory, see Fig. 7-9. The memory is arranged in groups of bits called *words*. In Fig. 7-9 there are four 4-bit words. Longer words are usually used. We have kept the bit length short to avoid cluttering the diagram. The location of a word in the memory is called its *address*. The x variable gives the word's address (or number). Note that we have numbered each memory element with a double number (e.g. 2,3). The first number indicates the word's address and the second the bit position. The x's select the desired word. For instance, suppose that we want to write on the x_3 word position. Then, a 1 is placed on the x_3 write line. Then, in the fourth row, the memory element 3,0 will set itself to the value of d_0. That is, the state of the flip-flop of the memory unit will be a 1 if d_0 is a 1, and will be a 0 if d_0 is a 0. Similarly, the memory element 3,1 will set itself to the state of input d_1 and memory element 3,2 will set itself to the state of d_2 and memory element 3,3 will set itself to the state of d_3. The memory elements in all the other rows will be unaffected since their write leads will have 0's on them.

If we want to read a word, then we place a 1 on the appropriate read lead. For instance, if we want to read the word corresponding to

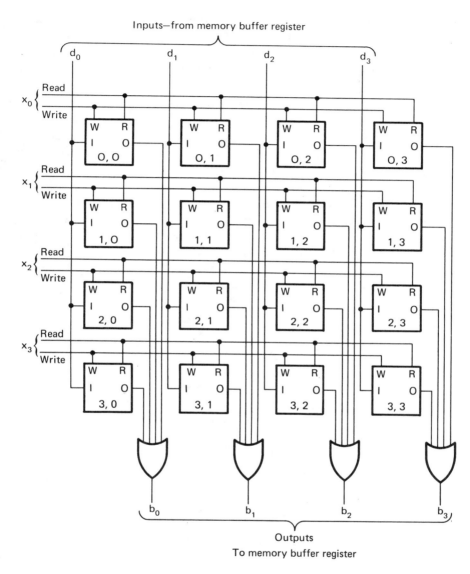

Figure 7-9 A semiconductor memory plane. The clock connections are omitted.

x_2, then x_2's read lead will have a 1 placed on it. The state of all the other read and write leads will be 0. Then, all the outputs except those in the x_2 row will be 0. In addition, b_0 will equal the state of memory element 2, 0, b_1 will equal the state of memory element 2,1, b_2 will equal the state of memory element 2,2 and b_3 will equal the state of memory element 2,3. Thus, we have obtained the desired output. These outputs will only be present while there is a 1 on the read line. Thus, the outputs are supplied to a register which stores them until they are needed. This is called the *memory buffer register*. Note that the readout is nondestructive. That is, reading the memory does not remove data from it.

When we want to write a word, the inputs d_0, d_1, d_2, and d_3 are usually stored in a register before they are transferred to the memory. This may be the same memory buffer register used to store the output. Note that since we do not read and write at the same time, we can use the same register for these two purposes.

Let us consider a circuit that can supply the x_0, x_1, x_2, and x_3 read and write signals. The word's address is usually supplied in binary form. For the four-word memory, we require two input bits, a_0 and a_1 which supply the numerical values of the subscripts of x, in binary form. That is, a_0 and a_1 supply the address of the word. We also assume that there are two lines that supply the read and write signals. These indicate whether a read or a write operation is to be performed. The control circuit which uses this information is shown in Fig. 7-10. One AND

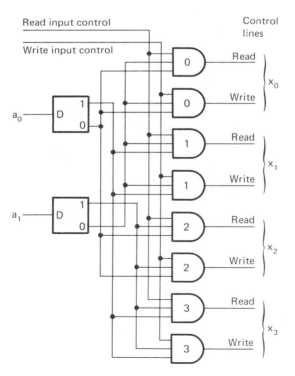

Figure 7-10 A control circuit for the memory of Fig. 7-9. The clock connections are omitted.

gate of each pair has an input which is from the read input control line while the other AND gate has an input from the write input control line. The two flip-flops apply 1's to the appropriate pair of AND gates. Thus, the desired lead will have a 1 on it while all the others will be 0.

There is often just a single control line. In this case the circuit is modified as shown in Fig. 7-11. Now, if the read/write signal is 1, the addressed word is written on. If the read/write signal is 0, the addressed word is read. Note that the address is stored in the two fkip-flops. , and these two flip-flops then consitute the *memory address register*.

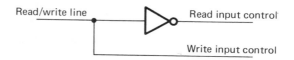

Figure 7-11 A circuit that can be used with Fig. 7-10 to provide a single read/write control line

This type of memory is called a *random access memory (RAM)* since each word can be read or written independently. For instance, if we want to read the fourth word, we do not have to read or write the first three. We shall discuss other types of memories subsequently.

7.3 READ-ONLY MEMORIES – DECODERS

At times, we work with memories which contain stored data which is not be changed. For instance, suppose that data is in binary form and we want to change it to another form. Note that the binary representation of decimal numbers can be considered to be a code. Other binary codes are, at times, used to represent decimal numbers. We shall discuss the reasons for this in Chapter 8. Suppose that data is represented in binary form and we want to change it to a different code. A memory could be encoded with the necessary information to convert from one code to the other. There would never be any reason to change this memory. Thus, the memory could be manufactured with the required information stored in it, and no writing provision would be made. In this section, we shall discuss these *read-only memories (ROM)*.

We have only considered one use of ROMs. Special programs that are repeated over and over are stored in them. These may be, for instance, instruction-decoders which are used in large computers or they may be used to control trigonometric function generators which are used in a desk calculator.

Let us consider a read-only memory in some detail. It is arranged in a rectangular array form and has a set of input leads. When a 1 is supplied to one of these leads, it outputs a single word which is determined by the design of the ROM. The output word depends upon which input lead is supplying the 1. Note that a ROM is arranged in essentially the same way as the RAM which was discussed in the last section. A word is addressed and the memory content of that word is output. There is no read/write control since a ROM can only be read. In a computer, the RAM and ROM are often treated as part of one memory with different addresses for the words of RAM and the words of ROM. A simple ROM memory is shown in Fig. 7-12. The *input* to the ROM memory is supplied to one of the *word lines*, the *output word* occurs in the *bit lines*. Suppose that a 1 is put on word line 0. (We assume that a 1 is a positive pulse and a 0 is zero volts.) The diodes *a* and *b* will conduct and connect bit lines 1 and 3 to the word line 0. Since this line has a 1 on it, the output word will be 1010. Now suppose that a 1 is put on word line 1. Then, the diodes *c, d* and *e* conduct and the output is 1101. Simi-

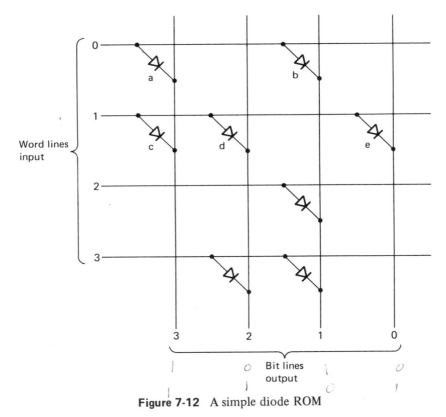

Figure 7-12 A simple diode ROM

larly, an input on word line 2 produces an output 0010, and an input on word line 3 produces 0110. We have just shown output lines. Usually, they supply values to a buffer memory register. The input may be supplied from a register.

We have considered a ROM using diodes. MOS devices are used in ROMs that require very little power. Junction transistors can also be used in high speed ROMs.

As an additional example, let us obtain a ROM which will convert from a binary representation to a different code called a *Gray code*. Gray codes were used in the Karnaugh maps. (We shall discuss the use of such codes in Sec. 8.2.)

Binary Number	Gray Code
000	000
001	001
010	011
011	010
100	110
101	111
110	101
111	100

Each binary number represents in input line. That is, each binary number causes a 1 to be placed on the appropriate word line (e.g. the line whose decimal number is equivalent to the binary number). Later in this section we shall discuss a device which supplies signals to the word lines. For the time being, let us just assume that the signals are available. The ROM is shown in Fig. 7-13. As an example, if the input signal is 100. then word line 4 will have a 1 placed on it. Then, the output will be coupled to bit lines 2 and 1. Thus, the output is 110.

We must now consider a circuit that will drive the appropriate word lines. This is called a *decoder*. Actually, we have already considered such circuits. For instance, Fig. 7-10 illustrated such a circuit. The decoder circuit for Fig. 7-13 is shown in Fig. 7-14. The flip-flops make up a memory which stores the input signal. The remainder of the circuit constitutes the decoder. Note that all three inputs to AND j, j=0,1,2,3, 4,5,6,7 will be 1 only when the binary number $a_2a_1a_0$ is equal to j.

We have shown an output line and AND gate for an input of 000. Actually, this is not needed since, see Fig. 7-13, there is no coupling

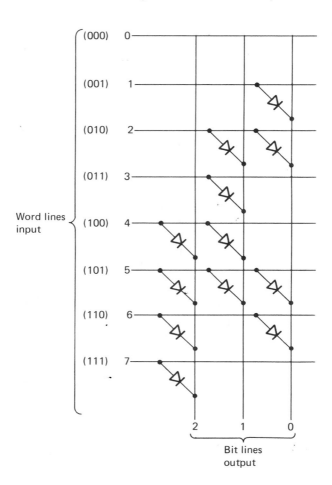

Figure 7-13   A ROM that converts three-bit binary numbers into a Gray code

between the word-zero line and any of the output lines. AND 0 is shown in Fig. 7-14 for completeness.

Use of ROMs to Generate Logic Functions

We can take a somewhat different viewpoint. Consider the 0-bit line of Fig. 7-13. That will have a 1 output if word line 1, 2, 5 or 6 has a 1 on it. Thus, bit line 0 can be considered to be a logic function of the input. Let us formalize this so that we can design the output of a specific bit line as a given function of the input variables. That is, we can use ROMs to generate a switching function. Let us realize some arbi-

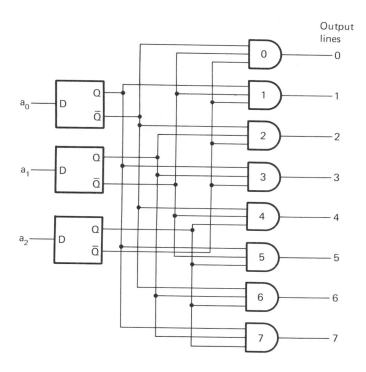

Figure 7-14 A driver circuit for the ROM of Fig. 7-13. The clock circuits are not shown.

trary functions. It is most convenient if these are in the nonminimized, canonical sum of products form. Suppose that we have

$$y_0 = x_1\bar{x}_2\bar{x}_3 + x_1\bar{x}_2 x_3 + \bar{x}_1 x_2 \bar{x}_3 + x_1 x_2 x_3 \tag{7-2a}$$

$$y_1 = \bar{x}_1 x_2 x_3 + \bar{x}_1 x_2 \bar{x}_3 + \bar{x}_1 \bar{x}_2 \bar{x}_3 \tag{7-2b}$$

$$y_2 = \bar{x}_1 \bar{x}_2 x_3 + x_1 x_2 \bar{x}_3 + x_1 x_2 x_3 + \bar{x}_1 \bar{x}_2 \bar{x}_3 + \bar{x}_1 x_2 \bar{x}_3 \tag{7-2c}$$

A ROM network which realizes this is shown in Fig. 7-15. Each input line represents one minterm. That is, the signal on the input line labeled $\bar{x}_1 \bar{x}_2 \bar{x}_3$ is 1 if $x_1 = 0$, $x_2 = 0$, $x_3 = 0$, etc. Then, there are diodes connecting each function to each of its required minterms. For instance, from the diodes connected to bit line y_1 we see that it will be 1 if $\bar{x}_1 \bar{x}_2 \bar{x}_3$ OR $\bar{x}_1 x_2 \bar{x}_3$ OR $\bar{x}_1 x_2 x_3$ are 1's. Thus, we have realized the desired function.

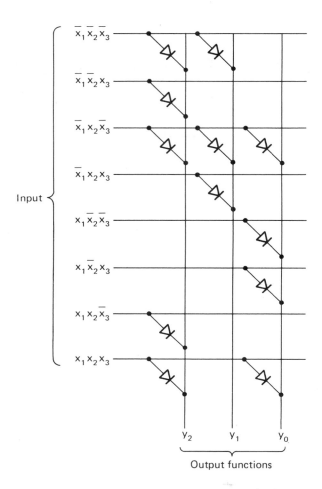

Figure 7-15　A ROM that realizes Eqs. (7-2)

Decoders

Now let us see how we can realize the driving functions. For each variable we need both the variable and its complement. These can be realized from a flip-flop circuit. Such a circuit is shown in Fig. 7-16. This is then used to drive the decoder used in Fig. 7-17. That is, the x_1 and $\bar{x}_1$ outputs of Fig. 7-16 are connected to the x_1 and $\bar{x}_1$ inputs of Fig. 7-17, etc. The outputs are all the functions we desire. This acts in essentially the same way as the decoder of Fig. 7-14.

This decoder is a simple one which uses only one level of logic (i.e. gates do not drive other gates). Thus, it is fast. In general, if there are n variables, a maximum of 2^n gates will be needed. Each gate has n

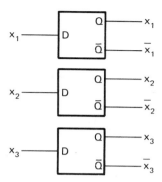

Figure 7-16 Flip-flops used to drive a three-variable decoder. The clock circuits are not shown.

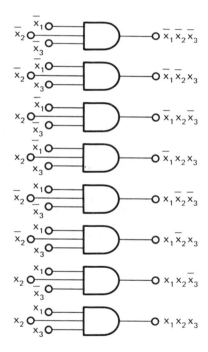

Figure 7-17 A simple one-level decoder that can drive the ROM of Fig. 7-15

inputs. Note that increasing the number of inputs increases the complexity of the gate. For this type of circuit there are, at most, $n \times 2^n$ inputs.

A multilevel decoder, called a *tree decoder*, is shown in Fig. 7-18. Each AND gate has only two inputs. We have shown a three-variable decoder. However, the basic layout can be used for more variables.

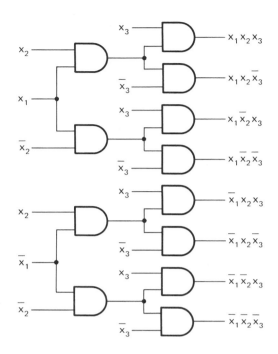

Figure 7-18   A three-variable, two-level, tree decoder

For instance, if there were four variables, then the first two levels would be as in Fig. 7-18. The third level would consist of 16 gates, in eight pairs. Each pair would have two inputs. One of the inputs of each gate in the pair would come from one of the outputs of Fig. 7-18. In addition, one gate in each pair would have an input from x_4, the other gate would have an input from $\bar{x}_4$. The number of AND gates required is

$$G = \sum_{k=2}^{N} 2^k \tag{7-3}$$

where N is the number of variables. The total number of inputs required is $2G$.

If N is a large number, then the total number of inputs will be less than with the single-level decoder. The single-level decoder will have fewer gates and will be faster. However, for a large number of variables, the single-level decoder becomes complex electronically.

Another decoder circuit that is used is the *matrix decoder*. This is also a multilevel decoder with two inputs per gate. The lines are arranged in a matrix fashion. In Fig. 7-19, we illustrate this for a three-variable decoder. There are six lines, with inputs $x_1 x_2$, $x_1 \bar{x}_2$, $\bar{x}_1 \bar{x}_2$, x_3 and $\bar{x}_3$. Then, AND gates are connected between every possible pair of lines. Note that this uses a total of 12 AND gates and 24 inputs. The corresponding tree decoder also used 12 gates and 24 inputs. The single-

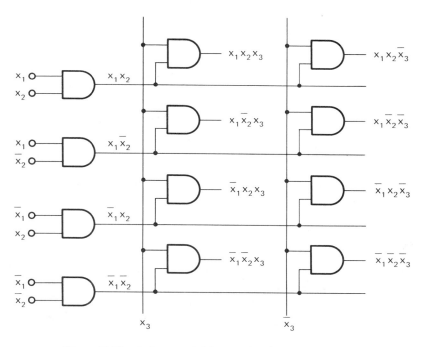

Figure 7-19 A three-variable, two-level, matrix decoder

level decoder used 9 gates with 24 inputs. It appears as though the single-level decoder is the simplest. However, when the number of gates increases, the matrix decoder will have many fewer inputs and gates than the tree decoder, and will have many fewer inputs than the single-level decoder. The line arrangement for a four-variable matrix decoder is shown in Fig. 7-20. Each line would be driven by a two-input AND gate. A two-input AND gate would be connected between each pair of lines. The outputs of these second-level AND gates would be the output of the decoder. If more inputs are used, these ideas are extended. For instance, if there were eight input variables, then these will be divided into sets of four each. Each pair would then be used as the inputs to a

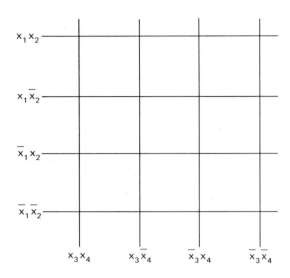

Figure 7-20 Line arrangements in a four-variable matrix decoder

four-variable matrix decoder. All possible pairs of outputs would then be combined using AND gates.

Programmable ROMs

Most ROMs are programmed during manufacture to specifications that are supplied by the user. However, there are ROMs that can be programmed by the user. They are fabricated like the ROM of Fig. 7-12 except that there is a link (e.g. diode) between *every* pair of crossing wires. In addition, a fusable wire is placed in series with each link. A fusable wire is one that will melt and open-circuit if a sufficiently large current is put through it. The user programs the ROM by placing a large current through those fusable lines where no coupling is desired. The links are then melted and the coupling is broken. Such ROMs are called programmable ROMs, PROMs. Once they are programmed, they cannot be reprogrammed.

Erasable ROMs

There are ROMs that can be programmed by the user and then erased and reprogrammed. The erasing is accomplished by exposing these ROMs to strong ultraviolet light. The ROM can then be reprogrammed. Normal light levels do not affect the ROM. The details of the electronics of such ROMs are given in the electronics circuit text cited at the end of the chapter. Those ROMs are called erasable ROMs or EROMs.

7.4 THE PROGRAMMABLE LOGIC ARRAY, PLA

The ROM can be used in conjunction with a decoder to generate switching functions. However, if the switching functions are complex, then the decoder circuits can become elaborate. There is an interconnection of ROMs that can be used which eliminates the need for decoders. This is called a *programmable logic array*, PLA.

Two different types of ROMs are used in this interconnection. The first is of the type discussed in the last section. This is called a sum ROM, for instance, consider Fig. 7-12. Bit line 2 will output a 1 if word line 1 OR word line 3 input a 1. Then, the bit line becomes a 1 if any of its associated lines becomes a 1. Thus, its output is a logical sum of its inputs. The second type of ROM used is called a product ROM. Its bit line becomes a 1 only if *all* its associated word lines are 1. Hence, the AND operation is applied here. A simple product ROM is shown in Fig. 7-21. If there is a 0 on an input line, it is converted to a 1 by the

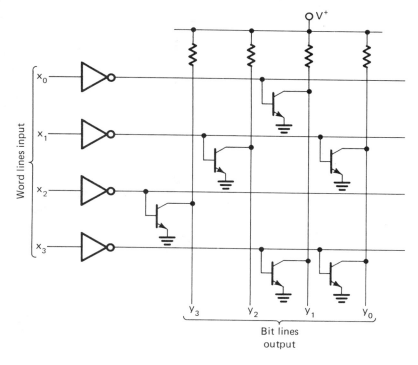

Figure 7-21 A product ROM

NOT gate. When the resulting voltage is applied to the base of a transistor, it turns it on. (We assume positive logic here.) This effectively connects the bit line to ground, reducing the output voltage to the level of a 0. If *any* transistor connected to a bit line is turned on, that bit line will output a 0.

Conversely, if the voltage on a word line is a 1, then the NOT circuit converts it to a 0. All the transistors whose bases are connected to that word line will be turned off. If *all* the transistors that are connected to a bit line are turned off, then the output of that bit line will be a 1. For instance, in Fig. 7-21 we generate the following functions:

$$y_0 = x_1 x_3 \tag{7-4a}$$

$$y_1 = x_0 x_3 = \bar{x}_1 \bar{x}_2 \bar{x}_3 + \bar{x}_1 x_2 \bar{x}_3 + \bar{x}_1 x_2 x_3 \tag{7-4b}$$

$$y_2 = x_1 \tag{7-4c}$$

$$y_3 = x_2 \tag{7-4d}$$

Now let us demonstrate how a product ROM and a series ROM can be interconnected to form a PLA which generates a set of switching functions. In Fig. 7-22 we indicate such a PLA which realizes Eqs. (7-2).

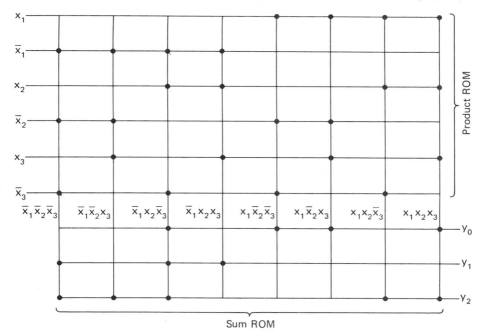

Figure 7-22 A PLA that realizes Eqs. (7-2)

We have symbolically indicated the ROMs by wires that cross each other. If there is a coupling, it is indicated by a dot. The upper ROM is a product ROM. Thus, it generates all the required minterms. The lower ROM is a sum ROM. It is connected so that it sums the proper minterms to obtain the desired functions. PLAs can be fabricated using LSI techniques and very complex logical functions can be realized using them.

In the last section we discussed PROMs that can be programmed by burning out fusable links. There are similar PLAs. They are called *field-programmable logic arrays*, FPLAs.

BIBLIOGRAPHY

Booth, T.L., *Digital Networks and Computer Systems*, Second Ed., Chap. 8, John Wiley and Sons, Inc., New York 1978

Chirlian, P.M., *Analysis and Design of Integrated Electronic Circuits*, Chap. 11, Harper and Row, Inc., New York 1981

Mano, M.M., *Digital Logic and Computer Design*, Chap. 7, Prentice Hall, Inc., Englewood Cliffs, NJ 1979

O'Malley, J., *Introduction to the Digital Computer*, Chaps. 4 and 9, Holt, Rinehart and Winston, Inc., New York 1972

PROBLEMS

7-1. Discuss the difference between series and parallel operation of registers.

7-2. Modify the shift register of Fig. 7-4 to use J–K flip-flops.

7-3. Repeat Prob. 7-2 using T flip-flops.

7-4. Design a three-bit register which is controlled by four input signals, c_0, c_1, c_2, c_3, in the following way:

$c_0 = 1$, enter right, shift left

$c_1 = 1$, enter left, shift right

$c_2 = 1$, set register to 101

$c_3 = 1$, enter sequence, 000 (clear)

all $c = 0$, no shift

Only one c will be nonzero at a time.

7-5. Modify the shift register of Prob. 7-4 so that parallel transfer can be used. An additional control signal will be required.

7-6. Obtain a semiconductor memory element which uses a D flip-flop.

7-7. Repeat Prob. 7-6 for a T flip-flop.

7-8. Modify Fig. 7-9 so that the inputs are taken from, and the outputs are supplied to, a single memory buffer register.

7-9. Describe the complete process of reading and writing for a semiconductor memory.

7-10. Obtain a ROM which converts four-bit numbers into ordinary binary in accordance with the following table.

Given Numbers	New Numbers
0000	0000
0001	0001
0011	0010
0010	0011
0110	0100
0111	0101
0101	0110
0100	0111
1100	1000
1101	1001
1111	1010
1110	1011
1010	1100
1011	1101
1001	1110
1000	1111

7-11. Obtain a ROM which realizes the following functions:

$$y_1 = x_1 x_2 x_3 + x_1 \bar{x}_2 x_3 + \bar{x}_1 \bar{x}_2 x_3$$
$$y_2 = x_1 x_2 \bar{x}_3 + \bar{x}_1 \bar{x}_2 \bar{x}_3 + \bar{x}_1 x_2 x_3 + x_1 \bar{x}_2 x_3$$
$$y_3 = x_1 x_2 x_3 + \bar{x}_1 x_2 \bar{x}_3 + x_1 x_2 \bar{x}_3 + \bar{x}_1 \bar{x}_2 \bar{x}_3$$

7-12. Repeat Prob. 7-11 for the functions

$$y_1 = x_1 x_2 + \bar{x}_1 x_3 + \bar{x}_2 x_3$$
$$y_2 = x_1 + x_2 \bar{x}_3$$
$$y_3 = x_1 + x_2$$
$$y_4 = x_1 + \bar{x}_2 + x_1 \bar{x}_3$$

7-13. Design the decoder for Prob. 7-11.

7-14. Repeat Prob. 7-13 for Prob. 7-12.

7-15. Design a one-level decoder for four input variables.

7-16. Repeat Prob. 7-15 for a tree decoder.

7-17. Repeat Prob. 7-15 for a matrix decoder. Show all gates.

8

Binary Codes

We have thus far discussed the components of digital systems. Before turning our attention to the microprocessor and the complete digital computer, we shall discuss procedures for supplying data in a form which the computer can process. Since digital systems handle all data in binary form, all information supplied to the computer must ultimately be converted into a binary code. These binary codes will be the subject of this chapter.

We shall start by considering numerical codes. Subsequently, we shall discuss how alphabetic data is handled.

8.1 BINARY CODED DECIMAL CODE (BCD CODE)

In digital computers, all signals are in the form of 0's and 1's. These lend themselves to binary arithmetic. Almost all large computers perform binary arithmetic. However, computer *users* work in the decimal system. In most computers, there are circuits which perform a conversion from decimal to binary when a number is supplied to the computer, and vice versa. However, the computer can only work with 0's and 1's. Thus, there must be some provision made for representing decimal numbers using bits. In addition, in some computers, the numbers are retained in decimal form within the compuer. The arithmetic operations are not as simple in this case but, if they are relatively few in number, then the elimination of the conversion of radix results in an overall saving. We shall now discuss procedures for representing decimal numbers using 0's and 1's. Each decimal digit is represented by a sequence of binary digits. That is, there is a *binary code* which represents each of 0, 1, 2, 3, 4, 5, 6, 7, 8, and 9. Such codes are called *binary coded decimal* (BCD) codes.

The BCD codes are special examples of *weighted* codes. In such codes, each bit is assigned a weight. A weight just means a numerical value here. For instance, a very common code is the 8421 code. Here, four bits are used to represent a decimal digit. If the first (rightmost) bit is 1, then it represents a weight of 1; if the second bit is 1, then it represents a weight of 2, etc. Thus, the number 1010 is equivalent to

$$8 + 0 + 2 + 0 = 10$$

Similarly, 1001 is equivalent to

$$8 + 0 + 0 + 1 = 9$$

Thus, Table 8-1 illustrates the 8421 BCD code.

Table 8-1 BCD Code Representation

Decimal Digit	*(8 4 2 1)*
0	0 0 0 0
1	0 0 0 1
2	0 0 1 0
3	0 0 1 1
4	0 1 0 0
5	0 1 0 1
6	0 1 1 0
7	0 1 1 1
8	1 0 0 0
9	1 0 0 1

Note that, in the 8421 BCD code, each decimal digit is represented simply by its binary equivalent.

As an example, we shall write 8624 using the 8421 BCD code. From Table 8-1, we have

$8 \rightarrow 1000$

$6 \rightarrow 0110$

$2 \rightarrow 0010$

$4 \rightarrow 0100$

Thus, the 8421 BCD representation of 8624 is

1000 0110 0010 0100 = 1000011000100100

The weighted code that we have just considered appears logical since each decimal digit is represented by its binary equivalent. However, it is not the only one that could have been used. For instance, another weighted BCD code that is sometimes used is the 2421 code. Here the weights are 2, 4, 2, and 1. For instance, 1101 is equivalent to $2+4+0+1 = 7_{10}$. Negative numbers can also be used as weights. For instance, the weights 8, 4, –2, and –1 are sometimes used. In this case, 1011 is equivalent to $8+0-2-1=5_{10}$. Table 8-2 represents the decimal representations for these two codes.

Table 8-2 Two Different BCD Codes

Decimal Digit	2 4 2 1 Representation	8 4–2–1 Representation
0	0 0 0 0	0 0 0 0
1	0 0 0 1	0 1 1 1
2	0 0 1 0	0 1 1 0
3	0 0 1 1	0 1 0 1
4	0 1 0 0	0 1 0 0
5	1 0 1 1	1 0 1 1
6	1 1 0 0	1 0 1 0
7	1 1 0 1	1 0 0 1
8	1 1 1 0	1 0 0 0
9	1 1 1 1	1 1 1 1

Then, 8624 in the 2421 representation is

1110 1100 0010 0100

Note that there is some freedom of choice here. For instance, the 2421 representation of 4 could be either 0100 or 1010. The choices made in Table 8-2 are standard ones which actually simplify arithmetic in those computers which perform arithmetic directly using BCD coded numbers. Although we have considered only three weighted BCD codes here, there are actually very many others.

There is another BCD code that has sometimes been used. It is formed by following this scheme: Three is added to each digit; the resulting base 10 number for each digit is then represented by its ordinary 8421 BCD representation. This is called an *excess* 3 code. For instance,

$$8476 \rightarrow 11 \ 7 \ 10 \ 9 \rightarrow 1011 \ 0111 \ 1010 \ 1001$$

The excess 3 code has the simple arithmetic feature discussed in conjunction with the 2421 and the 8421 codes.

8.2 REFLECTED CODES—GRAY CODES

Many times, digital systems are used in conjunction with analog ones. For instance, a digital computer may be used in a control system that varies the position of the rudder of a ship. Or, a digital system may be used in the guidance system of a space probe. The position of the shaft of the rudder or of the inertial guidance platforms of the space probe must be supplied to the computer. This is done by an analog-to-digital converter which converts the analog signal (e.g. position) to a binary representation. In such mechanical systems, the conversion is often a commutator connected to a rotating shaft. As an example, the commutator may have many rings, with each ring corresponding to a binary digit. Brushes make contact with the rings. Each ring is divided into segments which are insulated from each other, so that parts of the ring can be connected to a voltage which corresponds to a 0 and the remaining parts of the ring are connected to a voltage which corresponds to a 1. Suppose that there are four rings. Then, the position of the shaft can be encoded into a four-bit binary number. Suppose that the 8421 BCD codes is used here. Then, the shaft position can be divided into 10 parts. (If 8 rings are used, then the position can be divided into 100 parts, etc.) If ordinary binary representation is used, then 4 rings can represent 16 positions and *8* rings can represent 256 positions, etc. For instance, if we have four rings then, as the shaft rotates, the output voltage could vary in accordance with Table 8-2.

The shaft position is then given the appropriate numerical value. However, a problem can arise where when the position of the shaft is such that one or more brushes are at the junction(s) between segments of the rings simultaneously. In this case, large errors can arise. Let us illustrate this using an 8421 BCD code. Suppose that the position changes from 3, represented by 0011 to 4, represented by 0100.

Three bits must change *simultaneously*. Thus, three brushes must move across contacts on their respective rings simultaneously. However, because of mechanical tolerances, this change may not occur exactly simultaneously and errors can result. For instance, suppose that the second bit changes before the others. Then, the number 0000 results. This represents 0, which is not between 3 and 4 and is a serious error. This type of error can be avoided if adjacent numbers differ by only one bit. Then, there is no shaft position where the contacts to more than one segment must change simultaneously. A new BCD code must be used. (Note that it is not necessary to use BCD codes here.) A code which does this is called a *reflected* code or a *Gray* code. Since each successive number changes by only one binary digit then, as the position of the shaft is changed, only one commutator segment will ever have to be crossed at one time. Thus, we do not have the previously discussed problem of several bits having to change simultaneously. We have illustrated the use of a reflected code with a rotary shaft. However, such codes are often used with many types of analog-to-digital conversion. In addition, there are other applications. For instance, the coding used in generating a Karnaugh map was a reflected code.

A typical reflected code is shown in Table 8-3.

Table 8-3 A Reflected Binary Code

Decimal Number	Reflected Binary Number
0	0000
1	0001
2	0011
3	0010
4	0110
5	0111
6	0101
7	0100
8	1100
9	1101
10	1111
11	1110
12	1010
13	1011
14	1001
15	1000

Note the dashed line in Table 8-3. Except for the leftmost bit, there is a mirror symmetry (i.e. reflection) about this line. This is why this is called a reflected code.

Let us now discuss a general procedure for obtaining a Gray (reflected) code. We assume that we start with the ordinary binary representation of a number. First, place a 0 to the left of the number. Then, starting from the left, use each adjacent pair of bits and apply the exclusive OR (XOR) operation to them. The result of each of these operations yields a digit of the Gray code. Let us illustrate this with an example. We will obtain the Gray code for $7_{10} = 111_2$. Add a 0 to the left. This yields 0111. Then, the XOR operations are

$$0 \oplus 1 = 1$$

$$1 \oplus 1 = 0$$

$$1 \oplus 1 = 0$$

Thus, the Gray code is 100 or, equivalently, 0100. Similarly, the Gray code for $65_{10} = 1000001_2$ is 1100001.

To convert from a Gray code to an ordinary binary number, we can use the following procedure. Examine the number. Each digit in the binary code is either the same as the corresponding digit of the Gray code or its complement. To determine which, count the number of 1's to the *left* of the digit in question. If it is an even number, then the digit is unchanged. If the number of 1's to the left is odd, then take the complement. For example, if we have 1100001 in Gray code, then the first digit of the regular binary number remains a 1 since there is an even number (zero) of 1's to the left. The second digit must be changed since there is an odd number of 1's to the left in the original (Gray) number. Thus, we must change the 1 to a 0. Proceeding in this way, we obtain $1000001_2 = 65_{10}$, as we should. (Note that zero is treated as an even number here.)

8.3 ALPHANUMERIC CODES

We have thus far concentrated on numbers. However, computers often process data which contains letters or other symbols. For instance, lists of names can be alphabetized, or students' grades can be stored and printed by a computer. Also, computer programs are written in terms of words and symbols. Thus, the computer must be able to process all of these. All information in a digital computer is stored and processed in binary form (i.e. as 0's and 1's). Thus all data—numbers, letters, symbols, etc.—must have a binary representation or binary code. Various codes are used for this purpose.

Before discussing the codes, let us discuss some background material. Information is not simply processed within the computer. It must also be transmitted to it by the user and the answers must be transmitted from the computer to the user. At times, this information must be transmitted over great distances. For instance, the user may be transmitting his information, using a terminal and a modem, over telephone lines. The modem generates an apprrpriate digital code which can be transmitted over telephone lines. Even when the user is close to the computer, information must be transmitted, by means of a suitable code, to and from such input-output devices as terminals, card readers, etc. These are external to the central processor of the computer. The information is transmitted to these peripheral devices using a suitable code. We shall consider some codes that are in use. In Chapter 6 we considered the electronic circuits that are associated with coding and decoding (i.e. sequence-generating and sequence-detecting circuits). These codes provide for transmitting 26 letters, 10 digits, and various symbols. They are called *alphanumeric* codes since letters, as well as numbers and symbols, are transmitted. Two representative codes are given in Table 8-4. One is called the EBCDIC (Extended BCD Interchange Code). This code uses eight bits. The other is called the ASCII (American Standard Code for Information Interchange) code. Note that there are several versions of this code. A recent modification of this code is called ANSCII (American National Standard Code for Information Interchange).

In both of these codes, the numerical values, in binary, of the alphabetic characters is in increasing order. This is done to allow such things as alphabetizing to be easily done.

Table 8-4 Some Alphanumeric Codes

	ASCII Code	*EBCDIC Code*
A	100 0001	1100 0001
B	100 0010	1100 0010
C	100 0011	1100 0011
D	100 0100	1100 0100
E	100 0101	1100 0101
F	100 0110	1100 0110
G	100 0111	1100 0111

	ASCII Code	EBCDIC Code
H	100 1000	1100 1000
I	100 1001	1100 1001
J	100 1010	1101 0001
K	100 1011	1101 0010
L	100 1100	1101 0011
M	100 1101	1101 0100
N	100 1110	1101 0101
O	100 1111	1101 0110
P	101 0000	1101 0111
Q	101 0001	1101 1000
R	101 0010	1101 1001
S	101 0011	1110 0010
T	101 0100	1110 0011
U	101 0101	1110 0100
V	101 0110	1110 0101
W	101 0111	1110 0110
X	101 1000	1110 0111
Y	101 1001	1110 1000
Z	101 1010	1110 1001
0	011 0000	1111 0000
1	011 0001	1111 0001
2	011 0010	1111 0010
3	011 0011	1111 0011
4	011 0100	1111 0100
5	011 0101	1111 0101
6	011 0110	1111 0110
7	011 0111	1111 0111
8	011 1000	1111 1000
9	011 1001	1111 1001
blank	000 0000	0100 0000
.	010 1110	0100 1011
(	010 1000	0100 1101
+	010 1011	0100 1110
)	101 1101	0100 1111
{	101 1011	0101 1010
$	010 0100	0101 1011
*	010 1010	0101 1100
)	010 1001	0101 1101
;	011 1011	0101 1110

	ASCII Code	EBCDIC Code
,	010 1100	0110 1011
–	010 1101	0110 1101
=	011 1101	0111 1110

We have shown some symbols such as +, –, *, etc. here. Actually, provision for many more symbols is made. We have just given a representative list here. Note that the codes may vary from computer to computer. The ones given here are fairly representative.

These codes appear to be unduly long. For instance, the reader may ask why use seven or eight bits to transmit a decimal digit when BCD codes require only four bits. There are several reasons for this. The codes transmit much more information than just the 10 digits. For purposes of detection, it is desirable to have each symbol the same length. Another reason for having the *code word* (i.e. the number of bits for each symbol) longer than necessary is to enable us to detect or eliminate errors which occur during transmission. We shall discuss this next.

8.4 ERROR-DETECTING AND CORRECTING CODES

When data is transmitted from a remote location to a computer via telephone, or via other communcation channels, there is a reasonable probability that some errors will occur. For instance, noise introduced on the line may make a 0 appear as a 1 or vice versa. Thus, it is desirable to have some way of knowing if an error has occurred and, if possible, to correct it.

The usual transmission channels are not very noisy and the probability of an error in a single bit is usually low. Thus, the probability of an error in more than one bit in a code word is very low. Let us assume that this is so and discuss a simple procedure that will allow *single* errors to be detected. In this way, an error message can be generated and the word can be retransmitted.

We shall illustrate this with the 8421 BCD code given in Table 8-1. An extra bit called a *parity bit* is added to the code. This bit value is such that the number of 1's is always even (or odd). This is called an *even (odd) parity check*. The resulting code is called a *parity checking code*. For instance, an even parity checking code is illustrated in Table 8-5. Note that, in each case, the parity bit is such that the total number of 1's in the code word is even. (Note that zero is considered to be an even number.)

If we are to have error detection, then the code word must be in-efficient in the sense that more bits than necessary to simply transmit the information are used. For instance, if an error is to be detected, the change of a single bit must not produce another valid code word. If the change of a single bit produced a valid code word, then there would be no way of knowing that an error had occurred, since the changed word could have been the one transmitted. That is, to detect an error, it must result in a "word" that is not in the table. To be able to describe things such as this numerically, we define the *Hamming distance* between two code words as the number of bits as the number of bits which must be changed to convert one word into the other. The *minimum distance* of a code is the smallest distance between *any* two words of the code. For instance, in Table 8-1, the minimum distance is one, while in Table 8-5, the minimum distance is two.

Table 8-5 8421 BCD Code with Even Parity Check

Decimal Digit	*(8 4 2 1 p)*
0	0 0 0 0 0
1	0 0 0 1 1
2	0 0 1 0 1
3	0 0 1 1 0
4	0 1 0 0 1
5	0 1 0 1 0
6	0 1 1 0 0
7	0 1 1 1 1
8	1 0 0 0 1
9	1 0 0 1 0

If a code is to be such that a single error can be detected, then the *minimum distance must be at least two*. Again, if the minimum distance is one, then a single error can produce another valid code word and an error cannot be detected. If the minimum distance is three, then if two errors occur in a code word, the fact that an error has occurred can be detected, etc.

If the minimum distance is three, or more, then single errors can not only be detected, but they can also be corrected. That is, there is sufficient information present so that we not only know that an error has occurred, but we can also determine the digit in which the error has occurred. Once this is known, the errors can be corrected by replacing the erroneous digit by its complement, that is by changing a 0 to a 1 or vice versa. We shall now illustrate a code which is error-correcting if single errors occur. It is called a *Hamming code*. In these codes, several parity-checking bits are added. Parity is then checked for groups of bits. Consider a code which has four bits. The 8421 code is one such. Let us add three parity bits so that a code word is of the form

$$p_1 \quad p_2 \quad w_3 \quad p_4 \quad w_5 \quad w_6 \quad w_7 \tag{8-1}$$

The subscripts indicate the position of the bit in the word. Note that w_3, w_5, w_6 and w_7 would be the word if no parity checking were used. Then, the parity bits are adjusted so that there is an even number of 1's in the following groups of positions labeled c_0, c_1 and c_2.

$$c_2 \text{ positions } 4 \ 5 \ 6 \ 7 \tag{8-2a}$$

$$c_1 \text{ positions } 2 \ 3 \ 6 \ 7 \tag{8-2b}$$

$$c_0 \text{ positions } 1 \ 3 \ 5 \ 7 \tag{8-2c}$$

Note that only one parity bit is in each checking position. Thus, the parity bits do not interact with each other.

As an example, let us use the 8421 BCD code. The Hamming code is illustrated in Table 8-6.

Table 8-6 8421 BCD Code with Hamming Error Correction

Digit	Position 1	2	3	4	5	6	7
	p_1	p_2	8	p_4	4	2	1
0	0	0	0	0	0	0	0
1	1	1	0	1	0	0	1
2	0	1	0	1	0	1	0
3	1	0	0	0	0	1	1
4	1	0	0	1	1	0	0
5	0	1	0	0	1	0	1
6	1	1	0	0	1	1	0
7	0	0	0	1	1	1	1
8	1	1	1	0	0	0	0
9	0	0	1	1	0	0	1

When the word is received, the three parity checks are made and any error is detected and corrected.

For instance, suppose that 7 is incorrectly transmitted as 0001011, then the tests for c_2 will fail, c_1 will not fail and c_0 will fail. Thus, the error must be in position 5. Note that c_0 and c_2 have errors in position 5 and 7 in common. Thus, if c_0 and c_2 fail, the error could be either in position 5 or in position 7. However, c_1 did not fail. Therefore, the error is not in position 7 and the error is in position 5. Complementing the fifth bit, we obtain the correct value 0001111. Studying Eqs. (8-2), we obtain the following table for the position of failure. Here T indicates no failure and F indicates a failure.

Table 8-7 Position of Failure for Hamming Code of Table 8-6

Test c_2 c_1 c_0	*Digit of Failure*
T T T	None
T T F	1
T F T	2
T F F	3
F T T	4
F T F	5
F F T	6
F F F	7

In Table 8-7, if we consider the trues (no error) as 0's and the falses (error) as 1's, then the binary number resulting from the parity check indicates the position of the error. Note that transmission errors in the parity bits are also corrected.

The Hamming detecting procedure can be applied to codes with more than four (original) bits. In general, the parity checking bits are placed in positions 1, 2, 4, 8, 16, 32, 64, . . . , 2^n. The bits which are checked can be determined using the following scheme. p_1 checks all bits whose position (subscript) in binary has a 1 in the 2^0 position. For instance, 001, 101, 111 all have 1's in that position. Then, p_2 is a parity check for all bits whose binary representation has a 1 in the 2^1 position (i.e. 010, 111, 011) etc. Using this scheme, Hamming codes can be generated for words of any length.. Note that no parity bit checks any other parity bit. Thus, the parity bits are all independent. We have illustrated some error correcting codes. However, there are very many such codes that are used.

BIBLIOGRAPHY

Booth, T.L. *Digital Networks and Computer Systems*, Second Ed., Chap. 2, John Wiley and Sons, Inc., New York 1978

Kohavi, Z., *Switching and Finite Automatic Theory*, Chap. 1, McGraw-Hill Book Co. Inc., New York 1970

Mano, M.M., *Digital Logic and Computer Design*, Chap. 1, Prentice Hall, Inc., Englewood Cliffs, NJ 1979

O'Malley, J., *Introduction to the Digital Computer*, Chap. 5, Holt, Rinehart and Winston, Inc., New York 1972

Steinhart, R.F. and Pollack, S.V., *Programming the IBM System/360*, Chap. 2, Holt, Rinehart and Winston, Inc., New York 1970

PROBLEMS

8-1. Write the following numbers using an 8421 BCD code:

1796   437892   1264.36

8-2. Repeat Prob. 8-1 using a 2421 BCD code.

8-3. Repeat Prob. 8-1 using an 84–2–1 BCD code.

8-4. Write a table for a 74–2–1 BCD code.

8-5. Obtain a BCD code which uses different weights than those used previously.

8-6. Obtain a Gray code for numbers between 0 and 32.

8-7. Obtain the Gray code representation for 1785_{10} and 1236_{10}.

8-8. Obtain the ordinary binary representations of the following Gray code representations:

1010110   111011011

8-9. Write the ASCII representation of the word BOOK. If this is to be stored in a register which stores $4n$ bits ($n=1,2,\ldots$) what is the smallest value that n can have?

8-10. Repeat Prob. 8-8 using an EBCDIC representation.

8-11. Rewrite Table 8-5 but now use odd parity checking.

8-12. Modify Table 8-2 so that an even parity check can be made.

8-13. The following data are sent with 8421 BCD code with even parity checking. Which are in error? Assume that there is, at most, one error in each.

01011   10001   01111

8-14. The following are data sent with an 8421 BCD code with Hamming error correction. Identify the data sent, correcting any errors if necessary. Assume that there is, at most, one error in each.

1101001   1101010   1001101   0000001   1000000   1001111

8-15. Obtain the Hamming error correction code for the 2421 BCD code.

8-16. Describe how Hamming error correction would be applied to an 8-bit (uncorrected) code.

9

Basic Computer Arithmetic

In this chapter we shall consider the details of the arithmetic that is used in a digital computer. In Chapter 2 we discussed the basic ideas of addition using binary numbers. Here we shall consider all pertinent aspects of binary arithmetic and relate them to digital computers.

9.1 ACCUMULATORS–MODULAR ARITHMETIC

In Sec. 2.2 we discussed addition in the binary system. Let us now see how this is implemented. Addition is performed using an adder. The results are stored in a register which is usually called an *accumulator*. For instance, suppose that we have an 8-bit accumulator. Then it can store an 8-bit binary number. Uusually, one binary number is added to the accumulator, and the new stored (binary) number will be the sum of the original stored number plus the added number. If we want to add two numbers, the following procedure would be used. The accumulator would be cleared (set equal to zero). The first number would be put into the accumulator (stored there). Then, the second number would be added to the number stored in the accumulator. After this is done, the accumulator would store the sum of the two numbers. In Chapter 10 we shall consider details of the actual hardware that makes up the arithmetic circuits of a computer. In this chapter we shall consider the arithmetic details connected with obtaining the number stored in the accumulator register.

In the simplest case, the number stored in the accumulator is, as discussed, the sum of two numbers which have been added. However, problems can arise if the numbers to be added are so large that an overflow can result, see Sec. 2.2. For instance, an overflow will occur when we are working with an 8-bit accumulator (i.e. one that can store eight binary digits) if the sum of the two numbers which are added results in a 9-bit binary number. Since such overflows result in a loss of the most significant figures, see Sec. 2.2, they must be avoided. Provision is usually made to indicate that an overflow has occurred, so that the computer user is warned. We shall discuss this subsequently. Computation may stop when an overflow occurs. In the next section we shall see that, at times, this process can be used to advantage. However, for the time being, let us assume that overflow is undesirable.

Since the number of bits in a register or accumulator is limited, arithmetic using them assumes a special form, which we shall now discuss. We shall start with decimal numbers and then use binary numbers. Consider an accumulator which can store three decimal numbers. An example of such a "decimal accumulator" is an automobile odometer. Let us assume that it is set at zero and numbers are added to it, one at a time. The sequence of its readings will be

000
001
002
003
.
.
009
010
011
012   (9-1)
.
.
.

996
997
998
999
000
001
002
.
.
.

The accumulator can store numbers from 000 to 999. If the sum becomes greater than 999, then the *accumulator register cycles again*. This is a characteristic which is common to almost all accumulators including those found in computers. Thus, if the accumulator reads 041, this could actually represent any of the numbers

$$
\begin{array}{l}
41 \\
1041 \\
2041 \\
\vdots
\end{array}
\tag{9-2}
$$

(There could also be negative representations. However, we shall defer a discussion of these.) We shall now discuss an arithmetic which can be used not only to characterize registers and accumulators, but also to study arithmetic processes used with computers. We shall also use this arithmetic to find ways of using overflow for useful purposes.

Now let us define a mathematical process called *modular arithmetic* which is of fundamental importance in registers. Two integers M and N are said to be *congruent, modulo* k if M/k and N/k have the same remainder. In this case, we write

$$
M \equiv_k N
\tag{9-3}
$$

where k is called the *modulus*, and must be an integer. Let us consider an example.

$$
17 \equiv_3 29
\tag{9-4}
$$

Note that $17/3 = 5 + \frac{2}{3}$, and $29/3 = 9 + \frac{2}{3}$. In each case, the remainder is 2. Thus, 17 and 29 are congruent, modulo 3. Other terminology and symbols are sometimes used. For instance, in (9-3), M and N are also said to be *equivalent, modulo* k. Another symbolic representation for (9-3) is

$$
M \equiv N(\bmod k)
\tag{9-5}
$$

If Eq. (9-3) is satisfied, then we can write

$$
(M/k) - (N/k) = I
\tag{9-6}
$$

where I is an integer. Note that if M/k and N/k have the same remainder, then their difference will be an integer. Thus, Eq. (9-6) is a criterion which can be used to test if two numbers are congruent, modulo k. Now let us see how modular arithmetic applies to registers. Consider the register of (9-1) and the example (9-2). For the register, we demonstrated that the numbers, 41_{10}, 1041_{10}, 2041_{10}. . . would all produce the same reading. We can now express this using modular arithmetic.

$$41 \equiv_{1000} 1041 \equiv_{1000} 2041 \equiv_{1000} \cdots \tag{9-7}$$

Note that since $41/1000$, $1041/1000$, $2041/1000$, . . . all have the remainder 41, we can use modular arithmetic to characterize the reading of the register.

Let us now discuss binary registers (or accumulators). As an example, we consider a 3-bit binary accumulator register. (This contains fewer bits than those usually encountered.) Assume that the accumulator register is set at zero and that numbers are added to it one at a time.

$$
\begin{matrix}
000 \\
001 \\
010 \\
011 \\
100 \\
101 \\
110 \\
111 \\
000 \\
001 \\
010 \\
\vdots
\end{matrix}
\tag{9-8}
$$

These correspond to the decimal numbers 0, 1, 2, 3, 4, 5, 6, 7, . . . Now consider a specific number. A reading of 101 could correspond to 5, 13, 21, 29. . . (We shall again restrict ourselves to positive integers.) In terms of modular arithmetic, we have

$$5 \equiv_8 13 \equiv_8 21 \equiv_8 29 \equiv_8 \cdots \tag{9-9}$$

Note that, for each of these, $N/8$ yields a remainder of 5.

Let us generalize these results. Consider a register whose highest number is R. For instance, in the register of (9-8), the register ranges between 0 and 7. Thus, in this case, $R = 7$. We want the register to store a number N without overflow. That is, the value stored is to be the *true value* of N. Then,

$$N \leqslant R \tag{9-10}$$

where we consider that N and R are positive integers. If we are to avoid overflow, then relation (9-10) must be satisfied. Let us now consider this in terms of a binary register. If the register has b bits, then $R = 2^b - 1$. Thus, relation (9-10) becomes

$$N < 2^b - 1 \tag{9-11}$$

Relation (9-11) is valid for a binary radix accumulator register. At times, but usually not in computers, accumulators are constructed using other radices. For instance, an automobile odometer is a base 10 register. In general, if there are d digits in the accumulator, then $R = r^d - 1$, where r is the radix. Thus, we can generalize relation (9-11) to be

$$N < r^d - 1 \tag{9-12}$$

Thus far we have only considered positive numbers. Of course, almost all computers work with negative numbers as well. Thus, some provision must be made to store the sign of the number. The leftmost digit of the register is usually used for this purpose. In general, *a 0 or a 1 in the leftmost digit does not simply replace the sign.* For instance, in one system, 0111 would represent 7 while 1001 would represent –7. (Note $001_2 \neq 7_{10}$.) We shall consider such systems in the next section.

Now, if we have a k-bit register, only $k-1$ of the bits are available for the magnitude of the number. Thus, relation (9-11) becomes

$$|N| \leqslant 2^{k-1} - 1 \tag{9-13}$$

if there is to be no overflow. We shall modify this somewhat in the next section.

Addition

Let us assume that we are doing addition on a binary basis and that the accumulators or registers have k binary bits. Also assume that k_1 of these are available for storing the magnitude of the number. For instance, if the most significant bit is always 0 when the number is positive, then

$$k_1 = k - 1 \tag{9-14}$$

We shall again restrict ourselves to positive integers. The binary addition process follows that discussed earlier in this section. Since there are k bits, the modulus of our modular arithmetic system is 2^{k_1}. Now suppose that we add two integers N_1 and N_2 and that after adding them, the resulting total stored in the accumulator is T. There may be overflow. However, whether there is or not, the total stored in the accumulator is

$$T \equiv_{2^{k_1}} N_1 + N_2 \tag{9-15}$$

where T is the smallest positive integer which satsifies Eq. (9-15). Note that the accumulator reading cannot exceed $2^{k_1} - 1$. For instance, if $k_1 = 4$ and $N_1 = 6$ and $N_2 = 4$, then

$$T \equiv_{16} 6 + 4$$

In this case, the value read on the register will equal the true sum of 10. Of course, these numbers will be in binary form. We write them in base 10 for human convenience. Now suppose that $N_1 = 12$ and $N_2 = 8$. Then, the smallest integer value of T which satisfies this is 4. This will be the value stored in the accumulator. Note that an overflow has occurred in this case. This could also have been determined from relation (9-13).

Subtraction

We shall now consider subtraction. At the start we shall discuss the basic ideas and then shall apply these to computer subtraction. We will briefly consider radix 10 and then turn our attention to the binary case. Suppose that we subtract

2965_{10} minuend
$- 1432_{10}$ subtrahend

1533_{10} difference

Here each digit of the subtrahend is subtracted from the corresponding digit of the minuend to obtain the digits of the difference. Now suppose that a digit of the subtrahend is greater than the corresponding digit of the minuend. In this case, we "borrow" from the digit to the left. Let us illustrate this with the subtraction $6345 - 3726$. We write this as

```
 -1 -1
 ⌒ ⌒
 6 3 4 5              5 13 3 15
               →
- 3 7 2 6            - 3  7 2  6
─────────           ──────────────
 2 6 1 9             2  6 1  9
```

The first digit of the subtrahend, 6, is greater than the first digit of the minuend, 5. Thus, we borrow 10 from the next column. Column two now represents 30 while the first column is now 15. Then, $15-6$ yields 9. In the second column, we now have $30-20$ which is represented by 1 in the second digit of the answer. Now consider the third column. 300 is less than 700 and, thus, we borrow 1000 from the fourth column and the subtraction proceeds. Let us consider an additional example.

```
  -1-1-1
  ⌒⌒⌒
 6 3 4 5             5 12 13 15
               →
- 1 3 4 6            1  3  4  6
─────────           ──────────────
 4 9 9 9             4  9  9  9
```

Notice how the borrowing the first column causes borrowing to take place in the remaining columns.

We have discussed the subtraction of positive integers. It is assumed that the reader is familiar with the rules of arithmetic which relate to subtraction of negative numbers. Similarly, it is also assumed that the reader is familiar with the arithmetic operation that results if the subtrahend is larger than the minuend [i.e. $a-b = -(b-a)$].

Now let us work with subtraction on a binary basis. For instance,

$$
\begin{array}{r}
\overset{\overset{-1}{\frown}}{1\,0\,1\,1} \\
-\ 0\,1\,0\,1 \\
\hline
0\,1\,1\,0
\end{array}
\qquad \rightarrow \qquad
\begin{array}{r}
0\ 10\ 1\ 1 \\
-\ 0\quad 1\,.\,0\ 1 \\
\hline
0\quad 1\ 1\ 0
\end{array}
$$

In the first two columns, the digits of the subtrahend are not greater than those of the minuend and "simple subtraction" takes place. In the third column, we must borrow. That is, 4_{10} is greater than 0. Thus, 8_{10} is borrowed from the fourth column.

As a second example, let us consider the subtraction

$$
\begin{array}{r}
\overset{\overset{-1}{\frown}\overset{-1}{\frown}\overset{-1}{\frown}}{1\ 0\ 0\ 0} \\
-\ 0\ 0\ 1\ 1 \\
\hline
0\ 1\ 0\ 1
\end{array}
\qquad \rightarrow \qquad
\begin{array}{r}
0\ 1\ 1\ 10 \\
-\ 0\ 0\ 1\ 1 \\
\hline
0\ 1\ 0\ 1
\end{array}
$$

Since 1 is greater than 0, we borrow $2_{10}=10_2$ from the second column. However, the second column of the minuend was 0. Thus, we must borrow $4_{10}=100_2$ from the third column. This is also 0. Thus, we must borrow $8_{10}=1000_2$ from the fourth column. When several 0's are adjacent to each other, a borrow propagates through all of them.

For illustrative purposes, let us consider one final example where the minuend is less than the subtrahend. In this case, we shall subtract the minuend *from* the subtrahend and write the answer as a negative number. For instance,

$$
\begin{array}{r}
1010 \\
-\ 1111 \\
\hline
-\ 0101
\end{array}
\qquad\qquad
-\left\{
\begin{array}{r}
1111 \\
-\ 1010
\end{array}
\right\}
$$
$$
\qquad\qquad\qquad\qquad\ -\ 0101
$$

Now let us consider subtraction on the basis of registers and accumulators. Again, for clarity, we shall start the discussion by considering base 10. Consider a 3-digit accumulator. As 1 is added successively, we have

.
.
.
996
997
998

999
000
001
002
003
004   (9-16)
005
.
.
.
998
999
000
001
.
.
.

Suppose that we have 5 entered in the accumulator and that we sub-tract 2. The accumulator stores 3. Now suppose that we have 5−7. The accumulator is stepped back seven steps to 998. The answer should be −2. Thus, an underflow (i.e. the equivalent of an overflow when the number becomes too small) has resulted. Let us consider this on the basis of modular arithmetic.

$$-2/1000 = 0 - (-2/1000)$$

$$999/1000 = 1 - (2/1000)$$

Each of these has the same remainder (i.e. −2). Thus, we can say

$$-2 \equiv_{1000} 998 \tag{9-17}$$

Thus, on a modular basis, we have the correct answer. Actually, this is not a real underflow since we assume that −2 lies within the range of the register. This assumes that there is some provision for obtaining the correct sign. We shall consider such procedures in the next section.

Let us consider another aspect of the previous example. $998/1000 = 0 + 998/1000$. In this case, the remainder is not −2 but is 998. Consider Eq. (9-6). This states that two integers are equivalent, modulo k if their difference, divided by k is an integer. Then,

$$(998/1000) - (-2/1000) = 1 \tag{9-18}$$

Thus, the definition is satisfied. A true overflow can occur if the magnitude of the answe is too large. For instance, $999 - (-20) = 1019$ which is a true overflow.

Now let us consider these ideas using base 2 numbers. We shall again illustrate the ideas with a 3-bit register. Suppose it is cycled (i.e. as the number stored in the accumulator is incremented by 1, the new number is stored in the accumulator). The numbers stored are

.
.
.
100
101
110
111
000
001
010
011
100
101
110
111
000
001
010
.
.
.

Now let us perform $5 - 7$ which, in base 2, would be $101 - 111$. Cycling the register backward from 101, we obtain 110. On a modular basis, we

have

$$6 \equiv_8 -2 \tag{9-19}$$

Note that $6/8 = 1 - 2/8$. Thus, we have achieved the correct answer on a modular arithmetic basis. Again, an overflow appears to have resulted. However, as we shall see in the next section, this is actually *not* the case.

We have compared subtraction to cycling the accumulator backward. This procedure is convenient if the accumulator is an automobile odometer. However, in a digital computer, such backward cycling of the accumulator is awkward. Circuits can be built which can add and, similarly, other circuits can be built which can subtract, but these are different circuits. Circuits can be built which can both add and subtract but these are more complex than those circuits which only add (or subtract). We shall next consider a subtraction procedure which essentially uses only an adding circuit and, thus, reduces the complexity of the computer.

9.2 COMPLEMENT ARITHMETIC

To reduce the complexity of a digital computer, it would be desirable to utilize the same circuit to perform the operations of both addition and subtraction. In this section, we shall demonstrate how an adder and a register can also be used to perform subtraction by taking account of the "overflow properties" of the register which, in this case, we shall call the accumulator.

Let us illustrate the procedure that we shall use. Suppose we want to compute the difference of two numbers A and B. We assume that $R-1$ is the largest number that can be stored in the accumulator. The accumulator operations are thus defined modulo R. The difference that we want is

$$D = A - B \tag{9-20}$$

Now suppose that we form the number N in the following way:

$$N = R - B \tag{9-21}$$

Note that we have performed a subtraction here. However, it is a special subtraction in that the minuend is equal to R. We shall subsequently see that this *particular* subtraction is especially simple and does not require special subtractor circuits in the digital computer. Now suppose that we take the *sum* of A and N.

$$A + N = A + R - B$$

Since $R/R = 1$, an integer, we have

$$D \equiv_R A - B \equiv_R A + N \tag{9-22}$$

Thus, the sum $A + N$ and the difference D are congruent, modulo R. Hence, each produces the same result in the accumulator. Let us illustrate this using base 10. Suppose that $A = 80$ and $B = 6$ and $R = 1000$. That is, these are three decimal numbers. Then,

$$A - B = 80 - 6 = 74$$

Now

$$N = R - B = 994$$

$$A + N = 80 + 994 = 1074$$

$$1074 \equiv_{1000} 74$$

Thus, the accumulator will store the *correct* value of 74. Let us now consider an example in binary. We shall assume a 7-bit accumulator. Then,

$$A = 80_{10} = 1010000_2$$

$$B = 6_{10} = 0000110_2$$

Since there are seven bits in the accumulator, R will be an 8-bit number.

$$R = 128_{10} = 10000000_2$$

Then,

$$N = R - B = 10000000 - 0000110$$

Performing the subtraction, we obtain

$$N = 1111010$$

Hence,

$$A + N = 1010000 + 1111010 = 11001010$$

$$11001010_2 \equiv_{128} 1001010_2$$

This last statement is equivalent to

$$202_{10} \equiv_{128} 74_{10}$$

Thus, we obtain the desired result.

The process of subtracting $R - B$ is awkward in digital systems since R will equal 1 followed by a string of 0's. Thus, extensive carrying will have to be performed. However, we can simplify this. We shall illustrate this simplification with an example. Suppose that

$$R = 10000000$$

We can write this as

$$R = 1111111 + 0000001$$

When we take $R - B$ we can subtract B from 1111111 and then add 0000001 to the result. No carrying will be required in the subtraction since each digit of 1111111 will always be greater than the corresponding digit of B. Note that R will have one more digit than B. Remember that R is a number which is one greater than the maximum number stored by the accumulator. Let us consider the subtraction of a number from $111. . .111$, where each has the same number of bits. For example

$$
\begin{array}{r}
1111111 \\
- \ 1011010 \\
\hline
0100101
\end{array}
$$

Note that the answer is equal to the subtrahend with each digit reversed. That is 0's are replaced by 1's and vice versa. Note that this will always happen, independent of the number of binary digits. The subtrahend and minuend must have the same number of digits. If they do not, then we must fill in 0's to the left. For instance, if we have $11111_2 - 101_2$, we write

$$
\begin{array}{r}
11111 \\
- \ 00101 \\
\hline
11010
\end{array}
$$

Thus, to obtain $R - B$, we complement each of the digits of B, after filling in 0's to the left if necessary, and then add $00. . .01$ to the resulting number.

The process of reversing binary digits is not only simple for humans, it is also simple to perform in the computer. For instance, suppose that the accumulator register is composed of flip-flops. Not only will the stored number be available but the number with complemented binary digits will be available simultaneously. Thus, no circuitry need be added to reverse the bits. Hence, subtraction can be performed using adders without any special subtractors being needed.

We shall also demonstrate that the ideas just discussed will enable us to easily represent negative numbers. Before doing this, however, we shall formalize some of the previous discussion by defining some mathematical terms.

The r's Complement

We define the r's complement of a number M in radix r using the following relation

$$C_r = r^k - M \tag{9-23}$$

If we are working with a register then, in general, the maximum number stored is $r^k - 1$. Thus, if we have a k-digit binary register, i.e. k is the total number of bits, then the 2's complement is

$$C_2 = 2^k - M \tag{9-24}$$

For instance, in the previous discussion, $R - B$ represents the 2's complement of B. (The definition of a complement can be broadened on a modular arithmetic basis. However, this definition will suffice for our purposes.)

Thus, if we want to subtract two numbers, e.g. $A - B$, we can accomplish the same result by adding A and the 2's complement of B. This assumes that we are using modular arithmetic and binary digits, and that the exponent k in Eq. (9-24) is equal to the number of bits in the register.

Let us summarize the previous results. We use the following procedure to obtain the 2's complement of a binary number: (1) Write the number with the proper number of digits. (2) Complement the bits (i.e. replace 0's by 1's and vice versa); and (3) add 1. For example, if $k=7$, let us obtain the 2's complement of 101_2. Then, we have $101_2 = 0000101_2$. Complementing the bits and adding 1, we have

```
  1111010
+ 0000001
  1111011
```

Thus, 1111011 is the 2's complement we desire.

We can generalize this result to the r's complement. For instance, suppose we want to obtain the 10's complement of a number. From (9-23), we have

$$C_{10} = 10^k - M = 10000\ldots0 - M \tag{9-25}$$

This can be written as

$$C_{10} = 999\ldots999 - M + 1 \tag{9-26}$$

Use of 2's Complement to Represent Negative Numbers

The 2's complement is very useful if we want to represent a negative number. Indeed, we have essentially done this when we used the 2's complement to implement subtraction. We have mentioned that the leftmost bit can carry sign information. Let us now formalize this. If a number B is the negative of another number A then, using ordinary arithmetic, we have

$$A + B = 0 \tag{9-27}$$

Thus, we can express B uniquely in terms of A using ordinary arithmetic. In the case of compuers, we must work with modular arithmetic. Thus, if we have a k-bit register, the relation that corresponds to (9-27) is

$$A + B \equiv_{2^k} 0 \tag{9-28}$$

Now, there is *no* unique relationship between A and B since there is an ambiguity of 2^k. For purposes of example, assume that A is negative and B is positive. Now assume that the register has k bits. (If the magnitude of the number stored in the register is not to produce an overflow, then it must be less than $2^{k-1} - 1$, see Eq. (9-13). Let us assume that this is so.) Consider that B represents a number such that

$$|B| \leqslant 2^{k-1} - 1 \tag{9-29}$$

Then the 2's complement of B will be

$$C_2 = 2^k - B \qquad (9\text{-}30)$$

Now let us assume that B is the negative of another number A. Hence,

$$B = -A \qquad (9\text{-}31)$$

Then, substituting in Eq. (9-30), we obtain

$$C_2 = 2^k + A \qquad (9\text{-}32)$$

In terms of modular arithmetic we have $2^k + A \equiv_k A$. Hence,

$$C_2 \equiv_k A \qquad (9\text{-}33)$$

But C_2 is the 2's complement of B. Hence, on a modular basis, the 2's complement of B is equal to $-B$. Thus, in all operations involving modular arithmetic (e.g. using registers), we can replace a negative number by the 2's complement of its magnitude. Let us illustrate this with an example. Suppose that we have a 4-bit binary register. For positive numbers we used only three of the bits. That is, for positive numbers, the leftmost bit is 0. Now we shall use all four bits since negative numbers will also be considered. Consider -1. We represent it by the 2's complement of $+1$. Thus, we use

$$C_2 = 1110 + 0001 = 1111$$

Similarly, for -2, we use the 2's complement of $+2$. Hence,

$$C_2 = 1101 + 0001 = 1110$$

etc.

In Table 9-1 we represent the 2's complement of the negative numbers for -1 to -7.

**Table 9-1 4-Bit Binary Representation of Numbers from −7 to +7;
The 2's Complement is Used for the Negative Numbers**

Decimal Number	Binary Number
7	0111
6	0110
5	0101
4	0100
3	0011
2	0010
1	0001
0	0000
−1	1111
−2	1110
−3	1101
−4	1100
−5	1011
−6	1010
−7	1001

Note that, if the first digit is a 0, then the number is positive, while if the first digit is a 1, the number is negative. *Note that here we have not made a number negative by simply replacing the leading 0 by a 1.* Again note that if we have an accumulator or register using k bits, we can store an integer whose magnitude is equal to or less than $2^{k-1} - 1$. Finally, consider the negative numbers from −1 to −7. If the leftmost digit is ignored, then they are the binary equivalents of 1 through 7 in *reverse* order.

Note that there is a four-digit binary numbers that we have *not* used. It is 1000. This is the 2's complement of 8_{10}. Thus, using four binary digits, we can represent numbers from −8 to +7 using 2's complement. In general, when 2's complement is used with k bits, we can represent a number N in the range

$$2^{k-1} - 1 \geqslant N \geqslant -2^{k-1} \tag{9-34}$$

Note that there is one more negative number than positive number.

9.3 2's COMPLEMENT SUBTRACTION

We can now formalize our rules of subtraction. At the start of the last section, we demonstrated that subtraction could be accomplished by replacing the subtrahend by its 2's complement and adding. This demonstrates that negative numbers can be represented by their 2's complement. Let us extend our results. In the previous examples, we did not allow for the sign bit since, for simplicity, we only worked with positive differences. In general, we must include the sign bit to account for negative integers. Note that all the results are still valid. For instance,

$$11101 \equiv_{2^3} 1101 \equiv_{2^3} 101$$

This follows from the fundamental definition of binary numbers which is

$$a_n a_{n-1} \ldots a_2 a_1 a_0 = a_n 2^n + a_{n-1} 2^{n-1} + \ldots + a_2 2^2 + a_1 2^1 + a_0 2^0$$

where $a_j = 0$ or 1. If we represent this modulo 2^k, then all coefficients whose subscripts are k or higher can be ignored since they represent integral multiples of 2^k. Thus, if we have a 4-bit register and we want to subtract $A - B$ where

$$A = 0011_2 = 3_{10}$$

$$B = 0100_2 = 4_{10}$$

first we obtain the 2's complement of B

```
  1011
+ 0001
  ────
  1100
```

Then, adding this to A, we obtain

$$A - B = 1111$$

This represents -1_{10}, as it should, see Table 9-1.

Now let us consider another example. Suppose we want to subtract $A - B$ where

$A = 1111_2 = -1_{10}$

$B = 0100_2 = 4_{10}$

From the previous example, the 2's complement of B is 1100. Then, adding this to A, we obtain

$$\begin{array}{r} 1111 \\ +\ 1100 \\ \hline 11011 \end{array}$$

Note that there is an apparent overflow. However, from the previous discussion, we have

$$11011_2 \equiv_{2^3} 1011_2$$

That is, we can *ignore* the fifth digit, which has been caused by a carry. Hence,

$$A - B = 1011_2 = -5_{10}$$

The answer is correct. Note that the above was *not* an example of an overflow. The only time that a *true overflow* occurs is when the magnitude of the resulting number is too large, see Eq. (9-34). In this case, the allowable range is between $2^3 - 1$ and -2^3.

Actually, we can study the leftmost digit to determine if a true overflow has occurred. The only way that a true overflow can possibly occur is if we add two numbers of the same sign or subtract two numbers of opposite sign. Any other operation will result in an answer which is smaller in magnitude than the magnitude of at least one of the original numbers. Addition is always performed. When we want to subtract, we take the 2's complement and add. If we are actually adding two positive numbers, then the leftmost digit of each will be a 0. If their sum causes a 1 to be carried into the leftmost digit, then this is actually attempting to represent the magnitude and, thus, will result in a true overflow. On the other hand, if we add a positive to a negative number, then a true overflow cannot occur. In this case, if a 1 is carried into the leftmost digit, it will be added to a 1 (representing the negative number) which is already there. This will be carried out, i.e. there will be a 1 carried to the left of this "leftmost" column. In this case, this is not a true overflow and this carried digit is ignored. Now consider adding two negative numbers. In the 2's complement, the large magnitudes will have "smaller" representations (e.g. $-1_{10} = 1111_2$ while $-7_{10} = 1001_2$). An overflow can occur if we add two negative numbers whose magnitudes are large enough. Assume that we have one additional column to the left. If, as a result of addition, a number is carried into this extra column (i.e. an overflow appears to result), then we are actually adding two small magnitude negative numbers. (Note that they appear as large numbers if they are not considered on a 2's complement basis.) Thus, a true overflow does *not* result. On the other hand, when adding two negative numbers, if a carry into the leftmost column occurs without a carry out of this column, then a true overflow will occur. Then, summarizing the previous discussion, the only time that a true overflow occurs is when a carry into the leftmost column occurs *without* a carry out. Circuits can be built which sense when *both* of these "occur" (i.e. a carry in and no carry out). Thus, the user can be warned of an overflow and/or computation can be terminated.

These results can be extended. If we add several positive and negative numbers, then the result will be correct as long as the total sum does not overflow. This is true even if some of the partial sums *do* overflow. This is a characterisitic of modular arithmetic.

We have thus far considered integers. Now let us discuss numbers with fractional parts. Actually, we shall not discuss the problem completely here. In Sec. 9.7 we shall consider the representation of such numbers in more detail. However, the discussion here will provide some of the mathematical details. We start by assuming that the numbers all have an equal number of digits preceding the binary point and also that they have an equal number of digits following the binary point. If this is not true, then we can always add right or left 0's to accomplish it. For instance, suppose we want to perform the subtraction $A - B$ where $A = 1011.11_2$ and $B = 101.101_2$. We rewrite these as

$$A = 1011.110_2 \quad \text{and} \quad B = 0101.101_2$$

Now consider the binary number

$$A = a_n a_{n-1} \ldots a_2 a_1 a_0 . a_{-1} a_{-2} \ldots a_{-k} = a_n 2^n + a_{n-1} 2^{n-1} + \ldots + a_2 2^2$$
$$+ a_1 2 + a_0 2^0 + a_{-1} 2^{-1} + a_{-2} 2^{-2} + \ldots + a_{-k} 2^{-k} \quad (9\text{-}35a)$$

We can write this as

$$A = 2^{-k}[a_n 2^{n+k} + a_{n-1} 2^{n-1+k} + \ldots + a_2 2^{2+k} + a_1 2^{1+k} + a_0 2^k$$

$$+ a_{-1} 2^{k-1} + a_{-2} 2^{k-2} + \ldots + a_{-k} 2^0]$$ (9-35b)

Thus, we can represent A as a binary integer multiplied by 2^{-k}. We represent B in a similar way. If we add or subtract A and B, we can perform the operation on the integers and then multiply the result by 2^{-k}. Let us illustrate this with some examples. Suppose

$A = 0111.110_2$ and $B = 0101.101_2$

We then have

$2^3 A = 0111110_2$

$2^3 B = 0101101_2$

Now suppose we want $A - B$. Take the 2's complement of $2^3 B$. This yields 1010011_2. Adding this to $2^3 A$, we obtain

$2^3(A - B) = 10010001$

Note that we have carried an extra digit. Then, on the basis of modular arithmetic, we have

$2^3(A - B) = 0010001$

Dividing by 2^3 we have

$A - B = 0010.001$

Note that multiplying and dividing by 2^3 just involves shifting the binary point first three digits to the left and then three digits to the right. If this step were omitted in the previous example then, when we obtain the 2's complement, we just reverse the digits and then add 0000.001_2.

Most computers use the 2's complement system for handling subtraction or, equivalently, negative numbers. However, there are other systems that can be and are, at times, used. We shall consider them briefly.

9.4 1's COMPLEMENT REPRESENTATION

When we take the 2's complement of a negative number, we perform two operations. First, all the digits are reversed (i.e. 1's become 0's and vice versa). Next we add 1. We can form a different complement by omitting the second step. That is, we reverse the digits but do not add 1. Now, instead of forming $2^k + N$, we form $2^k - 1 + N$, where N is a negative number. This is referred to as taking the 1's complement. The 1's complement is simpler to obtain but calculations with it are somewhat more difficult. We shall illustrate this subsequently. Let us consider how numbers would be represented in 1's complement. To do this, we shall write a table which corresponds to Table 9-1.

Table 9-2 4-Bit Binary Representation of Numbers from –7 to +7; The 1's Complement is Used for the Negative Numbers

Decimal Number	Binary Number
7	0111
6	0110
5	0101
4	0100
3	0011
2	0010
1	0001
0	0000
−0	1111
−1	1110
−2	1101
−3	1100
−4	1011
−5	1010
−6	1001
−7	1000

One complexity can immediately be seen. There are two 0's. To designate this, one is labeled –0 and the other +0. They yield equivalent results when used in arithmetic. Because of the extra zero, Eq. (9-34) becomes

$$|N| \leqslant 2^{k-1} - 1$$

We now consider addition and subtraction. Note that subtraction is performed by adding the complement of the subtrahend. Thus, we need only speak of addition. If two positive numbers are added, the result will be correct. However, if a positive and a negative number are added, the result will be in error by 1 if the result is positive. Arithmetic performed modulo 2^k will be correct if 2's complement is used. However, this is not true when 1's complement is used. We can consider that the 2's complements of the negative numbers were obtained "incorrectly." However, in many cases, this error is self-correcting. Note that, in 1's complement, negative numbers are represented by a "code," see Table 9-2, which is *not* correct modulo 2^k. If a positive and a negative number are added and the result is negative, then no correction is necessary since the code corrects for it. For instance, when the number is printed, it would be decoded properly. That is, in both 2's complement and 1's complement representation of negative numbers, decoding is used before the numbers are printed output to be read by humans. Compare Tables 9-1 and 9-2. The representations for negative numbers differ by one. This corrects for the "error" that was previously introduced.

When two negative numbers are added, there will be an error when the numbers are decoded. Note that each number will be "in error" by one. Thus, the sum will be "in error" by two. The decoding corrects for an error of one. Thus we see that, at times, one must be added to the answer, while at other times, this is not necessary. If the cases when one must be added are studied, it can be seen that this occurs if, and only if, a one is carried beyond the "leftmost" column. If this occurs, then one must be added to the sum. Let us consider this. Add 5 and -2. Then,

$$
\begin{array}{r}
0101 \\
+ \ 1101 \\
\hline
10010
\end{array}
$$

There is a 1 carried "beyond" the leftmost column. Then, add 1. Thus, the result is

$$0011_2 = 3_{10}$$

which is correct. The leftmost carried 1 is omitted since we worked on the basis of modular arithmetic and a 4-bit register. (Note that accumulators provide an indication if a 1 is carried beyond the leftmost bit.)

It is sometimes considered that the discarded digit is carried around and added to the answer. This is called an *end around carry*. Let us illustrate this with the previous example.

$$\begin{array}{r} 0101 \\ +\ 1101 \\ \hline 1\ 0010 \\ \underline{\ 1} \\ 0011 \end{array}$$ end around carry   (9-36)

Note that the end around carry is a symbolic representation of the two steps (discarding one digit and then adding 1). Actually, the end around carry is not difficult to implement in a computer since the carry bit is available. Thus, with 1's complement arithmetic we must always perform the end around carry. (Note that true overflow must be guarded against here just as in the case of 2's complement.)

Let us compare 1's complement with 2's complement. The 1's complement is easier to obtain since we only have to reverse the digits. However, 1's complement has disadvantages. The end around carry must be performed. There are also two 0's in the 1's complement which require us to keep track of more data. We have considered the 1's complement in far less detail than the 2's complement. This is because 2's complement arithmetic is very widely used, while 1's complement is not.

9.5 SIGNED MAGNITUDE REPRESENTATION

A representation which is easy for people to use is one where the leftmost digit represents the sign and the remaining digits represent the magnitude. Thus, 0001 would equal 1_{10} while 1001 would equal -1_{10}. Similarly, $0011_2 = 3_{10}$ and $1011_2 = -3_{10}$. Thus, humans need only remember the positive integers. However, addition and subtraction with these numbers is more difficult. For instance, if a positive and a negative number are added, the result will not be correct, modulo 2^k. Nor can it be made correct simply by adding 1. Thus, this system introduces complexities and is not often used for addition and subtraction. The signed magnitude representation is used, at times, in special-purpose computers such as digital filters, when multiplication is to be performed.

9.6 THE EXCESS 2^{k-1} REPRESENTATION

Let us consider another representation that is used at times. Assume that we have a register or accumulator with k digits and that the largest magnitude we encounter is 2^{k-1}. Hence, if N is a number within the range of the accumulator, then $2^{k-1}+N$ will always be a nonnegative number. Consider a 4-bit register where all the positive numbers are represented by 3-bit numbers. That is, the first bit is 0. Adding 2^4 will just change the leading 0 to a 1. For instance,

$$0110_2 = 6_{10} \tag{9-37a}$$

$$1110_2 = 6_{10} + 2^4_{10} \tag{9-37b}$$

Now consider a negative number. To obtain the 2's complement, we take $2^k + N$, see Eq. (9-32). However, now we take $2^{k-1} + N$. In binary, the difference will be in the leftmost digit. Since it is 1 in the 2's complement case, it will be 0 in the excess 2^{k-1} representation. For instance, see Table 9-2, the 4-bit 2's complement representation of –5 is 1011; in excess 2^3 representation it is 0011.

In summary, the excess 2^{k-1} representation is closely related to the 2's complement representation. In fact, they are the same except that the leftmost digit is reversed (e.g. a leading 0 represents a positive number in 2's complement and a negative number is excess 2^{k-1} representation).

9.7 FLOATING-POINT NUMBERS

Often, especially in scientific or engineering work, we encounter numbers of very large or very small magnitude. These are usually represented using a multiplier which is an integral power of 10. For instance, we write 1,630,000 as 1.83×10^6. Also, 0.214×10^{-16} represents .0000000000000000214. This type of notation is called *floating-point notation*, since the position of the decimal point is allowed to vary, or float. That is, changing the exponent effectively changes the location of the decimal point. Floating-point notation can be contrasted with *fixed-point notation*. Here the location of the decimal point is fixed. For instance, numbers with no fractional part are fixed point since they have a fixed number of digits (zero) after the binary point.

Let us consider some terminology. If we have the number 0.23×10^{14}, then 0.23 is called the *fractional part* while 14 is called the *exponent*. (At times, the fractional part is also called the *mantissa*.)

Floating-point numbers are also used in digital computers. Of course, there, binary rather than decimal representation is used. Let us illustrate this.

Instead of using powers of 10_{10}, the binary system uses powers of $10_2 = 2_{10}$. (We shall consider exceptions to this subsequently.) For instance, we can write the number $64_{10} = 2_{10}^7$ as

$$2_{10}^7 = 0.1_2 \times 2^8 \tag{9-38}$$

The computer binary representation would be

$$0.1_2 \times 10_2^{1000_2}$$

Often, to avoid confusion for humans, the exponent is expressed in decimal form. The $10_2 = 2_{10}$ is also often expressed in decimal terms. Let us consider some additional examples. Here we shall express the exponents in binary form.

$$0.10110 \times 2^{100} = 1011.0 \tag{9-39a}$$

$$0.1011011 \times 2^{10000} = 0.1011011 \times 2^{16} = 1011011000000000. \tag{9-39b}$$

$$0.1011 \times 2^{-100} = 0.1011 \times 2^{-4} = 0.00001011 \tag{9-39c}$$

The same terminology is used for base 2 as for base 10. For instance, in Eq. (9-39b), 0.1011011_2 is the fractional part while 10000_2 is the exponent. We have written the number with a 0 preceding the binary point. In computers, this is not done since it would be wasteful of storage space to store a known 0. For example, it is thus conventional to write Eq. (9-39b) as

$$.1011011 \times 2^{10000} \tag{9-40}$$

Now consider the leftmost digit of the fractional part. It is a 1. It is conventional to make this a 1. Note that we can make the first digit *either* a 0 *or* a 1 by properly changing the exponent. Let us demonstrate this. A binary number representation of a fraction is

$$.a_{-1}a_{-2}\ldots a_{-k} = a_{-1}2^{-1} + a_{-2}2^{-2} + \ldots + a_{-k}2^{-k} \tag{9-41}$$

Note that the a_j's are either 1 or 0. Consider that, starting from the binary point, many of the digits are 0's. Assume that a_{-j} is the first nonzero digit (i.e. $a_{-j} = 1$). In that case, Eq. (9-41) can be rewritten as

$$.a_{-1}a_{-2}\ldots a_{-k} = [a_{-j}2^{-1} + a_{-(j+1)}2^{-2} + \ldots + a_{-k}2^{-(k-j+1)}]2^{-(j-1)} \tag{9-42}$$

Hence, by appropriately adjusting the exponent, we can always make the leading digit a 1. (The one exception to this is if all the digits are 0, that is, the number represented is zero.) Let us illustrate this with an example.

$$.00101 = .101 \times 2^{-2}$$

The number $.101 \times 2^{-2}$ is said to be *scaled*. Thus, the fractional part is written as a binary fraction which lies between 0.5_{10} and 1_{10}. (i.e. $0.5 \leqslant N < 1$) We shall see subsequently that scaling allows us to store the greatest number of significant figures in a given amount of storage.

The fractional part and the exponent can each be either positive or negative. Thus, in order to store a floating-point number, we must store both two numbers *and* their signs. All the representations used in the previous section can be used here. To understand something of the storage of numbers, let us digress and consider some elementary definitions of storage.

A register stores a given number of binary digits, or bits. Groups of these are called *bytes*. Often, many of the registers in a computer are of the same length. This is termed a *word*. For instance, a typical word could consist of two (eight-bit) bytes or 16 bits. In microprocessor-based systems, a byte is typically either 4, 8, or 16 bits in length.

In many computers, the word length is fixed at an integral number of bytes. Such a word is illustrated in Fig. 9-1 where each byte is eight bits. There, the word consists of 16 bits. Most computers are organized so that the word length is fixed. For instance, we can speak of a computer's having 32-bit words. Such fixed word lengths are usually used when scientific calculations are performed. On the other hand, there are computers where the word length is variable. That is, the memory is organized in such a way that the user can specify the number of bytes in a word and these need not all be of the same length.

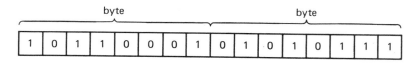

Figure 9-1 A 16-bit register. Eight-bit bytes are illustrated.

For purposes of the discussion of floating point numbers, let us assume that all words are the same length. When we considered integers, the word length determined the maximum magnitude of the number, see Eq. (9-34). In the case of floating-point numbers, the magnitude is determined in large part by the exponent. In this case, the problem of determining the largest magnitude is somewhat more complex. Let us consider an example to illustrate this. We shall assume that the word length is 32 bits. This is much longer than the word length of most microprocessors used by computers. This will be discussed in the next section. We must store both the exponent and the fractional part. Either of these may be positive or negative. Therefore, there are two signs which must be stored. Whatever system is used to store negative numbers (e.g. 2's complement, etc.), one bit is used to store the sign. Thus, 2 of the 32 bits are used to store the signs. Typically, 6 bits will be used to store the "magnitude" of the exponent and the remaining 24 bits are used to store the "magnitude" of the fractional part. Let us consider the implications of this. Considering that the maximum positive number that can be stored for the exponent is

$$2^6 - 1 = 63_{10}$$

Then,

$$2^{63} = 9.223 \times 10^{18}$$

This represents a large number. For instance, if this multiplies a fractional number between 0.5 and 1.0, the result can be a very large number. Conversely, if the exponent is negative, multiplication by 2^{-63} can result in an extremely small number. Note that in 2's complement there can be one more negative number than positive number represented. Thus, the preceding figure for negative numbers would be 2^{-64}. Hence, the range of allowed numbers, when floating-point (i.e. scientific) notation is used, is very large.

In some computers, the floating-point notation is not written to base 2 but to base 8 or base 16. The numbers are, of course, stored in binary. Let us illustrate this. Suppose a hexadecimal base is used. If the exponent stored is positive with a binary representation of $000011_2 = 3_{10}$, then the "exponent component" of the number would be

$$16^3 = 4096_{10}$$

If six bits are used for the exponent, then the largest number stored for the exponent is, as before, $2^6 - 1 = 63$. Thus, the largest numerical value of the exponent portion is

$$16^{63} = 7.23 \times 10^{75}$$

Therefore, by using the hexadecimal system, the largest positive number becomes much greater than that obtainable when the binary system is used. Remember that the actual stored numbers are in binary form. Scaling is somewhat more complex in this hexadecimal system.

Now let us consider the fractional part. In the previous example, we stored this in 24 bits. This represents a fraction. That is, the binary point is at the left of the digits. If we consider the 24-bit binary number, we must shift the binary point 24 digits to the left. This is equivalent to dividing by 2^{24}. Thus, the largest value that this fraction can take on is $(2^{24}-1)/2^{24} = .99999994$. The fractional part will always be scaled by adjusting the exponent so that its leading binary digit is 1. Since the largest number of 9's in the decimal number is 7, then 24 bits lead to no more than seven significant figures in the fractional part. Remember that the scaling keeps the number N in the range $0.5 \leqslant N < 1$. The problem of significant figures can also be studied by considering the following. The smallest increment in the binary fraction is $1/2^{24} = .596 \times 10^{-7} = .0000000596$. There are seven 0's. Thus, again, we can only have seven signficant figures. That is, the smallest decimal step we can take is $.596 \times 10^{-7}$. Hence, the fractional part in this example can have up to seven signficant figures. For example, we could not distinguish betwen .12461281 and .12461282 because we could not take small enough steps. At times, a hexadecimal (or octal) basis is used for the fractional part. The details of such use are similar. Note that the digits of the register are always binary.

Let us illustrate these ideas. Assume that we use decimal registers since we are more familiar with numbers to base 10. Suppose that we can store two digits for the exponent and four for the fractional part. In addition, there are two signs stored. Then, the exponent value can range between 10^{-99} and 10^{99}. Now let us represent the number 1439. We can write this as 0.1439×10^4. Now let us represent 15963. Only four digits can be stored in the fractional part. The last digit will then be lost. Thus, we have 0.1596×10^5 as the stored number. This could represent 15960, 15961, . . . , 15969. For this reason, we say that this register has a *precision* of four significant figures. Note that, because of the exponent, we can represent very large numbers but, because of the register size, only to four signficant figures. For instance, 1,596,421 and 1,596,287 would be represented identically as $.1596 \times 10^7$. In computer registers, the same concepts of precision apply. Of course, the numbers are stored in binary. The precision of a computer is usually rated in decimal terms for the convenience of the human users. Typically, there may be seven or eight significant figures. At times, this is not enough. Most computers, and their associated software, are such that additional registers can be used to store additional digits of the fractional part. This usage is termed *double precision*, since the number of available significant figures is increased. Usually, it is (approximately) doubled. We shall consider this in the next section.

Note that we have not indicated any storage location for the binary point. This is because the binary numbers are scaled so that the number immediately to the right of the binary point is a 1. (All numbers to the left of the binary point are 0's.) Thus, there is no reason to store the location of the binary point since its position is fixed at the left of the number. Scaling also optimizes precision. For instance, $.1011011_2 \times 2^8$ = $.0010110111_2 \times 2^{10}$. However, if we can only store seven bits, then the second number would become $.0010110 \times 2^{10}$. This represents a loss of significant figures.

Let us now consider how arithmetic is performed with numbers written in floating-point notation.

Addition and Subtraction

When numbers are added or subtracted, their radix points are aligned. Equivalently, when floating point notation is used, we must adjust the exponents so that they are equal. We assume that the binary fractional parts are scaled (i.e. a 0 to the left of the binary point and a 1 as the first digit to the right of the binary point). Now let us consider addition. Suppose we want to add $.1011011 \times 2^3$ and $.1101111 \times 2^6$. Assume, for purposes of simplicity, that there are seven significant figures. Then, to align the binary points, we write the addition as

$$.1101111 \times 2^6$$
$$+ \ .0001011011 \times 2^6$$
$$\overline{.1111010011 \times 2^6}$$

There is a problem here. To accurately represent this sum, 10 significant figures are needed. However, only seven can be stored. Thus, the above addition would actually become

$$.1101111 \times 2^6$$
$$+ \ .0001011 \times 2^6$$
$$\overline{.1111010 \times 2^6}$$

Thus, the answer is *inaccurate*. This loss of accuracy is a roundoff error (other roundoff errors were discussed in Sec. 2.1). When many calculations are performed, the accumulated roundoff error can become very large. At times, double precision calculations must be used to prevent large errors even though extreme accuracy is not needed in the final answer.

Multiplication and Division

When numbers in scientific notation are multiplied or divided, we follow the rules for exponential notation. For instance, if

$$A = M_a 2^{E_a} \tag{9-43a}$$

$$B = M_b 2^{E_b} \tag{9-43b}$$

where M_a and M_b are the fractional parts and E_a and E_b are the exponents, respectively, then

$$A \cdot B = M_a M_b 2^{E_a + E_b} \tag{9-44}$$

That is, when two numbers are multiplied, we multiply the fractional parts and add the exponents.

Similarly,

$$A/B = (M_a/M_b) 2^{E_a - E_b} \tag{9-45}$$

When two numbers are divided, we divide the fractional parts and subtract the exponents.

Note that multiplication and division do not require that the binary points be aligned. Thus, we do not have this complexity and the associated roundoff error. However, roundoff error can result. For instance, consider the multiplication $A \cdot B$ where

$$A = .1011 \times 2^4$$

$$B = .1001 \times 2^9$$

$$A \cdot B = .011100011 \times 2^{13}$$

This would be represented as

$$A \cdot B = .11100011 \times 2^{12}$$

Note that we had four significant figures in each of A and B. Assume that this is all that can be stored. Then, the above product would actually be stored as

$$A \cdot B = .1110 \times 2^{12}$$

Again, a roundoff error results. Similarly, roundoff errors occur with division. This will be discussed in greater detail in Chapter 10.

All this discussion of roundoff error does not imply that computer calculations are inaccurate. Actually, many significant figures can be stored and very accurate results can usually be obtained. However, the possibility of roundoff error should always be considered.

There is another factor which should be discussed when roundoff error is considered. The data used in many scientific calculations is obtained from measurements. These usually contain only a limited number of significant figures. It is unreasonable to assume that calculations involving these figures should result in more significant figures than are supplied. Thus, to be realistic, if calculations tend to "introduce more significant figures," as in the case of the previous multiplication, then the excess significant figures should be rounded off.

Some computers have special floating-point hardware (i.e. special circuits that perform floating-point arithmetic automatically). For instance, when multiplication is performed, the exponents would automatically be added without having to provide special instructions. This hardware gives high-speed operation. However, many computers do not have it. In face, most microcomputer-based systems do not have floating point hardware. Software, i.e. programs, is written which directs each step of the floating-point operation. The use of the software simplifies the computer hardware. However, it results in slower execution of floating-point operations.

9.8 USE OF MULTIPLE WORDS IN NUMERICAL REPRESENTATIONS AND CALCULATIONS

In microprocessor-based computers and even in larger computers, the word-lengths are often too small to adequately represent numbers, see Sec. 9.7. For instance, we may not be able to represent large enough integers. Suppose that we have a microprocessor that uses 8-bit words. That is, the accumulator register holds eight bits, the adder can add two 8-bit numbers, etc. In this case, the largest integer that we can work with appears to be, see Eq. (9-12), $2^7 - 1 = 127$. However, we are really not limited in this way. A single number can be represented by more than one word. For instance, consider that we have 8-bit words and are representing numbers using 2's complement. Now suppose that two words are used to represent a number so that 16 bits are available. Now 15 bits can be used to represent the magnitude, and we can work with much larger integers, $2^{15} - 1 = 32767$. Similar statements can be applied to floating-point operations.

The use of more than one word to represent a number leads to complications. For instance, if we have 8-bit words and 16-bit numbers, then each number must be stored in two different memory locations, two different 8-bit registers must be used as an accumulator register, etc. In general, only an 8-bit adder will be available. Thus, the computations must be done in two stages. Let us illustrate this with some calculations. We shall work in decimal for simplicity. Suppose that we have two-digit numbers and a one-digit adder and one-digit registers. If we want to add 47 + 32 we would first add 7 + 2 = 9. Then, we would add 4 + 3 = 7. Thus, the result of 79 would be obtained and stored in two registers.

The previous example was not complicated because there was no carrying. For instance, suppose we want to add 47 + 35. The first addition adds 7 + 5 = 12. The result 2 is stored. Then we add 4 + 3 = 7. Since there was a carry, the second result must be incremented by 1 to obtain 8. When these calculations are performed using a computer, provision must be made to include such carries.

In an actual computer, the computations would be in binary. Again, we shall assume that a single word is eight bits. Now we must store the number in two registers. This two-register pair takes the place of the single accumulator register. Microprocessors are fabricated so that they can work with pairs of registers to facilitate such operations. We shall describe these operations in detail in Chapter 10. For the time being, consider the following: Microprocessors have instructions that result in the contents of one 8-bit register being added to the contents of another 8-bit register. This is an ordinary 8-bit operation. On the other hand, there are also instructions that cause a 16-bit number, stored in a pair of 8-bit registers, to be added to another 16-bit number stored in another pair of 8-bit registers. The sum is then stored in the second pair of 8-bit registers which functions as an accumulator register.

In this chapter we have considered various types of arithmetic operations. They can all be extended so that each number can be represented by more than one word of storage. This is often termed *double precision*. This terminology is somewhat ambiguous. For instance, in some computers, single-precision numbers are stored in two register pairs. In such cases, double precision would utilize four words of storage.

BIBLIOGRAPHY

Booth, T.L., *Digital Networks and Computer Systems*, Second Ed., Chap. 4, John Wiley and Sons, Inc., New York 1978

Hill, F.J. and Peterson, G.R., *Introduction to Switching Theory and Logical Design*, Second. Ed., Chap. 2, John Wiley and Sons, Inc., New York, 1974

Mano, M.M., *Digital Logic and Computer Design*, Chap. 8, Prentice Hall, Inc., Englewood Cliffs, N.J. 1979

O'Malley, J., *Introduction to the Digital Computer*, Chap. 8, Holt, Rinehart and Winston, Inc., New York 1972

PROBLEMS

9-1. Determine four integers which are congruent, modulo 3, to 7.

9-2. Repeat Prob. 9-1 for modulo 2.

9-3. A binary register has 32 bits. What is the largest positive number that can be stored in this register?

9-4. Repeat Prob. 9-3 but now assume that one bit of the register is used as a sign bit.

9-5. We want to perform the addition $729_{10} + 1436_{10}$ in a binary accumulator. How many bits are required?

9-6. Repeat Prob. 2-18 but now subtract the numbers.

9-7. Repeat Prob. 2-19 but now subtract the numbers.

9-8. Repeat Prob. 2-20 but now subtract the numbers.

9-9. Perform the subtraction of Prob. 9-6 but now use the procedure of Sec. 9-2.

9-10. Obtain the 10's complement of the following numbers:
123_{10}, 146_{10}, 937.12_{10}

9-11. Obtain the 2's complement of
10101, 110101, 101.101

9-12. Use the 2's complement to perform the subtraction of Prob. 9-6.

9-13. Demonstrate that, when using accumulators, a negative number can be represented by its 2's complement.

9-14. Extend Table 9-1 for numbers ranging from 15 to –15. Use five bits.

9-15. Add the following two numbers which are given in 2's complement:

0110,   1011

9-16. Can −16 be included in the table of Prob. 9-14? If the answer is yes, include it.

9-17. In your own words, demonstrate how the leftmost digit can be used to determine if a "true overflow" has occurred in a register.

9-18. Use the 2's complement to perform the following subtraction:

$$\begin{array}{r} 1011.0101 \\ - \quad 101.1101 \\ \hline \end{array}$$

9-19. Obtain the 1's complement of the numbers in Prob. 9-11.

9-20. Extend Table 9-2 for numbers ranging from 15 to −15. Use five bits.

9-21. Can −16 be included in the table of Prob. 9-20? Compare this answer with that of Prob. 9-16. What is the difference?

9-22. Add the following numbers which are given in 1's complement representation.

0110,   1011

9-23. Write a table which gives numbers ranging from 7 to −7 using signed magnitude representation.

9-24. Repeat Prob. 9-23 but now use the excess 2^3 representation.

9-25. Write the following numbers in floating point binary:

126,   1479, 138×10^5, 1.427×10^3

9-26. The following pairs of numbers represent the fractional part and the exponent (in base 2) of a number. Express the number in base 10. The exponent is given in 2's complement. Assume that the fractional part is positive.

.1011,   0110
.1101,   1011
.101,   111110
.1011,   0111110101

9-27. A register has 64 bits. Twelve bits are used to store the exponent of a number and the remaining 52 bits are used for the fractional part. What is the range of numbers that can be stored? What is the precision?

9-28. Add the following two numbers. Assume that the register can store seven bits for the fractional part. What is the roundoff error?

$.1101101 \times 2^6$
$.1011011 \times 2^5$

9-29. Repeat Prob. 9-28 for

$.1101101 \times 2^{16}$
$.1101101 \times 2^3$

9-30. Repeat Prob. 9-28 for

$.1011101 \times 2^{-8}$
$.1011101 \times 2^{-10}$

9-31. Perform the following multiplication. Assume that only seven significant figures can be stored for the fractional part.

$(.1011011 \times 2^9)(.1011111 \times 2^{16})$

9-32. Repeat Prob. 9-31 for

$(.10111 \times 2^{-9})(.1011101 \times 2^5)$

9-33. Discuss the use of multiple-precision arithmetic.

10

The Arithmetic Logic Unit — ALU

In this chapter we shall discuss the arithmetic logic unit, the ALU, that is used to perform computations. The basic arithmetic operations of addition and subtraction will be discussed. (It is assumed that the reader is familiar with all of the introductory material discussed in Chapter 9.) The operations of multiplication and division will then be developed. Other mathematical and logical operations shall also be considered.

The material presented in this chapter lays the groundwork for the discussion of the complete microprocessor-based digital computer which will be given in the next chapter. Actually, it would help you to understand the material discussed in this chapter if some ideas of complete digital computers were known before the material of the present chapter were discussed. Thus, here we shall briefly consider some ideas of the elements of the digital computer to be discussed in greater detail in Chapter 11.

Computers can be considered to consist of four basic units: the *central processing unit* (CPU), the *control unit* (CU), the *main memory unit* (MMU), and various input-output (IO) devices. The computations are actually performed in the central processing unit. These operations are directed by the control unit, which also directs the operation of the entire computer. The actual operations are performed by the arithmetic logic unit, the ALU, which is part of the CPU. The CPU, when fabricated on a single chip, is called a *microprocessor*. The main memory unit is a random access memory, see Chapter 7. Its function is to store the data and the results of computation. In addition, such things as the program which directs computations are also stored in the main memory unit. All data is input and/or output using input-output devices such as floppy disk drives and printers. In this chapter we shall consider important aspects of the ALU. In particular, we shall consider arithmetic and logical operations as well as the hardware which can be used to perform them. We shall also discuss a notation which is used to describe the computer operations. In the next chapter, we shall discuss the complete microprocessor-based computer.

10.1 ADDITION

The fundamental arithmetic operations in a digital computer are addition and subtraction. The other arithmetic operations are performed using the ideas of addition and subtraction as well as the circuits associated with them. Actually, subtraction can be considered to be a special case of addition. In this section we shall consider addition circuits. It is assumed that the reader is familiar with Sec. 3.4 where the full and half-adders were discussed.

We shall start by reviewing the *full adder*. This is a circuit which adds two binary digits and a carry digit and yields their sum plus a carry digit. From Eq. (3-26), we have

$$
\begin{array}{r}
c_{j-1} \\
a_j \\
b_j \\
\hline
c_j s_j
\end{array}
$$

where a_j, b_j, and c_{j-1} are the digits to be added, s_j is the sum and c_j is the carry digit. In Eqs. (3-32) we expressed c_j and s_j as

$$s_j = a_j \oplus b_j \oplus c_{j-1} \tag{10-1a}$$

$$c_j = a_j b_j + b_j c_{j-1} + c_{j-1} a_j \tag{10-1b}$$

Remember that we shall be adding two N-bit binary numbers. Here we have considered the addition of their j^{th} digits, plus the carry resulting from the addition of the $j\text{-}1^{\text{st}}$ digits. A half-adder sums two binary digits but does not make provision for the input of a carry term from the previous digit. Realizations for half and full adders are given in Figs. 3-19 through 3-21. A block diagram representation of a full adder is given in Fig. 10-1.

Now let us discuss a circuit that adds two 3-bit binary numbers, see Fig. 10-2. Actual adders add many more bits. However, all the funda-

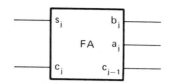

Figure 10-1   The block diagram of a full adder

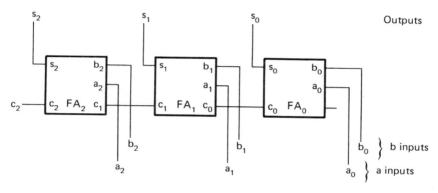

Figure 10-2   A 3-bit adder

mental principles can be described using three bits without cluttering the circuit with excessive detail.

The circuit uses three full adders. We assume that the a_j and b_j inputs are stored in registers during the addition process. Note that the inputs to each adder are the ones called for in Eqs. (10-1). Thus, the sum digits will be correct. FA_0 adds the rightmost bits. Thus, there is no carry input for this adder. Hence, it could be a half adder. The carry digit from the leftmost adder is not used. (It could be used, and we shall consider this subsequently.)

Carry Propagation

The diagram of Fig. 10-2 seems to imply that, if we apply the inputs a_0, a_1, a_2, b_0, b_1, b_2, then the output will be obtained after the time interval required by a full adder to produce its output. Actually, this is not the case. FA_1 does not have all its input signals until it has received the c_0 input from FA_0. Thus, if it takes Δt seconds for a full adder to respond to its input signal, FA_1 cannot complete its response until $2\Delta t$ seconds have elapsed. Thus, FA_2 will not receive its carry input for $2\Delta t$

seconds. Thus, the final sum will not be achieved until $3\Delta t$ seconds have elapsed. If the register adds n bits, then the delay will be $n\Delta t$ seconds. This delay is said to occur because of *carry propagation*.

At times, such delays are acceptable. However, it is often desirable to speed up the process. One way of doing this is called *carry look-ahead* where the carry bits are calculated without waiting for the adder to do so. Let us illustrate this. From Eqs. (10-1) we have

$$c_0 = a_0 b_0 \tag{10-2a}$$

$$c_1 = a_1 b_1 + b_1 c_0 + c_0 a_1 \tag{10-2b}$$

$$c_2 = a_2 b_2 + b_2 c_1 + c_1 a_2 \tag{10-2c}$$

We would like to express c_1 and c_2 in terms of the a's and b's only. Substituting Eq. (10-2a) into Eq. (10-2b), we obtain

$$c_1 = a_1 b_1 + b_1 b_0 a_0 + a_1 a_0 b_0 \tag{10-3a}$$

Substituting Eq. (10-3a) into Eq. (10-2c), we obtain

$$c_2 = a_2 b_2 + b_2 a_1 b_1 + b_2 b_1 b_0 a_0 + b_2 a_1 a_0 b_0$$
$$+ a_2 a_1 b_1 + a_2 b_1 b_0 a_0 + a_2 a_1 a_0 b_0 \tag{10-3b}$$

In Fig. 10-3 we have illustrated a 3-bit adder which utilizes Eq. (10-3a) to eliminate the effect of carry propagation to the second adder. To eliminate it in a 4-bit adder, we would also implement Eq. (10-3b). (Note that the adders themselves can be simplified since they do not have to output the carry bit.)

If this implementation is carried out, the carry propagation effect is eliminated. However, in large adders, the number of gates required, and the associated gate interconnection problem becomes very large. There are other implementations which use fewer gates, but these are not as fast. These are discussed in the Rhyne reference cited in the bibliography at the end of this chapter.

In the remainder of this book, we shall draw adder circuits as shown in Fig. 10-2. However, there may be gates which are not shown that are added to eliminate the effects of carry propagation.

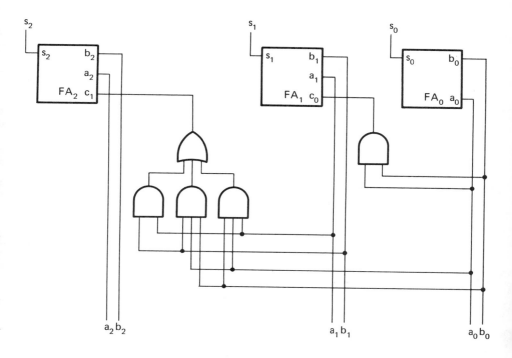

Figure 10-3 A 3-bit adder that does not suffer from carry propagation

A Simple Arithmetic Logic Unit — ALU

In most computer circuits, we do not just have a simple adder, but an adder and register combined. The register is often called the accumulator. We shall label it register A. The number in the register is added to another number which is supplied by an external circuit. Their sum is placed in register A. Thus, the original number in register A is lost since it is replaced by the sum. This device is a simple *arithmetic logic unit*, ALU. Actually, ALUs do much more than we have indicated here. Thus, the descriptions of this section will only apply to a simple ALU. In Secs. 10-2 and 10-4 we shall consider the ALU in more detail.

We want to implement a circuit which has a register which stores a binary number and which then accepts another number. After computation has been completed, this register should store the sum of the two numbers. Consider Fig. 10-4. At the start, assume that the ADD and CLEAR signals are 0. Thus, no clock pulses are supplied to the D flip-flops and the lower inputs of ANDs 0, 1, and 2 are 1's. The D flip-flops

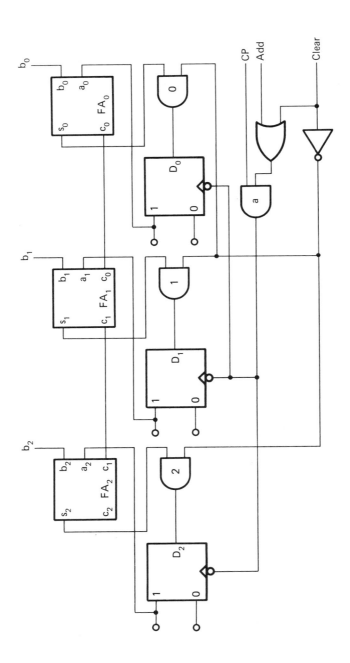

Figure 10-4 A simple ALU

constitute the A-register. Their outputs are made available to the rest of the computer. Assume that there is a binary number stored in the register which will be supplied to the a_j inputs of each full adder. The b inputs to the full adder will be supplied from an external source. Thus, after sufficient time has elapsed, the full adder will produce the sum $s_2 s_1 s_0$. These digits will be applied to the inputs of AND 2, AND 1, and AND 0, respectively. Thus, these can be applied at the D_2, D_1, and D_0 inputs, respectively, of the flip-flops. However, the flip-flops will not change state since no clock pulse is applied to their clock pulse terminals. We assume that the clock is running and generating pulses. However, AND a prevents the clock pulses from being applied. Now suppose that the ADD signal is given (for one clock pulse). That is, a 1 is placed on the ADD line. After the next clock pulse is applied, the state of the flip-flops will become equal to their respective D inputs, which are s_2, s_1 and s_0. The number now stored in the register will be the sum of the two original numbers as desired.

If a signal is put on the clear line, the one input to each of AND 0, 1 and 2 becomes 0. Hence, D_0, D_1 and D_2 are 0. At the next clock pulse, all flip-flops will set themselves to state 0. Hence, we have cleared the register. Note that the next $b_2 b_1 b_0$ added will be added to the 000 in the register giving a new stored $b_2 b_1 b_0$.

The B-register which supplies b_0, b_1, and b_2 is often on the same microprocessor chip as is the accumulator. We have only shown single-word arithmetic here. If double-word (double precision) operations are performed, then this discussion should be modified in accordance with Sec. 9.8. In particular, each register must be replaced by a pair of registers. For instance, in Fig. 10-4 six flip-flops, in three pairs, would be used to store the sum.

Sequential Adders

We have discussed parallel adders. That is, information is input to each full adder simultaneously (with the exception of carry propagation). Thus, to add n bits we require n full adders. Now let us consider the serial or sequential adder. Only one full adder is used. One bit of each number is input at a time and the resulting sum and carry bit are stored. Then, the next bit of each number is added using the last stored carry bit and the new sum and carry bits are stored. Note that the carry bit need only be stored until the addition using it is performed. Thus, only one register is needed to store the carry bit. Let us consider a simple ALU which performs sequential addition, see Fig. 10-5. The storage is a shift register of the type shown in Fig. 7-2b. AND

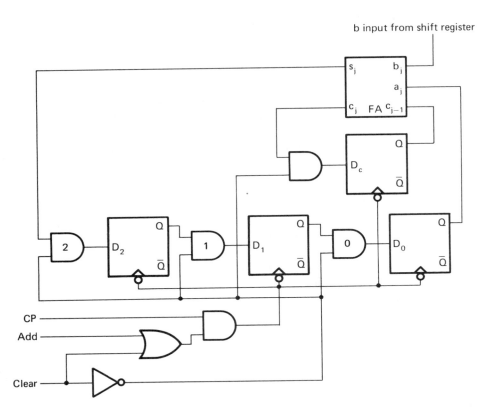

Figure 10-5 A serial adder

gates 0, 1, and 2 are included to allow the register to be cleared. This is essentially the same procedure as discussed in connection with Fig. 10-4. At the start, the shift register will store one of the numbers to be added. When the addition is completed, it will store the sum.

Now let us assume that an ADD signal is present and that the CLEAR signal is 0. The b input to the full adder is also assumed to come from a shift register. Prior to the first clock pulse, a_0 and b_0 are supplied and $c_{j-1} = 0$. The adder then computes s_0 and c_0. It is assumed that all this is completed before the next clock pulse arrives. Note that s_0 is calculated before this clock pulse. Thus, it has not, as yet, changed the state of flip-flop D_2.

After the next clock pulse, the following takes place: The shift register shifts to the right. Thus, a_1 is stored in D_0 and s_0 is input to D_2. Then, b_1 and a_1 are input to the adder and c_0 becomes the output of flip-flop D_c. The adder then computes s_1 and c_1. At this clock pulse, s_0 becomes the state of the leftmost flip-flop, D_2, of the shift register prior to addition. Note that the spacing between clock pulses must also be long enough to allow all this to occur.

Proceeding in this manner, after two more clock pulses, the contents of the shift register will be $s_2 s_1 s_0$. The ADD signal is removed and the sum is stored in the accumulator.

10.2 ALU OPERATIONS

We have considered simple ALUs and seen some of the operations that they can perform. In this section we shall discuss other operations which we would expect from ALUs. In the next section we shall see how these can be used to perform other arithmetic operations. Finally, we shall discuss a circuit for a complete, simple ALU. In Chapter 11 we shall see how this relates to the central processing unit.

We must now introduce some notation that will be used in our discussion of ALUs and of the arithmetic operations which they perform. Registers will be indicated by capital letters. For instance, A represents register A. The number stored in a register, i.e. its *contents*, will be indicated by parentheses; (A) represents the contents of A.

If information in register A is transferred to register B, we write this as

$$(A) \rightarrow B \tag{10-4}$$

This means that, after the next clock pulse, the contents of B will be the same as the contents of A.

The operations of arithmetic are indicated by the following symbols:

Addition	+	
Subtraction	−	(10-5)
Multiplication	*	
Division	/	

Thus, we can represent addition using an adder whose accumulator is register A will receive a number from register B as

$$(A) + (B) \rightarrow A \tag{10-6}$$

That is, the contents of registers A and B are added and the sum is placed in register A. If we want to divide the contents of register B by the contents of register C and place the result in register E, we would write

$$(B)/(C) \rightarrow E \qquad (10\text{-}7)$$

In addition to the arithmetic operations, we can perform logical operations. To avoid confusion with the arithmetic symbols, we shall use the alternative notation:

AND $\qquad \wedge \qquad$ (10-8a)

OR $\qquad \vee \qquad$ (10-8b)

XOR $\qquad \oplus \qquad$ (10-8c)

For instance, we can write

$$(A) \vee (B) \rightarrow C \qquad (10\text{-}9)$$

This means that the first locations of A and B are compared, using an OR operation, and if either or both is a 1, then a 1 is put in the first location of register C. Otherwise, a 0 is placed there. This is repeated with each location in turn.

Another notation that we shall use is that (A_i) represents the contents of the ith location of register A. (A_{ij}) represents the contents stored in the i to j locations of register A. For instance, if the contents of register A is 10011010, then $(A_0)=0$ and $(A_{02})=010$. Then, we can also represent Eq. (10-9) as

$$(C_i) = (A_i) \vee (B_i) \quad i = 0, 1, 2, \ldots \qquad (10\text{-}10)$$

We can also speak of signals. For instance, in Fig. 10-5, the output signal could be the voltage on the 1-lead of flip-flop D_0. We shall denote this by a lower case letter, e.g., y. The signal at a particular time will be written as (y) or $y(t_0)$. If we want to designate that a particular signal is the result of the addition of the contents of two registers, we would write

$$(A) + (B) = y \qquad (10\text{-}11)$$

For instance, if y were the output signal of a serial adder, then it would consist of an appropriate sequence of 0's and 1's. Note that y can be either a series of bits or a single bit depending upon the circumstances.

In addition to the arithmetic and logical operations, there are certain operations that indicate the condition of the variable in the A-register.

This information is stored in a register called the *flag register, F*. Each bit of the flag register contains different information. For instance,

$$(F_0) = 1 \tag{10-12a}$$

if there has been a carry (borrow) to (from) the leftmost bit in the previous addition (subtraction)

$$(F_0) = 0 \tag{10-12b}$$

if there has not been a carry (borrow) to (from) the leftmost bit in the previous addition (subtraction)

Note:

(F_j) is the contents of the jth bit of the flag register. $\tag{10-13}$

Information concerning the value of (A) is also given:

$(F_1) = 1$ if (A) is negative $\hspace{2cm}$ (10-14a)
$(F_1) = 0$ if (A) is positive $\hspace{2cm}$ (10-14b)
$(F_2) = 1$ if $(A) = 0$ $\hspace{3cm}$ (10-15a)
$(F_2) = 0$ if $(A) \neq 0$ $\hspace{3cm}$ (10-15b)
$(F_3) = 1$ if the parity of (A) is even $\hspace{1cm}$ (10-16a)
$(F_3) = 0$ if the parity of (A) is odd $\hspace{1.2cm}$ (10-16b)

In order to implement the mathematical operations, we require the ALU to perform certain operations in response to control signals generated by the control unit. We have already discussed some of these operations in this chapter and in Chapter 6. However, we shall list a more complete set of them here.

Table 10-1 Elementary ALU Operations

Operation	Abbreviation	Control Designation	Symbol
Addition	ADD	c_1	$(A) + (B) \rightarrow A$
Clear	CL	c_2	$0 \rightarrow A$
AND	AND	c_3	$(A) \wedge (B) \rightarrow A$
OR	OR	c_4	$(A) \vee (B) \rightarrow A$
XOR	XOR	c_5	$(A) \oplus (B) \rightarrow A$
shift right	SR	c_6	$(A_{j+1}) \rightarrow A_j$ $j=0,...,n-1, (A_n) = 0$
shift left	SL	c_7	$(A_{j-1}) \rightarrow A_j$ $j=0,...,n, (A_0) = 0$

Operation	Abbreviation	Control Designation	Symbol
complement	COMP	c_8	$(\bar{A}) \rightarrow A$
increment	INCR	c_9	$(A) + 1 \rightarrow A$
negative check	NC	c_{10}	$(F_1) = 1$ if (A) represents a negative number
zero check	ZC	c_{11}	$(F_2) = 1$ if (A) represents a 0

We assume that there are two registers, A and B, associated with the ALU. We shall discuss this in greater detail in the next chapter.

Most of these have been considered previously. For instance, the shift right and shift left cause the entire contents to be shifted as in a shift register. In the complement operation, each element of the register is replaced by its complement (e.g. 011011 is replaced by 100100). In the increment, 1 is added to the contents of the register.

The ALU has a set of command leads, one for each of the desired operations. When a 1 is placed on that lead, then, after the next clock pulse, the accumulator performs in accordance with the given instruction. A block diagram of a simple ALU is given in Fig. 10-6.

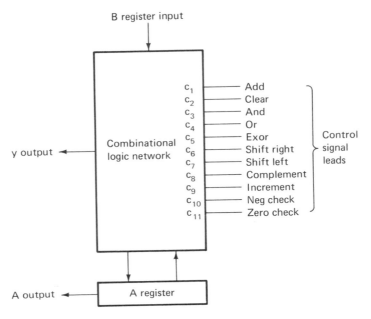

Figure 10-6 A block diagram of a simple ALU

Suppose that we want to perform the following operations which, as we shall see, are related to the 2's complement subtraction. Take the complement of the contents of (A). Then, add 1 to this. Finally, take the resulting number and add it to another binary number which is in B and then store the final result in A. The sequence of operations to be performed is

$$(\overline{A}) \rightarrow A$$
$$(A) + 1 \rightarrow A$$
$$(A) + (B) \rightarrow A$$

$$(10\text{-}17)$$

Suppose that, originally,

$$(A) = 0100$$

and

$$(B) = 0011$$

The results of operations (10-17) are

$$(\overline{A}) = 1011$$
$$(A) + 1 = 1100$$
$$(A) + (B) = 1111$$

In order to perform the sequence of operations, 1's must be put on the proper control signal lines at the appropriate times. This is accomplished by the control unit. For instance, in order to carry out operations (10-17), we would have to use control lines c_1, c_8 and c_9. The first set of control signals would put a 1 on line c_8, with all the other control lines 0. The next set of control signals would put a 1 on control line 9 with all other control lines 0. Finally, the third set of control signals would put a 1 on control line 1 with all other control lines 0. These signals would have to be timed properly. This will be discussed in greater detail in Chapter 11.

10.3 ARITHMETIC OPERATIONS – SUBTRACTION, MULTIPLICATION AND DIVISION, BCD–BINARY CONVERSION

There are many techniques that can be used to implement the operations of subtraction, multiplication and division. We shall only consider a few of them to establish some fundamental ideas. The text by Stein and Munro cited in the bibliography contains many such procedures.

Subtraction

We shall use 2's complement arithmetic in subtraction. We assume that the reader is familiar with the discussion in Secs. 9.2 and 9.3.

Now suppose that we want to subtract $(C) - (D)$, where C and D are two registers whose contents are available to the ALU. We form the 2's complement of (D) by complementing each of its bits and then adding 1 to the result. Then, the 2's complement of (D) is added to (C). Their sum is the desired result of the subtraction. Then, using the notation developed in the last section, we have the following algorithms, or set of operations, that are used to obtain the subtraction $(C) - (D)$.

$$0 \rightarrow A \tag{10-18a}$$

$$(D) + (A) \rightarrow A \tag{10-18b}$$

$$(\bar{A}) \rightarrow A \tag{10-18c}$$

$$(A) + 1 \rightarrow A \tag{10-18d}$$

$$(C) + (A) \rightarrow A \tag{10-18e}$$

We assume that the minuend is stored in the C register and that the subtrahend is stored in the D register. Let us consider these steps. In the first step, (10-18a), we clear the accumulator. In (10-18b) we add the contents of D into the accumulator register. In (10-18c) we take the complement of the contents of the accumulator, and in (10-18d) we add 1 to the contents of A. Now (A) is the 2's complement of the original (D). In the last step we take the sum $(C) + (A)$ which is the desired result.

Let us use Table 10-1 to obtain the command sequence that can be used to implement (10-18). This is

$$c_2 \tag{10-19a}$$

$$c_1 \tag{10-19b}$$

$$c_8 \tag{10-19c}$$

$$c_9 \tag{10-19d}$$

$$c_1 \tag{10-19e}$$

Note that we assumed that the minuend and subtrahend were stored in registers C and D, respectively. At the appropriate time, the contents of D must be supplied to the input of the accumulator and then, at a subsequent time, the contents of C must be so supplied. It is assumed that the computer circuit allows this to be done. We shall consider this in Chapter 11.

Multiplication

Now let us consider the multiplication of two numbers contained in registers C and D. We shall assume that they are both positive binary numbers. Multiplications algorithms exist for multiplication using 2's complement numbers. However, we shall discuss the multiplication of two positive binary numbers since it is simpler conceptually.

Let us start by considering the binary multiplication of two positive integers as would be done with pencil and paper. Then we shall see how this can be implemented on the computer. Suppose that we want to multiply 1101 by 1011. Then, following the usual rules for multiplication, we have

$$
\begin{array}{ll}
1101 & \text{multiplicand} \\
\underline{1011} & \text{multiplier} \\
1101 & \text{first partial product} \\
1101 & \text{second partial product} \\
0000 & \text{third partial product} \\
\underline{1101} & \text{fourth partial product} \\
10001111 & \text{product}
\end{array} \qquad (10\text{-}20)
$$

We multiply the multiplicand by the first digit to obtain the first partial product. Next, we multiply the multiplicand by the second digit of the multiplier, shift the result one digit to the left and add the result to the first partial product, etc. When we implement this in the computer, we could shift the partial products to the left. However, it is more convenient if we consider the fourth (last) partial product as the reference. Then, the third partial product is shifted one digit to the right, the second partial product is shifted two digits to the right, and the first partial product is shifted three digits to the right.

Now let us consider the digital computer implementation of this. In order to perform multiplication we could use a lengthy sequence of instructions of the type shown in Table 10-1. However, the multiplication is much faster if we assume that several additional registers are available. Let us consider this. We shall assume that we have an extra 1-bit register associated with register A. It is called a *link* register, L. We must also have the usual register B, called a storage register. In addition, we assume that we have one other register MQ, called the multiplier-quotient register. (This will also be used for division.) Both B and MQ will be assumed to store four bits. In this example we shall multiply two 4-bit numbers. For this example, A has only four bits. (The number of bits is kept small to simplify the example.)

The MQ, L and A registers are arranged into a single register such that data can be shifted from one to the other. If both registers are cleared, then we have

$$\overbrace{\begin{matrix} L \end{matrix}} \quad \overbrace{\begin{matrix} A \end{matrix}} \quad \overbrace{\begin{matrix} MQ \end{matrix}}$$

$$0 \quad 0\,0\,0\,0 \quad 0\,0\,0\,0 \qquad\qquad (10\text{-}21a)$$

Now suppose that the multiplicand 1101 in example (10-20) is stored in the B register. Assume that the multiplier is also stored. For the time being, let us not say where it is stored, but we shall assume that we can use its digits for control. The multiplier is 1011. Now, since the rightmost digit of the multiplier is a 1, we add the contents of the B register to the A register, i.e. this carries out the mathematical operation of multiplying the contents of the B register by 1 and then adding. Doing this we obtain

$$\overbrace{\begin{matrix} L \end{matrix}} \quad \overbrace{\begin{matrix} A \end{matrix}} \quad \overbrace{\begin{matrix} MQ \end{matrix}}$$

$$0 \quad 1\,1\,0\,1 \quad 0\,0\,0\,0 \qquad\qquad (10\text{-}21b)$$

Now shift the contents of the combined L-A-MQ register one digit to the right. This yields

$$\overbrace{\begin{matrix} L \end{matrix}} \quad \overbrace{\begin{matrix} A \end{matrix}} \quad \overbrace{\begin{matrix} MQ \end{matrix}}$$

$$0 \quad 0\,1\,1\,0 \quad 1\,0\,0\,0 \qquad\qquad (10\text{-}21c)$$

The next digit of the multiplier (second from right) is 1. Thus, we again add the contents of the B register to the A register. This yields

$$\overbrace{\begin{matrix} L \end{matrix}} \quad \overbrace{\begin{matrix} A \end{matrix}} \quad \overbrace{\begin{matrix} MQ \end{matrix}}$$

$$\begin{array}{cccc}
0 & 0\,1\,1\,0 & 1\,0\,0\,0 \\
 & 1\,1\,0\,1 & \\
\hline
1 & 0\,0\,1\,1 & 1\,0\,0\,0
\end{array} \qquad (10\text{-}21d)$$

Note that the L register stores the "overflow" bit. Actually, this is not an overflow since we will next shift to the right.

Next, shift the L–A–MQ register to the right to obtain

L $\quad$ A $\qquad$ MQ

0 1 0 0 1 1 1 0 0 $\qquad\qquad\qquad$ (10-21e)

The next digit of the multiplier (third from right) is 0. Thus, we add 0 to the A register, leaving the results unchanged. Now shift to the right again. This yields

L $\quad$ A $\qquad$ MQ

0 0 1 0 0 1 1 1 0 $\qquad\qquad\qquad$ (10-21f)

The last (fourth from right) digit of the multiplier is 1. Thus, we add the contents of the B register to the A register to obtain

L $\quad$ A $\qquad$ MQ

1 0 0 0 1 1 1 1 0 $\qquad\qquad\qquad$ (10-21g)

Finally, shift to the right. This yields

L $\quad$ A $\qquad$ MQ

0 1 0 0 0 1 1 1 1 $\qquad\qquad\qquad$ (10-21h)

which is the result of the multiplication.

The process of shifting to the right and adding that we have performed exactly follows the steps used in multiplying with pencil and paper.

We assumed that the multiplier was stored in some unspecified register. Actually we can use the MQ register for this purpose. Note that, from the start until after the first step is performed, [(10-21b)], the MQ register is empty. After the first step, we no longer need the rightmost digit of the multiplier. Thus, we can shift the contents of the MQ register to the right as before. Therefore, we could store the multiplicand in the MQ register. The rightmost digit of MQ controls the multiplication that generates the first partial product of (10-21b). When we shift to the right, the rightmost digit of the multiplier is lost and the second digit of the multiplier becomes the rightmost digit. Now it can control the multiplication for the second partial product. Therefore, we can continue in this way. Then, for the example just considered, the sequence will be

0 0 0 0 0 1 0 1 1	(10-22a)
0 1 1 0 1 1 0 1 1	(10-22b)
0 0 1 1 0 1 1 0 1	(10-22c)
1 0 0 1 1 1 1 0 1	(10-22d)
0 1 0 0 1 1 1 1 0	(10-22e)
0 0 1 0 0 1 1 1 1	(10-22f)
1 0 0 0 1 1 1 1 1	(10-22g)
0 1 0 0 0 1 1 1 1	(10-22h)

In this example we have assumed that we could link three registers together (i.e. L, A, and MQ) so that we shift from one to the other. In the typical microprocessor, this would probably require several instructions. Such operations will be discussed in the next chapter.

In the example of multiplication just performed, two 4-bit binary numbers were multiplied and the result was an 8-bit number. It may seen as though this is an overflow. If the remainder of the computer circuitry is such that we can only work with 4-bit binary integers, then it is indeed an overflow. That is, the multiplier and/or multiplicand are too large. Now suppose that we multiply

```
   0011
   0010
  ─────
   0000
   0011
  0000
 0000
 ──────
 0000110
```

In this case the four rightmost digits, which will be stored in the MQ register after the multiplication is completed, constitute the complete answer and an overflow does not result even though there is an 8-digit answer. Note that, in the case of integers, it is the rightmost digits, which are stored in the MQ register, with which we are concerned.

In the case of floating-point numbers, we can often tolerate a loss of the rightmost bits. Suppose that we multiply 0.1101×2^a and 0.1011×2^b. The exponents a and b must be added. Now consider the actual multiplication of the fractional parts.

```
   0.1101
   0.1011
  _____
    1101
   1101
  0000
 1101
_____
.1001111
```

Note that the steps are the same as for the integer multiplication. The leftmost digits represent the most significant figures. Thus, if we can only save four digits, roundoff error results, and we get

.1000

as the answer. This is stored in the A register. The rounded off answer would be

$0.1000 \times 2^{a+b}$

The detained of the store are further discissed in Sec. 9.7.

We have considered that the extra digits constitute an overflow as in the case of ir tegers or must be rounded off in the case of floating-point numbers. However, the number stored in the combined A–MQ register is correct. It is possible to control the computer (i.e. to write programs) which can use this data, in which case, overflow or roundoff need not occur. For instance, suppose that we have an 8-bit number and the adders can add only 4-bit numbers. If we want to add the 8-bit numbers, then we can use the procedure discussed in Sec. 9.8.

We have considered parallel multiplication. Serial multiplication circuits also exist which utilize only one adder in the accumulator.

Division

Now let us consider the process of division. The operation is carried out using a minor modification of the circuitry used for multiplication. Let us start by considering a pencil and paper division following the ordinary rules of long division. We shall use 2's complement for subtraction. Let us divide 010001111 by 01011.

```
01011 | 010001111
```

The first step would be to see if 1000 is greater than 1011. Obviously, it is not. However, the computer is not aware of this and would try it out. Thus, a first trial would be

$$
\begin{array}{r}
1 \\
\hline
01011 \;|\; \overline{010001111} \\
-\,01011 \\
\hline
\end{array}
\qquad (10\text{-}23\text{a})
$$

In 2's complement we would have (the digits to the left of the dotted line respresent "sign bits")

$$
\begin{array}{c}
0\;|\;1\;\;0\;\;0\;\;0 \\
①|\;0\;\;1\;\;0\;\;1 \\
\hline
1\;|\;1\;\;1\;\;0\;\;1
\end{array}
$$

The leftmost digit, to the left of the dotted line, is a 1. Thus, the result is negative. If we ignore the circled number, we can see that there is no carry into, nor existing 1 in, the fifth position. Thus, a 1 must appear in the sign bit position, and the result is negative. Then, the choice of the first digit of the quotient as 1 is too large. Thus, the next step in the division is (omitting the sign bits)

$$
\begin{array}{r}
1 \\
\hline
1011 \;|\; \overline{10001111} \\
-\,1011 \\
\hline
\end{array}
\qquad (10\text{-}23\text{b})
$$

In 2's complement, the subtraction becomes (including all bits)

$$
\begin{array}{c}
0\;\;1\;\;0\;\;0\;\;0\;\;1 \\
+\,(1\;\;1)\,0\;\;1\;\;0\;\;1 \\
\hline
(1\;\;0\;\;0)\,0\;\;1\;\;1\;\;0
\end{array}
\qquad (10\text{-}23\text{c})
$$

This is positive and, thus, the choice of quotient digit is correct. Note that, to be completely correct, we must include the encircled digits when we use the 2's complement. It would be convenient not to have to add these bits. If we do not, then we have

$$
\begin{array}{c}
1\;\;0\;\;0\;\;0\;\;1 \\
0\;\;1\;\;0\;\;1 \\
\hline
1\;\;0\;\;1\;\;1\;\;0
\end{array}
$$

Now there is a 1 introduced into the fifth position of the answer. Thus, we know that the result will be positive since, if we had included the sign bit, this would have been a 0. Note that, if we discard this 1, then we will have the correct answer. Thus far, then, we have

$$
\begin{array}{r}
01 \\
1011\ |\ \overline{10001111} \\
+0101 \\
\hline
01101
\end{array}
$$

Continuing this way, we obtain

$$
\begin{array}{r}
011 \\
1011\ |\ \overline{10001111} \\
+0101 \\
\hline
01101 \\
+0101 \\
\hline
(1)0010
\end{array}
$$

The fifth position has a 1 which indicates, as before, that the result of the subtraction is positive. Again, we discard it and proceed. Thus, we have

$$
\begin{array}{r}
01101 \\
1011\ |\ \overline{10001111} \\
+0101 \\
\hline
01101 \\
+0101 \\
\hline
001011 \\
0101 \\
\hline
(1)0000
\end{array}
$$

Again, the 1 in the fifth position is disregarded. Thus, there is no remainder. When humans divide, we can look and see if the result of the subtraction would be negative. However, the computer cannot do this.

Now let us consider the computer operation. The previously developed multiplier is used with a different set of instructions. The combined A and MQ registers are used to store the dividend. The B

register is used to store the divisor. (Note that the dividend can have essentially twice the number of bits as can the divisor. The discussion of this in regard to multiplication applies here.)

After each subtraction, we shift in one more digit of the dividend. This is equivalent to shifting the dividend to the left. Note that this will also eliminate the encircled terms in the example.

As we shift to the left, 0's will be stored at the right side of the MQ register. We can use these "empty spaces" to store the digits of the quotient. Then, when the division is complete, the quotient will be stored in the MQ register.

Now let us carry out the previous example illustrating the contents of the L–A–MQ register. At the start, we have

$$L \quad A \qquad MQ$$
$$0 \; \overbrace{1 \; 0 \; 0 \; 0} \; \overbrace{1 \; 1 \; 1 \; 1} \hspace{3cm} (10\text{-}24\text{a})$$

Note that the A register has five bits. Actually, the fifth bit will be in the link register. To avoid confusion, we shall show this bit encircled. Note that it can be used to check that the results of each subtraction are positive. Thus, we shall only have four bits in the A register. After implementing (10-23a), we have

$$L \quad A \qquad MQ$$
$$Ⓞ \; \overbrace{1 \; 1 \; 0 \; 1} \; \overbrace{1 \; 1 \; 1 \; 1}$$

Note that the link bit indicates a negative result. (Remember that we have omitted the sign bit from the B register.) Hence, the divisor is added back, giving

$$L \quad A \qquad MQ$$
$$① \; \overbrace{1 \; 0 \; 0 \; 0} \; \overbrace{1 \; 1 \; 1 \; 1}$$

If we clear the link, we obtain the original expression

$$L \quad A \qquad MQ$$
$$Ⓞ \; \overbrace{1 \; 0 \; 0 \; 0} \; \overbrace{1 \; 1 \; 1 \; 1}$$

Now, the contents of the L–A–MQ register are shifted to the left. (Note that the above clearing of the link bit may seem a useless step, since the link bit is lost. However, in implementing the procedure, the link bit can be used to determine the next bit of the quotient. Thus, it is desirable that it be correct.) Then, after shifting, we have

$$\underbrace{\textcircled{1}}_{L}\ \underbrace{0\ 0\ 0\ 1}_{A}\ \underbrace{1\ 1\ 1\ 0}_{MQ}$$

$$(10\text{-}24\text{b})$$

Note that the previous link bit is 0, which is the correct rightmost bit of the MQ register. Next, subtraction is performed and succeeds. Thus, we obtain

$$\underbrace{\textcircled{1}}_{L}\ \underbrace{0\ 1\ 1\ 0}_{A}\ \underbrace{1\ 1\ 1\ 0}_{MQ}$$

$$(10\text{-}24\text{c})$$

Since the subtraction succeeded, the term in the quotient is 1. This can be obtained from the link bit. Thus, a 1 is put at the right input of the MQ register. Now the contents of the L–A–MQ register are shifted to the left. This yields

$$\underbrace{\textcircled{0}}_{L}\ \underbrace{1\ 1\ 0\ 1}_{A}\ \underbrace{1\ 1\ 0\ 1}_{MQ}$$

$$(10\text{-}24\text{d})$$

After the next subtraction, we have

$$\underbrace{\textcircled{1}}_{L}\ \underbrace{0\ 0\ 1\ 0}_{A}\ \underbrace{1\ 1\ 0\ 1}_{MQ}$$

$$(10\text{-}24\text{e})$$

Since the subtraction succeeded, i.e. there is a 1 in the L register, a 1 is put at the right input of the MQ register. Thus, after shifting left, the result it

$$\underbrace{\textcircled{0}}_{L}\ \underbrace{0\ 1\ 0\ 1}_{A}\ \underbrace{1\ 0\ 1\ 1}_{MQ}$$

$$(10\text{-}24\text{f})$$

The next subtraction fails. Then, after adding back the divisor, clearing the link, and shifting, we obtain

$$\underbrace{\textcircled{0}}_{L}\ \underbrace{1\ 0\ 1\ 1}_{A}\ \underbrace{0\ 1\ 1\ 0}_{MQ}$$

$$(10\text{-}24\text{g})$$

After the next subtraction, we have

$$\underset{\underbrace{\text{L}}}{L} \quad \underset{\underbrace{\quad\quad}}{A} \quad \underset{\underbrace{\quad\quad}}{MQ}$$
$$①0\ 0\ 0\ 0\ 0\ 1\ 1\ 0 \qquad\qquad (10\text{-}24\text{h})$$

The subtraction succeeds. Thus, after the next shift, we have

$$\underset{\underbrace{\text{L}}}{L} \quad \underset{\underbrace{\quad\quad}}{A} \quad \underset{\underbrace{\quad\quad}}{MQ}$$
$$0\ 0\ 0\ 0\ 0\ 1\ 1\ 0\ 1 \qquad\qquad (10\text{-}24\text{i})$$

Thus, the quotient is in the *MQ* register and the remainder (0) is in the *A* register.

BCD–Binary Conversion

The numerical data supplied to a computer usually comes from a terminal, disk, or tape as a sequence of decimal digits. These are often encoded by the associated electronic circuits into a BCD code, see Sec. 8.1. Arithmetic can be performed using numbers encoded in this way. At times, as in desk calculators or in some small computers, this is done. However, for a digital computer, it is much more efficient if standard binary numbers are used. Thus, there must be a conversion from BCD to binary. This can be done by a simple combinational logic circuit. However, there would be millions of possible inputs, corresponding to each number that could be entered, and an equal number of binary outputs. Thus, the number of gates required would be prohibitive. We shall now discuss a procedure that can be used to convert without requiring extensive electronics.

In Eqs. (2-8) to (2-11) we illustrated a procedure for obtaining a binary number from a decimal number by successively dividing by two. The remainders yielded the binary number. It is assumed that the reader is familiar with this procedure. Let us illustrate it by obtaining 43 in binary form. Then,

$$
\begin{aligned}
43/2 &= 21 + 1/2 \\
21/2 &= 10 + 1/2 \\
10/2 &= 5 + 0/2 \\
5/2 &= 2 + 1/2 \\
2/2 &= 1 + 0/2 \\
1/2 &= 0 + 1/2
\end{aligned}
\qquad\qquad (10\text{-}25)
$$

Now take the remainders in reverse order. Thus, we have

$$43_{10} = 101011_2$$

Note that division by 2 is particularly simple in binary. The number is simply shifted one bit to the right. The remainder is simply the right-most bit. Now we shall consider how this can be implemented on the computer. We cannot simply divide since the arithmetic circuits are set up for binary numbers, not for BCD numbers. Let us assume that an 8-4-2-1 BCD binary code is used. Then, although *each* decimal digit is encoded in binary, the complete number is not. For instance, 43 will be encoded as

0100 0011

Since each digit is encoded in binary, we can use the division circuits to divide *each set of four bits* by 2. Let us see how this would work. We shall use the decimal representations of the digits.

$$4/2 = 2 + 0/2 \qquad\qquad 3/2 = 1 + 1/2$$

Note that the 4/2 actually represents 40/2. The next step is

$$2/2 = 1 + 0/2 \qquad\qquad 1/2 = 0 + 1/2$$

Next we have

$$1/2 = 0 + 1/2 \qquad\qquad 0/2 = 0 + 0/2$$

Thus, there is a remainder of ½ in the tens column. This is actually $10/2 = 5$. This must now be added into the ones column. Thus, we have $0 + 5 = 5$.

0   5

The next steps are then

$$5/2 = 2 + 1/2$$

$$2/2 = 1 + 0/2$$

$$1/2 = 0 + 1/2$$

Again we consider the remainders in the ones column and obtain

$43_{10} = 101011_2$

A block diagram for the implementation of this is shown in Fig. 10-7.

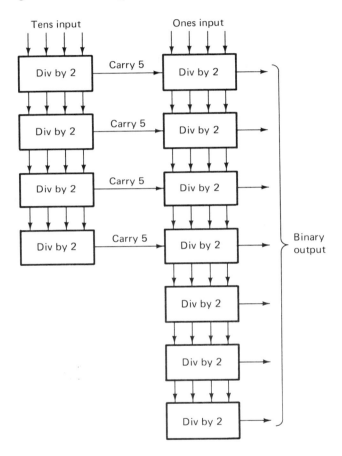

Figure 10-7 A block diagram which implements binary conversion of two digits of an 8421 BCD

We have illustrated this procedure using two digits. It can, of course, be extended. Remember that each carry adds 5 to the next "column" to the right. The electronics must account for this.

When the results of the computation are obtained, we often want the results to be printed out as a decimal number, that is, as a sequence of decimal digits. In this case, the binary number is converted into a BCD number. Each digit is then used to activate a printer. The converse process can then be used to convert from binary to 8-4-2-1 BCD. For instance, consider

$$101011_2$$

We perform the following operations. (Note that we work from left to right in this case.)

$$
\begin{array}{llll}
1 \times 2 = 2 & 2 + 0 = 2 & \text{(Note: 0 is second digit from the left)} & \\
2 \times 2 = 4 & 4 + 1 = 5 & \text{(Note: 1 is third digit from the left)} & \\
5 \times 2 = 10 & 10 + 0 = 10 & \text{(Note: 0 is fourth digit from the left)} & (10\text{-}26) \\
10 \times 2 = 20 & 20 + 1 = 21 & \text{(Note: 1 is fifth digit from the left)} & \\
21 \times 2 = 42 & 42 + 1 = 43 & \text{(Note: 1 is sixth digit from the left)} &
\end{array}
$$

Thus we can, by means of multiplication by 2 and addition, obtain the decimal number. The actual details of the implementation are the converse of the preceding ones. Remember that we obtain each digit encoded in 8-4-2-1 BCD. Note that multiplication of a binary number by 2 only involves shifting to the left by one bit.

10.4 ALU CIRCUITRY

We shall now consider the circuit of an arithmetic logic unit which performs the operations of Table 10-1. Some of these operations have already been considered. For instance, addition and clearing have been implemented by the circuit of Fig. 10-4. To easily obtain more control, the accumulator register often use J-K flip-flops instead of D flip-flops. To avoid excessively cluttered drawings, we shall draw only one stage of the ALU. Such a stage using J-K flip-flops, which performs the ADD operation is shown in Fig. 10-8. Note that the J-K flip-flop and the NOT gate function as the D flip-flop of Fig. 10-4. We introduce the ADD control signal c_1 differently here. When $c_1 = 0$, then

$$J_i = K_i = 0 \tag{10-27}$$

and the flip-flop does not change its state. When $c_1 = 1$, then

$$J = S_i \tag{10-28a}$$

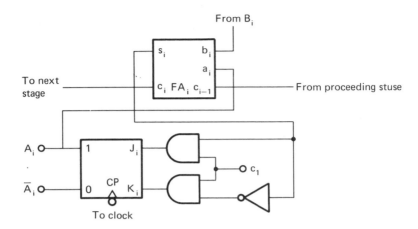

Figure 10-8 A portion of an ALU which performs an addition when $c_1 = 1$

$$K = \overline{S}_i \qquad (10\text{-}28\text{b})$$

and the flip-flop sets itself to S_i. The point labeled c_1 is where the ADD command signal is applied.

To clear the flip-flop, we use the circuit of Fig. 10-9. When the clear signal c_2 is given

$$J_i = 0 \qquad (10\text{-}29\text{a})$$

$$K_i = 1 \qquad (10\text{-}29\text{b})$$

Thus, the state of the flip-flop becomes 0 after the next clock pulse. We assume that only one command will ever be given at a time. Thus, we can use OR gates to apply the various signals to the flip-flops. We shall illustrate this subsequently. The β_{2i} is a connection point which we shall discuss later.

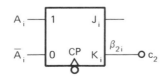

Figure 10-9 The circuit used to clear stage i of the ALU when $c_2 = 1$

Logical AND

To perform the logical AND, we use the circuit of Fig. 10-10. Note that, if $\bar{B}_i$ is not available, then a NOT gate must be included to obtain it from B_i. Note that if c_3 is a 1, and A_i and B_i are not *both* 1's then, after the next clock pulse, $A_i = 0$. We could draw truth tables and design these sequential circuits using the procedures of Sec. 6.5. However,

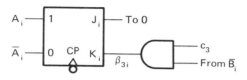

Figure 10-10 The circuit used to perform the logical AND operation with stage *i* of the ALU when $c_3 = 1$

these circuits are so simple that this step can be omitted. Note that, in this case

$$J_i = 0 \tag{10-30a}$$

$$K_i = \bar{B}_i \tag{10-30b}$$

Logical OR

The logical OR is implemented as shown in Fig. 10-11. In this case, when c_4 is a 1, we have

$$J_i = B_i \tag{10-31a}$$

$$K_i = 0 \tag{10-31b}$$

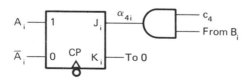

Figure 10-11 The circuit used to perform the logical OR operation with stage *i* of the ALU when $c_4 = 1$

Logical XOR

The logical XOR is implemented as shown in Fig. 10.12. In this case, when $c_5 = 1$

$$J_i = K_i = B_i \qquad (10\text{-}32)$$

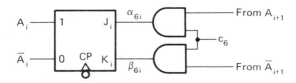

Figure 10-12 The circuit used to perform the logical XOR operation with stage i of the ALU when $c_5 = 1$

Thus, the flip-flop functions as a T flip-flop. Hence, if $A_i = 1$, and $B_i = 0$ or if $A_i = 0$ and $B_i = 1$, after the next clock pulse, we will have $A_i = 1$. If both A_i and B_i are 1's then, after the next clock pulse, $A_i = 0$. If both A_i and B_i are 0's, then the state will not change.

Shift Right

To shift right we take the output from the flip-flop to the left (i.e. the $i+1^{st}$ flip-flop) and set the ith flip-flop to its state. To do this, see Fig. 10-13, with $c_6 = 1$, we set

$$J_i = A_{i+1} \qquad (10\text{-}33a)$$

$$K_i = \overline{A}_{i+1} \qquad (10\text{-}33b)$$

For the leftmost flip-flop we connect the A_{i+1} input to 0 and the $\overline{A}_{i+1}$ input to 1. This can be done by open circuiting the A_{i+1} input and connecting the $\overline{A}_{i+1}$ input to c_6.

Figure 10-13 The circuit used to shift right with stage i of the ALU when $c_6 = 1$

Shift Left

This basically follows the shift right circuit, except that A_{i-1} replaces A_{i+1}. In this case, it is the rightmost (i.e. first position)A_0 whose A_i lead is open circuited and whose $\overline{A}_{i-1}$ lead is connected to c_7. The circuit is shown in Fig. 10-14.

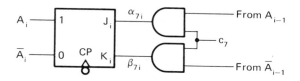

Figure 10-14   The circuit used to shift left with stage *i* of the ALU when $c_7 = 1$

Complement

To obtain the complement, when $c_8 = 1$, we change the state of the flip-flop. To do this, we connect it as a T flip-flop, and make its inputs equal to c_8. This is shown in Fig.10-15.

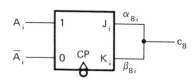

Figure 10-15   The circuit used to complement with stage *i* of the ALU when $c_8 = 1$

Increment

When we increment, 1 is added to the contents of the register. Thus, it is essentially an addition operation. In Fig. 10-16, we have combined both the operations of addition and incrementation. Note that when c_1 is applied, we essentially hve the circuit of Fig. 10-8. If $c_1 = 0$, then the B inputs are disconnected by the AND gate. If $c_9 = 1$, then the ADD command is essentially given except that now all the B inputs are cut off. Effectively $(B) = 0$. However, $c_9 = 1$ and it is connected to the carry input of the first full adder. (We assume that all the adders are full adders.) Thus, we add 1.

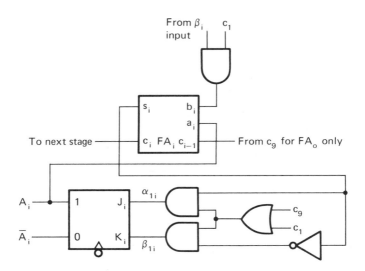

Figure 10-16 The portion of the ALU which performs an addition when $c_1 = 1$ and increments when $c_9 = 1$

Carry Check

The carry check is performed during an addition. The F_0 bit of the flag register acts as an additional (leftmost) bit of the A register. Hence, if there is a carry, $(F_0) = 1$.

Negative Check

We assume that 2's complement arithmetic is being used. Then, the number in the accumulator is negative if the number stored in the leftmost register is a 1. Thus, in Fig. 10-17, when $c_{10} = 1$, the output y will be a 1 if the stored number is negative. If it is positive or 0, then

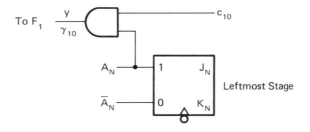

Figure 10-17 Negative check in ALU. If stored number is negative then $y = 1$ when $c_{10} = 1$. This number is stored in F_1 of the flag register.

$y=0$. This quantity is stored as one bit of the flag register, see Sec. 10-2. We assume that it is stored in the 1-bit position (i.e. F_1).

Zero Check

If the stored number is 0, then

$$\bar{A}_0 = \bar{A}_1 = \ldots = \bar{A}_N = 1 \tag{10-34}$$

In the circuit of Fig. 10-18, when $c_{11}=1$, y will equal 1 if Eq. (10-34) is satisfied. This is then put in F_2.

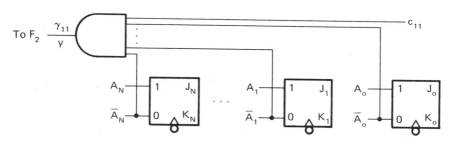

Figure 10-18 Zero check in ALU. If stored number is 0, then $y=1$ when $c_{11}=1$. This number is put in F_2.

Complete ALU

Now let us consider the complete ALU. Note that we assume that the controller is such that only one control signal of c_1 through c_{11} will be 1 at one time. Thus, to implement the complete accumulator, we can break all the inputs to the J-K flip-flop at the points marked with an α or a β in Fig. 10-19. Each set of α inputs for each stage is used as the input to an OR gate, which is connected to the J_i input. A similar statement can be made for the β points and the K inputs.

Thus, we break the α_{ni} and β_{ni} connections and insert the OR gates as shown in Fig. 10-19. Note that each flip-flop of the accumulator has a pair of OR gates inserted as shown. (Note that there is no α_{9i} or β_{9i}, since two functions were realized in Fig. 10-16.) Also note that there is one circuit of the type shown in Fig. 10-19a for *each stage* of the accumulator (i.e. a total of $N+1$).

Let us conclude by discussing some definitions. The device we have developed is called and ALU and its register is called an accumulator. However, other terminology is used at times and this situation can be confusing. For instance, an ALU is sometimes called an accumulator,

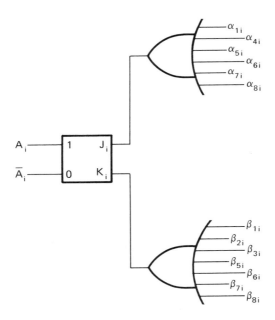

Figure 10-19 Diagram showing the interconnection of Figs. 10-8 and 10-18 to produce complete ALU. OR gates are inserted in series with each flip-flop input. Note α_{2i}, α_{3i} and β_{4i} are missing since these represent D-inputs.

or sometimes a *true accumulator*. There is also a much simpler device which will just add two numbers. In general, this is no longer used in digital computers. However, it is used in some desk calculators. Such a device is called an *accumulating register,* or a *working register.* We shall conform with modern usage and refer to an accumulator as a particular register of the ALU.

10.5 INTEGRATED CIRCUIT IMPLEMENTATION OF COMPUTATION OPERATIONS

All of the computational operations that we have discussed, e.g. addition, multiplication, BCD-binary conversion, are implemented electronically on a single chip using integrated circuits. We shall consider some representative ones here. Remember that these are only a small sample of the available integrated circuits. We shall not consider microprocessors here since they shall be discussed in the next chapter.

In Fig. 10-20a, the logic circuit diagram of a type MC1059 full adder is illustrated. This chip contains two full adder circuits. Thus, 2-bit numbers can be added. The carry output of the first adder is directly connected to the carry input terminal of the second adder. In addition to the sum and carry outputs, other logic outputs are given at appropriate terminals. These extra terminals increase the versatility of the device. The truth table is given in Fig. 10-20b. Its schematic diagram is shown in Fig. 10-20c. Only one-half of the circuit is shown since the

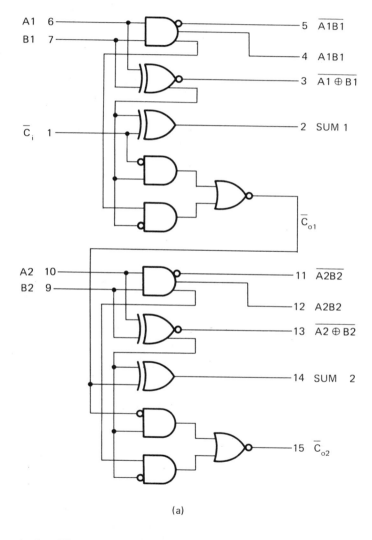

(a)

Figure 10-20 The type MC1059 binary adder. (a) logic circuit diagram; (b) truth table; (c) schematic diagram (Courtesy Motorola Semiconductor Products, Inc.)

TRUTH TABLE

Inputs			Outputs			
A1	B1	$\overline{C}_i$	A1 · B1	$\overline{A1 \oplus B1}$	S1	$\overline{C}_{o1}$
0	0	0	0	1	1	1
1	0	0	0	0	0	0
0	1	0	0	0	0	0
1	1	0	1	1	1	0
0	0	1	0	1	0	1
1	0	1	0	0	1	1
0	1	1	0	0	1	1
1	1	1	1	1	0	0
A2	B2	$\overline{C}_{o1}$	A2 · B2	$\overline{A2 \oplus B2}$	S2	$\overline{C}_{o2}$
0	0	1	0	1	0	1
1	0	0	0	0	0	0
0	1	0	0	0	0	0
1	1	0	1	1	1	0
1	1	1	1	1	0	0
0	1	1	0	0	1	1
1	0	1	0	0	1	1
0	0	0	0	1	1	1

(b)

Figure 10-20 (Continued) part (b)

other half is essentially the same. We have shown a 2-bit adder here. Actually, chips are available which will add two 16-bit (and larger) binary numbers.

As another example of an integrated-circuit-implemented computational operation, consider the type SN54145 BCD-to-decimal decoder/driver, see Fig. 10-21. The input to this device is a BCD number. This causes one of the appropriate output lines to be energized. The output lines are numbered 0 to 9. The energized one corresponds to the input BCD number. These outputs can activate relays or turn lamps on for display purposes. The logic circuit diagram is shown in Fig. 10-21a. Figures 10-21b and c give the equivalent input and output circuits. The truth table is given in Fig. 10-21d.

In this chapter we have considered some ideas of computation circuits. In the next one we shall discuss their use in the complete digital computer.

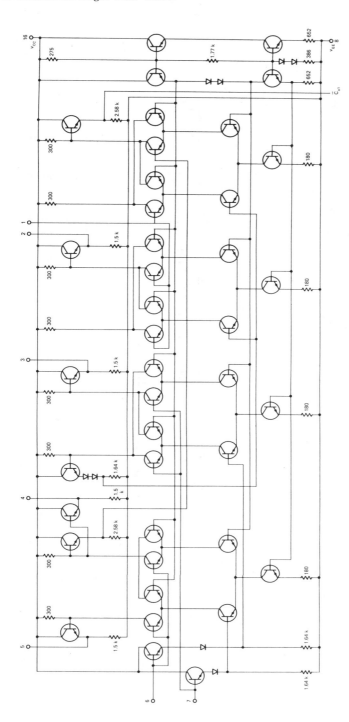

Figure 10-20 (Continued) part (c)

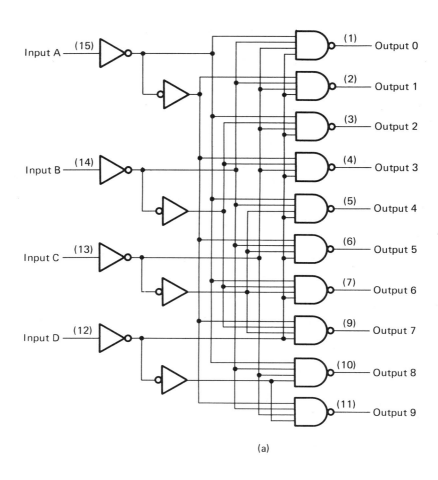

(a)

Figure 10-21 The type SN54145 BCD to decimal decoder/driver. (a) logic circuit diagram; (b) typical input circuit; (c) typical output circuit; (d) function table (Courtesy Texas Instruments, Inc.)

EQUIVALENT OF ALL INPUTS

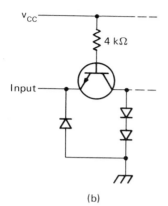

(b)

TYPICAL OF ALL OUTPUTS

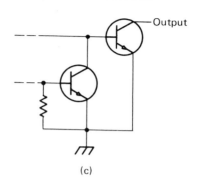

(c)

FUNCTION TABLE

No.	Inputs				Outputs									
	D	C	B	A	0	1	2	3	4	5	6	7	8	9
0	L	L	L	L	L	H	H	H	H	H	H	H	H	H
1	L	L	L	H	H	L	H	H	H	H	H	H	H	H
2	L	L	H	L	H	H	L	H	H	H	H	H	H	H
3	L	L	H	H	H	H	H	L	H	H	H	H	H	H
4	L	H	L	L	H	H	H	H	L	H	H	H	H	H
5	L	H	L	H	H	H	H	H	H	L	H	H	H	H
6	L	H	H	L	H	H	H	H	H	H	L	H	H	H
7	L	H	H	H	H	H	H	H	H	H	H	L	H	H
8	H	L	L	L	H	H	H	H	H	H	H	H	L	H
9	H	L	L	H	H	H	H	H	H	H	H	H	H	L
INVALID	H	L	H	L	H	H	H	H	H	H	H	H	H	H
	H	L	H	H	H	H	H	H	H	H	H	H	H	H
	H	H	L	L	H	H	H	H	H	H	H	H	H	H
	H	H	L	H	H	H	H	H	H	H	H	H	H	H
	H	H	H	L	H	H	H	H	H	H	H	H	H	H
	H	H	H	H	H	H	H	H	H	H	H	H	H	H

(d)

Figure 10-21 (Continued) part (b), (c) and (d)

BIBLIOGRAPHY

Hellerman, H., *Digital Computer System Principles*, Chap. 1, McGraw-Hill Book Co., New York 1967

Kohonen, T., *Digital Circuits and Devices*, Chap. 4, Prentice Hall, Inc. Englewood Cliffs, N.J. 1972

Mano, M.M., *Digital Logic and Computer Design*, Chap. 5, Prentice Hall Inc., Englewood Cliffs, N.J. 1979

O'Malley, J., *Introduction to the Digital Computer*, Chaps. 8 and 11, Holt, Rinehart and Winston, Inc., New York 1972

Rhyne, V.T., *Fundamentals of Digital System Design*, Chap. 8, Prentice Hall, Inc., Englewood Cliffs, N.J. 1973

Stein, M.L. and Munro, W.D., *Introduction to Machine Arithmetic*, Addison-Wesley Publishers, Inc., Reading, Mass. 1971

PROBLEMS

10-1. Derive Eqs. (10-1).

10-2. Describe the difference between a full adder and a half adder.

10-3. Design a 4-bit adder that does not suffer from the effect of carry propagation.

10-4. Modify the circuit of Prob. 10-3 so that fewer gates are used. Carry propagation through one gate is now allowed.

10-5. Modify the circuit of Fig. 10-4 so that J–K flip-flops are used.

10-6. Repeat Prob. 10-5 but now use T flip-flops.

10-7. Repeat Prob. 10-5 for the circuit of Fig. 10-5.

10-8. Repeat Prob. 10-7 but now use T flip-flops.

10-9. Discuss and illustrate the notation introduced in Sec. 10-2.

10-10. An ALU uses the control designations of Table 10-1. The initial values that are stored in the accumulator register A and in the B register that supplies values to the accumulator are

$(A) = 0000$
$(B) = 0010$

What are the contents of the A register after the following sequence of commands is given?

c_1

c_1

$$c_1$$

$$c_8$$

$$c_9$$

Assume that the command signals are appropriately synchronized with the clock pulses.

10-11. Repeat Prob. 10-10 if the initial value of (A) is 0011.

10-12. Repeat Prob. 10-10 for the sequence

$$c_2$$

$$c_1$$

$$c_1$$

$$c_4$$

$$c_3$$

$$c_8$$

$$c_9$$

$$c_6$$

10-13. Repeat Prob. 10-12 if the initial value of (A) is 0011.

10-14. What modifications would have to made in the subtraction algorithm if 1's complement subtraction were used?

10-15. Describe in detail the operation of multiplication of two binary numbers on a computer.

10-16. Repeat Prob. 10-15 for division.

10-17. Discuss the difference between integer and floating-point multiplication.

10-18. Discuss the difference between integer and floating-point division.

10-19. Discuss overflow and/or its lack in multiplication and division.

10-20. Extend the diagram of Fig. 10-7 to a 3-digit BCD conversion.

10-21. Obtain a block diagram for the decimal to 8-4-2-1 BCD conversion. Assume two decimal digits.

10-22. Repeat Prob. 10-21 for three decimal digits.

10-23. Design a circuit which converts each digit of an 8-4-2-1 BCD code to 2-4-2-1 BCD code.

10-24. Implement the gate circuit of Fig. 10-8 using a D flip-flop.

10-25. Repeat Prob. 10-24 for the circuit of Fig. 10-9.

10-26. Repeat Prob. 10-24 for the circuit of Fig. 10-10.

10-27. Repeat Prob. 10-24 for the circuit of Fig. 10-11.

10-28. Repeat Prob. 10-24 for the circuit of Fig. 10-12.

10-29. Repeat Prob. 10-24 for the circuit of Fig. 10-13.

10-30. Repeat Prob. 10-24 for the circuit of Fig. 10-14.

10-31. Repeat Prob. 10-24 for the circuit of Fig. 10-15.

10-32. Repeat Prob. 10-24 for the circuit of Fig. 10-16.

10-33. Discuss the flag register.

11

The Microprocessor-Based Digital Computer

In this chapter we shall discuss a complete digital computer. The structure of the computer will be typical of those utilizing microprocessors. However, the ideas can be applied to all digital computers. Its ALU can be considered to be based upon the ALU discussed in Chapter 10. We shall start by considering the general configuration of the computer. The transfer of information among the parts of the computer and the control of the sequencing of those operations that a computer must do when computation is performed are of fundamental importance. We shall consider them and discuss machine language and assembly language programming. The interfacing of the computer with the "external world" will also be considered.

11.1 THE GENERAL ORGANIZATION OF THE MICROPROCESSOR-BASED COMPUTER

In this section we shall consider, in a rather broad, qualitative way, the general organization of the digital computer. We shall see that some of the most important aspects of the computer have already been considered in detail. We shall discuss the others in this chapter.

Fundamentally, a digital computer can be described by the block diagram of Fig. 11-1, which illustrates four basic units. There is an arithmetic logic unit (ALU) where all the computations are actually performed. In a simple computer, this could consist of an ALU of the type discussed in Sec. 10.4, plus additional registers. We have developed all the basic types of circuits which could be used in this unit. The operations of the ALU are controlled by a *control unit* (CU). In addition to controlling the operations of the ALU, the CU must control the

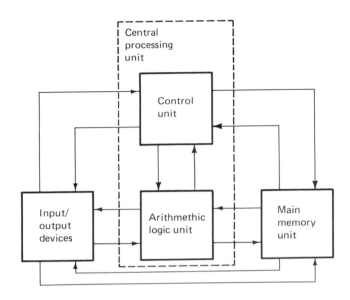

Figure 11-1 Block diagram of the basic computer units

transfer of information from the memory to the ALU and in the reverse direction. Transfer of data to and from the input and output devices may also be controlled by the control unit. The fabrication of the ALU and CU on a single chip is called a *microprocessor*. In general, the combination of the ALU and the CU is called a *central processing unit*, or CPU. When the ALU and the CU are combined in this manner, the computer is called a *microcomputer*.

The *main memory unit* (MMU) is a random access memory of the type discussed in Chapter 7. Semiconductor memories are almost always used with microcomputers. The main memory is capable of storing a large number of words. A single word can be selected and read or written in turn. These procedures were discussed in detail in Chapter 7. In this chapter we shall be concerned with the details of data storage in the memory.

A computer is useless unless data and other information, such as programs, can be supplied to it and unless the computed information can be extracted from it. This is the function of the input/output devices. A typical input device is the terminal, where the user types the information on a typewriter-like keyboard. Contacts on the keys are connected with logic circuits which generate coded signals. For instance, the codes of Table 8-4 are typical of the ones generated.

These are transmitted to the computer. Further code conversion usually takes place in the computer. For instance, numerical data may be first decoded to BCD and then to binary, see Sec. 10-3. (Note that these codes are usually arranged so that the conversion to 8-4-2-1 BCD is very simple.) Other forms of input devices are tape and disk drives where the information is stored on magnetic films. There are also special-purpose input devices. For instance, many "cash registers" now directly input information to a computer, which can then keep track not only of cash, but also of inventory, etc.

The output device could be the same terminal that was used for input. In this case, encoded signals are sent to the terminal which, with its logic circuits, decodes the signals and applies current to electromagnetic devices such that the appropriate character is printed. Alternatively, the information may be displayed on a video screen. Line printers are also used. In addition, output may be to the tape or disk drives. Of course, the input/output devices are much slower than the computer itself.

At times, the computer does not receive or transmit printed information. For instance, the input may be electric signals from gyroscopes which indicate the position, altitude, etc., of an airplane. The output may be signals which control the airplane's throttle, airlerons, etc. There are many other forms of input/output devices. We have just discussed a representative sample.

The first large digital computers occupied entire rooms. However, now, using LSI (large scale integration), a small computer can be built on several chips. This is true of the computer that we shall discuss in this chapter. Of course, some of the peripheral devices such as magnetic tape drives or terminals are still large. However, the actual electronics of the computer can occupy a very small volume.

11.2 MICROPROCESSOR ARCHITECTURE

We shall now discuss the architecture of a typical microprocessor. The microprocessor that we shall use will not be a specific one but will combine the characteristics of most representative microprocessors. In the examples of this chapter, we shall work with a microprocessor that utilizes 8-bit words. There are 4-, 8-, and 16-bit microprocessors available. The ideas that we discuss here are applicable to all of them. The data is transmitted between the microprocessor and RAM along eight wires plus a ground wire. This combination is termed a *data bus*. There is another group of wires that is used to address the memory. This is called the *memory address bus*. We shall assume that 16 bits can be transmitted along this bus (i.e. there are 16 wires plus ground). Sixteen bits allow us to address 2^{16} = 65,536 words of memory. There are also interconnections within the microprocessor chip itself which allow for the routing of data. This is called the *internal data bus*. There are other wires that carry such things as the power supply voltages and various control signals. These, plus the data bus and the memory address bus constitute the *system bus*.

Let us now indicate the architecture of the microprocessor with which we shall work, see Fig. 11-2. The ALU consists of the logic unit

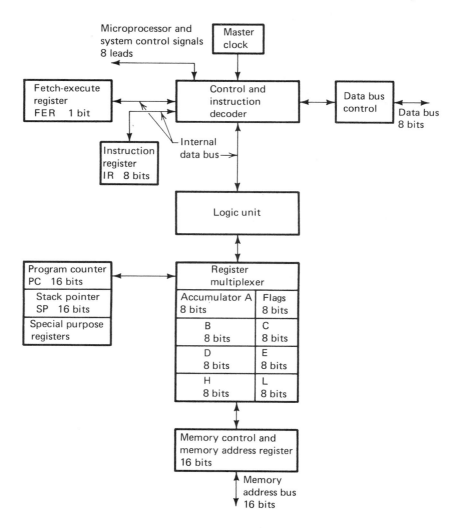

Figure 11-2 Block diagram of an 8-bit microprocessor

and the register multiplexer and the registers below it. The A-register is normally used as the accumulator, but this may not always be the case. The multiplexer allows the switching of the registers. For instance, we could add the contents of the *H*, *L*, or *D* register to the A-register. The other registers, *FER, IR, PC,* and *SP* are special purpose registers that we shall discuss subsequently. The remainder of the block consists of the control unit. Before we consider it and the other elements of the block diagram, we shall discuss how instructions are provided to the microprocessor.

11.3 MICROPROCESSOR INSTRUCTIONS

When a program is executed, the instructions and data will be stored in the RAM. Let us assume that the instructions are stored in memory locations 0 to 200 and are executed in order. We shall first discuss the structure of the instructions.

The simplest instruction just involves the data in the A-, B-, C-, D-, E- or H-, and L-registers. For instance, we could add the contents of the H-register to the A-register and store the result in the A-register. Since we assume that we are working with a microprocessor that uses 8-bit words, an 8-bit word, i.e. an 8-bit binary number, would be used as such an instruction. Note that the instruction is decoded by the control unit using combinational logic circuits.

The instruction is called an *operation code* or an *op code*. Such an 8-bit instruction is illustrated in Fig. 11-3a. If we have eight bits in an op code, then $2^8 = 256$ different instructions can be accommodated. This is usually sufficient for microprocessor applications. Nevertheless, at times, more than one word is required for an instruction. For instance, suppose that we want to add an 8-bit number to the number stored in the accumulator. Now 16 bits are required, 8 bits for the op code and 8 bits for the number we want to add to the accumulator. This is diagrammatically illustrated in Fig. 11-3b. Two consecutive memory locations are required to store this instruction. Note that the *operand* is a number upon which some operation is to be performed.

Some instructions require even more words. For instance, suppose that we want to add a number stored in a particular memory location to the contents of the A-register. Now three words are required, one for the op code, and the other two for the memory locations. This is illustrated in Fig. 11-3c. This instruction must be stored in three consecutive memory words. Note that, in Fig. 11-3c, we indicate that the second word stores the most significant memory bits and that the third word stores the least significant memory address bits. With some micro-

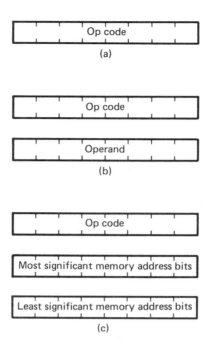

Figure 11-3 Typical microprocessor instructions. (a) one word; (b) two words; (c) three words

processors, the converse is true, and the least significant bits are stored in the second word while the most significant bits are stored in the third word. In general, all the words of an instruction will be stored in consecutive memory locations. However, we shall see that not all instructions need to be executed in sequence.

When we deal with large computers, the words are structured differently. For instance, a large computer may use 32-bit words. In such cases, the op codes and memory locations may all be included in a single instruction word.

11-4. THE CONTROL UNIT

We shall now discuss the execution of instructions by the microprocessor. Both data and instructions are stored in the RAM. There is no way of looking at the number stored in a given memory location to determine if it stores instructions or data. Hence, there must be some means of keeping track of this. The 1-bit *fetch-execute register*, FER, see Fig. 11-2, is used for this purpose. When (FER)= 1 (i.e. the contents of the FER is 1), then the computer treats the information that it is fetching from memory as an instruction. If (FER)=0, then the microprocessor treats the information that it is fetching from the memory as data.

Suppose that we have a program stored in memory locations 0 to 300. (For convenience, we will use hexadecimal numbers to describe the memory locations in this discussion.) There is a register called the *program counter* (PC). This register keeps track of the memory location where the next instruction is stored.

There are circuits in the computer that initialize the PC. For instance, when a start button is pressed, (PC) is set equal to 0. Suppose that a 1-word instruction is stored in that memory location. The contents of the FER will initially be set to 1. The instruction will then be fetched and stored in the *instruction register*, IR. Next, (PC) will be incremented by 1; now (PC)=1 so that the next instruction fetched would be that in memory location 1. However, before this is done, the instruction stored in the IR would be implemented. That is, the number stored in the IR would be decoded and the necessary control signals, see Sec. 10-4, would be generated so that the instruction would be carried out.

Suppose that the next instruction resides in memory locations 1, 2, and 3. The op code is in memory location 1 while the (memory) address of an operand is in locations 2 and 3. For instance, the instruction may call for the adding to register A of the number stored in the memory location whose address is stored in memory locations 2 and 3. The first word in memory address 1 would be put into the IR. The two 8-bit words stored in memory locations 2 and 3 would be put into the 16-bit memory address register. The (FER) would be set to 0. Now the data whose address is indicated by (MAR) would be fetched and added to the A-register. Note that the quantity fetched is treated as data since (FER)=0. The op code would now cause PC to be incremented by 3. The (FER) would be set to 1 and next the instruction in location 4 would be fetched and the process repeated.

We have thus far assumed that the instructions are executed in order. However, there are instructions called *branching instructions* that can vary the (PC). For instance, there is a branch-on-zero instruction. This is a 3-word instruction which affects the (PC) only if the value stored in the accumulator is 0. For instance, suppose that this instruction is in locatins 4, 5, and 6. The op code in memory location 4 is fetched and put into the IR and implemented. If (A)≠0 then, in essence, this instructions is ignored. The (PC) is set equal to 7 and the next instruction (that stored in memory location 7) will be executed. On the other hand, if (A)=0, then the branch-on-zero instruction is not "ignored." The (PC) will be set to the address specified by the second and third words of the branch-on-zero instruction. Suppose that memory locations 5 and 6, when combined as a 16-bit word, specify 401. Then, the value of the (PC) will be set to 401 and this will be the location of the next instruction to be fetched. After this instruction is executed, the (PC) will be incremented by 1 (or 2 or 3) as before so that the next instruction to be executed will be the one in location 402, etc. and thus the program will proceed along a new branch.

In Fig. 11-2 we have a set of leads labeled "microprocessor and system control signals." These are used for various purposes. One would be for the read/write memory control. Let us consider another control lead application. There are control leads that can initiate what is called an *interrupt routine*. Suppose that the microprocessor is monitoring a chemical process and that it uses a number of transducers that read temperature, pressure, and acidity. Suppose that the temperature and acidity are continuously monitored and that various readings cause appropriate actions to be taken. The heaters could be turned on or off, more acid could be added, etc. The pressure transducer only indicates an emergency situation. It causes no action to be taken unless the pressure becomes too high, in which case vents are opened and heaters are turned off. (We shall consider an actual program which takes such action in Sec. 11-6.) The program can be written so that the pressure transducer is not normally monitored. However, the microprocessor has an interrupt lead. If a 1 is placed on this lead, then the contents of the program counter are set to a new location, called the interrupt location. The operation now proceeds along a new branch. We say that the program has been interrupted and that an *interrupt routine* has been initiated. The program branches to this interrupt routine and carries out the appropriate steps. With some microprocessors, the location of the interrupt routine is fixed, for instance, when an interrupt occurs, the contents of the program counter are changed so that (PC)=1000. The user must write the appropriate programming starting at location 1000 to take care of the interrupt condition. The last instruction(s) in this routine cause the program to resume from where it was interrupted. We shall consider examples of this in Sec. 1106.

Let us see how the computer can resume its operation. Suppose that (PC)=50 when the interrupt is received. This value plus the contents of any of the registers that may be changed by the interrupt routine must be saved. Thus, all of the data is saved and the program can be resumed. All of this information must be stored in the main memory. There is a set of memory locations called a *stack* that is used for this purpose. The stack is just a set of continuous memory locations reserved by the programmer for this purpose. The stack is defined by the *stack pointer* register, SP. An initial value is put into SP [e.g. (SP)=2000]. Suppose that the interrupt routine changes the values of the A-, F-, H-, and L-registers. When the interrupt is received, the (PC) is stored in the memory location indicated by SP. Actually, PC is a 16-bit register so that two memory locations are used. To explain this, let us first define some notation. If M(X) is the memory location whose value is stored in the 8-bit register X, then the most significant bits of the (PC) are stored in M(SP). The value of the SP is then incremented (decremented) by 1 and the least significant bits of the (PC) are stored in M(SP). Note that the new (SP) is one more than the original (SP). Now (SP) is again incremented by 1. This is called *pushing* the value of (PC) onto the stack. Suppose that the interrupt routine changes the values of the A-, F-, H-, and L-registers. The interrupt routine must start with four instructions that push the contents of A, F, H, and L onto the stack. These values are now saved. The next instructions in the interrupt routine service the interrupt (e.g. open the vents, turn off the heaters). When the interrupt routine is complete, the final instructions of the interrupt routine restore the data from the stack. The following would be done: M(SP) is put in L, (SP) is decremented by 1 and an instruction to put M(SP) in H is executed, etc. Thus, L, H, F, and A are restored to their former values. This is called *popping* data off the stack. Finally, the original contents of PC is popped off the stack and put back into PC. (This requires two steps.) The program proceeds from the point where it was interrupted. Note that data is pushed on and popped off the stack in opposite order. We shall consider typical stack instructions in Sec. 11-6.

There are two types of interrupts, *maskable* and *nonmaskable*. A nonmaskable interrupt is as described. If a nonmaskable interrupt is received, the operation jumps to the interrupt routine and there is no way of stopping this from happening. On the other hand, maskable interrupts are controlled by the program. There is an instruction (op code) called an *interrupt disable* that will cause the microprocessor to ignore any maskable interrupt. That is, once this instruction is executed, the microprocessor ignores any maskable interrupt. There is another op code called an *interrupt enable* that causes the processor to react to a maskable interrupt. If, in a program, an interrupt disable op code is encountered, all maskable interrupts will be ignored until the interrupt enable op code is executed. Now the processor will respond to maskable interrupts. Note that any maskable interrupts given before the interrupt enable op code is executed will be lost. The micropro-cessor has different external leads for maskable and nonmaskable interrupts.

11.5 MICROPROCESSOR SIGNALS AND TIMING

We shall now discuss some of the external signals that are supplied to microprocessors or generated by them and see how this must be timed. Manufacturers furnish timing diagrams and data that supply such information.

One external signal that is supplied to the microprocessor is the master clock signal. Manufacturers specify the maximum clock fre-quence. If this maximum frequency is exceeded, then the micropro-cessor's circuits will not function properly. The master clock is often operated at a frequency that is considerably less than the maximum because the speed of the peripheral equipment is slower than that of the microprocessor, or simply because great speed is not necessary. Since the microprocessor is a synchronous device, manufacturers specify timing diagrams in terms of the master clock signals.

To illustrate a timing diagram, we shall consider the signals sent between the microprocessor and the memory when data is to be fetched from the memory. We assume here that, among the microprocessor's external control leads, there are the following two. First, there is a memory request lead, MREQ. This generates a MREQ signal that indicates to the memory that data is to be read or written. The second lead is the read/write lead, called READ. When MREQ is high (i.e. a 1 in positive logic), the memory is requested. When READ is high, the memory is to be read; when READ is low, data is to be written to the memory.

In Fig. 11-4 we illustrate a timing diagram for the fetch cycle. For simplicity, timing diagrams of this type are often drawn showing instantaneous rises and falls of signals but, of course, a finite time is required here. Note also that the actual clock, MREQ, and READ signals are actually indicated on the diagram, but only the presence or absence of address and data signals is indicated. The actual 8 data bus signals and 16 memory address bus signals are not shown.

The fetch cycle starts at the time marked 0 on the diagram and continues for four clock periods. Slightly after the first clock pulse, the MREQ line becomes high. This enables the memory circuits so that they can receive and transmit data. Somewhat before this, see Fig. 11-4, the address signals are put on the address bus. This timing sequence is used to allow the address signals to stabilize before the memory is enabled. Note that the actual data is not transmitted from the memory to the microprocessor until about one and one-eighth clock periods after the memory has been enabled and addressed. This is because memory circuits are relatively slow. This time is required for proper addressing.

Note that the fetch cycle continues for several clock cycles after the data has been transmitted. This provides time for putting the instruction in the instruction register, decoding it, and for the generation of the required control signals.

There is another external lead that can alter timing. It is called a $\overline{\text{WAIT}}$ lead. An external device such as memory or a terminal may not be ready to receive or transmit data. The external device must send a signal to the microprocessor to indicate whether or not it is ready. The $\overline{\text{WAIT}}$ lead is used for this purpose. If the $\overline{\text{WAIT}}$ line is high, then the external device is ready. The bar over $\overline{\text{WAIT}}$ indicates that an actual request to wait occurs when the lead is caused to go low by the external device. For instance, in the previous example, when the microprocessor caused MREQ to become high, if the memory is not ready, it would cause $\overline{\text{WAIT}}$ to go low. Now all signals to the microprocessor remain

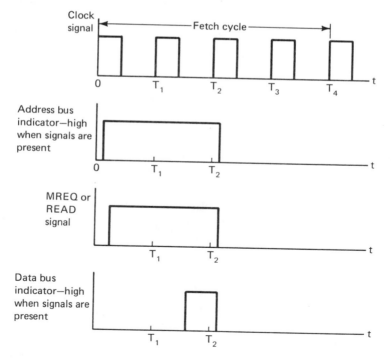

Figure 11-4 Timing diagrams for an instruction-fetch cycle

unchanged until the $\overline{\text{WAIT}}$ signal becomes high. Suppose that this takes four clock cycles. Then, the fetch cycle would last for an additional four cycles.

$\overline{\text{WAIT}}$ signals are most often used by slow peripherals such as printers and terminals. For instance, there is a 1-bit register associated with the printer. If it is ready to receive data, then this signal is low; if the printer is not ready to receive data, then this signal is high. This register controls the $\overline{\text{WAIT}}$ signal if the microprocessor indicates to the printer that its services are requested. In general, the computer is very much faster than a printer. Thus, when data is to be printed, the computer would have to wait between every letter. To avoid slowing down the computation, many printers have small memories called *buffers* which store the data to be printed. Thus, data can be input to the buffer at a rapid rate. Data is stored in the buffer until the printer prints it. This speeds the operation. If the buffer becomes full, then a $\overline{\text{WAIT}}$ signal is generated. The use of a buffer allows the printer to print while the computer is performing other operations.

We have not considered all possible signals here, but have discussed representative ones. The important interrupt signals were considered in the last section. Note that the words high and low as used in this discussion may vary from microprocessor to microprocessor. For instance, with one microprocessor, a line may have to be made high to achieve a result, while with another microprocessor, the line may have to be made low.

11.6 PROGRAMMING THE MICROPROCESSOR

In this section we shall present some of the basic ideas of microprocessor programming. It is not our purpose to teach programming but to discuss the fundamental ideas of microprocessor programming. The microprocessor is directed by a sequence of instructions called op codes, see Sec. 11.4. This collection of instruction is called a *machine language program*. In this section we shall discuss the op codes needed to write some simple programs and then develop some representative programs.

An op code is simply a binary number. For instance, 1011 0011 could be the op code that causes the contents of the B-register to be added to the A-register. We assume here that we are working with the 8-bit microprocessor of Fig. 11-2. It is difficult for people to remember op codes, so that mnemonics are usually used to represent them when programs are written. For instance, it is much easier to remember that

ADDA, B means "add the contents of the B-register to the accumulator" than to remember the meaning of 1011 0011. When the actual program is entered into the memory, the binary code must, of course, be used. However, when the machine language program is written, it is simpler to use mnemonics, so that the programmer can more easily understand and check the program.

We shall now consider some typical microprocessor op codes. Again, rather than writing binary numbers that are difficult to remember, we shall use mnemonics. However, they stand for numerical op codes.

The first set of op codes that we shall consider have to do with loading a number into a specified location. We start with the 3-word instruction

LOAD,M   (11-1a)
0010 1101   (11-1b)
0000 1111   (11-1c)

This causes the 8-bit contents of the accumulator (i.e. the A-register) to be loaded into the memory location specified by the second and third words. For instance, (11-1) causes the contents of the accumulator to be loaded into memory location 0010 1101 0000 1111 = $2B0F_{16}$. Note that it is conventional to use hexadecimal notation here. Of course, the memory locations given in (11-1b) and (11-1c) are just typical. Any valid memory location could be used.

The 3-word instruction

LOAD,A   (11-2a)
1101 0101   (11-2b)
0111 0001   (11-2c)

causes the contents of memory location 1101 0101 0111 0001 to be loaded into the accumulator. Note that this is not addition. Any previous data stored in the accumulator is lost.

There are instructions that allow the movement of data from one register to another. There is a 1-word instruction

LOAD,a,b (11-3)

where a and b are either A, B, C, D, E, H, or L. This causes the contents of register b to be put into register a. For instance

LOAD,B,A   (11-4)

causes the contents of the B-register to become the same as that of the A-register.

The last load command that we shall consider consists of two words and is called an *immediate load*. The specified number is loaded into the specified register. Its form is

LOADI,*a*   (11-5a)
1001 1001   (11-5b)

where *a* stands for A, B, C, D, E, H, or L. For instance,

LOADI,B   (11-6a)
1111 0001   (11-6b)

causes the number $1111\ 0111 = F1_{16}$ to be stored in the B-register. The previous contents of the B-register is lost.

Now we shall consider some arithmetic and logical instructions. We shall start with the following 1-word op code.

ADDA,*b*   (11-7)

Here *b* stands for B, C, D, E, H, or L. For instance

ADDA,H   (11-8)

causes the contents of the H-register to be added to the contents of the accumulator (A-register). The old contents of the A-register is lost.

The following are some op codes that perform logical operations. The op code

ANDA,*b*   (11-9)

where *b* stands for B, C, D, E, H, or L, performs the logical AND, for instance

ANDA,D   (11-10)

causes the contents of the A- and D-registers to be compared on a bit-by-bit basis using the AND operation. The result is stored in the A-register. For instance if, prior to the execution of (11-10)

$$(A) = 1001\ 1100 \qquad (11\text{-}11a)$$
$$(D) = 1001\ 0110 \qquad (11\text{-}11b)$$

then after execution we would have

$$(A) = 1001\ 0100 \qquad (11\text{-}11c)$$

and the contents of the D-register will be unchanged.

Two similar instructions involve the OR and XOR operations.

$$ORA,b \qquad (11\text{-}12)$$
$$XORA,b \qquad (11\text{-}13)$$

The following command causes a register to be incremented by 1. That is, 0000 0001 is added to the value stored in the indicated register. The old value is lost.

$$INC,a \qquad (11\text{-}14)$$

Here a stands for A, B, C, D, E, H, or L. Similarly,

$$DEC,a \qquad (11\text{-}15)$$

causes the contents of a to be reduced by 1.

The contents of the A-register is replaced by its complement, on a bit-by-bit basis, using the following instruction:

$$COMP \qquad (11\text{-}16)$$

If $(A) = 1001\ 1100$ before the execution of (11-4) then, after its execution, $(A) = 0110\ 0011$.

We shall work with two branching instructions. The first is called an *unconditional branch* and consists of the 3-word instruction:

$$JUMP \qquad (11\text{-}17a)$$
$$0011\ 0010 \qquad (11\text{-}17b)$$
$$0111\ 1111 \qquad (11\text{-}17c)$$

When this instruction is executed, the contents of the PC register is set to the memory location specified in the second and third words. In this case, after (11-17) is executed, the instruction in memory location 0011 0010 0111 1111 = $327F_{16}$ will be fetched and executed.

A conditional branching instruction is like an unconditional branch except that branching only occurs if some specified condition is met. Usually, this condition is specified by a bit of the flag register. We shall use a branch on zero.

JUMP,Z	(11-18a)
0000 1111	(11-18b)
1011 0101	(11-18c)

Suppose that these three words are stored in memory locations 0006_{16}, 0007_{16}, and 0008_{16}. If (A)=0 (i.e. the contents of the accumulator is 0) then (PC) will be set to 0000 1111 1011 0101 = $0FB5_{16}$ and the program will branch there. On the other hand, if (A)$\neq$0 then this instruction will be ignored. The next instruction fetched will be in memory location 0009_{16}.

There are various rotational commands that can be used to implement such things as multiplication and division, see Sec. 10.3. For instance, the instructions

RLA	(11-19)

and

RRA	(11-20)

rotate the A-register to the left and to the right, respectively. This is diagrammatically illustrated in Fig. 11-5a for the RRA.

Another set of rotational commands include the carry bit in the rotation.

RL,*a*	(11-21)
RR,*a*	(11-22)

Here *a* stands for A, B, C, D, E, or L. These commands are illustrated in Figs. 11-5b and c. Note that by working with instructions such as (11-21) and (11-22) we can shift bits up and back between registers and implement the operations discussed in Sec. 10.3. There are other operations that involve shifting without rotation, for instance

SRA	(11-23)
SLA	(11-24)

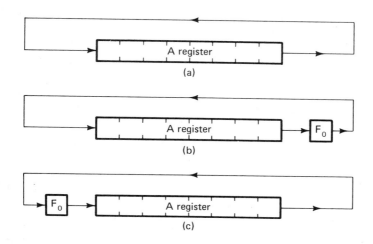

Figure 11-5 (a) Illustration of the RRA instruction; (b) illustration of the RR,a instruction; (c) illustration of the RL,a instruction

These shift the contents of the accumulator to the right or left. Typically there are other instructions that include the carry bit in such shifting operations.

The following are instructions that will be used in conjunction with the stack, see Sec. 11.5. The first instruction is

PUSH,*a*   (11-25)

Here *a* is either A, B, C, D, E, H, L, or PC. This pushes the contents of the specified register on the stack and increments (SP) by 1 unless PC is the register that is pushed. Since PC is a 16-bit number, it occupies two memory locations and, in this case, the stack pointer is incremented by 2. The instruction

POP,*a*   (11-26)

is the converse of PUSH. The value stored in the memory location pointed at by the current stack pointer location is put into the register designated by *a* and the stack pointer is decremented by 1, except if *a* represents the PC, in which case, the stack pointer is decremented by 2.

LOSPHL   (11-27)

puts the contents of the pair of registers HL into the SP register. This instruction is used by the programmer to establish the intial value of the stack pointer.

The instruction

RES   (11-28)

causes the program to resume at the value specified by (PC). We shall assume that, when an interrupt is received, PUSH,SP is automatically implemented so that the old (SP) is saved. Then the program branches to the interrupt routine.

A final instruction is

STOP   (11-29)

This stops computation.

Now let us consider some programs that illustrate the use of these instructions. We shall start by discussing a control application where the microprocessor is used to control a chemical process. Suppose that we want to keep the temperature of the process constant. A sealed vat containing the chemical is heated by an electric heater as shown in Fig. 11-6. A temperature transducer monitors the temperature. This transducer provides an electrical signal that is a function of temperature. If it is too low, the heater is to be turned on; if the temperature is too high, then the heater is to be turned off. If the pressure becomes too high, then the pressure transducer causes an interrupt, the safety vent is opened and the heater is turned off. The system remains in this condition until the high-pressure condition has passed, at which point normal operation proceeds. Note that, for illustrative purposes, we have oversimplified the process. For instance, we have not included controls to keep the level of the chemicals constant, etc.

Now let us consider the program. We shall assume here that the transducer can be read by simply observing memory locations. For instance, memory location 1111 0001 0000 0000 = $F100_{16}$ will represent the heater transducer. If 0000 0000 is in this memory location, then the temperature is too low, while if 1111 1111 is in this location then the temperature is too high.

In a similar way we shall assume that 1111 0010 0000 0000 = $F200_{16}$ is the pressure transducer location. If its contents are 0000 0000 the pressure is satisfactory. If the contents are 1111 1111, then the pressure is too high. We also assume that the pressure transducer has a second set of control leads. These close momentarily when the pressure becomes

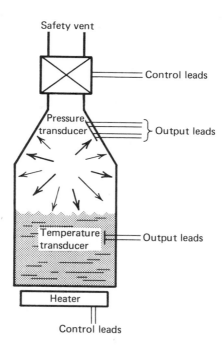

Figure 11-6 A single chemical process to be controlled

too high. Then they open and do not close again unlesss the pressure drops to normal and then they reset. These contacts activate the nonmaskable interrupt.

We shall also assume that the numbers stored in certain memory locations control the operation of heaters and the safety vent. For instance, memory location 1111 1001 0000 0000 = $F900_{16}$ will be the memory location for the heater control. If its contents is 0000 0000 then the heater will be off, while if its contents is 1111 1111, then the heater will be on. Similarly, 1111 1010 0000 0000 = $FA00_{16}$ is the memory location for the safety vent control. If its contents is 0000 0000 then the safety vent will be open, while if its contents is 1111 1111 then the safety vent will be closed.

It may seem unrealistic to treat memory locations as transducer outputs or heater or vent controls. However, in the next section we shall demonstrate that this can easily be done. A program that carries out the desired process control is: (11-30)

Memory Location (hexadecimal)	Instruction	Comments
0	LOADI,H ⎫	
1	0000 0000 ⎪	set initial
2	LOADI,L ⎬	(SP)=0096_{16}
3	1001 0110 ⎭	
4	LOSPHL	
5	LOAD,A	
6	1111 0001 ⎫	temperature transducer
7	0000 0000 ⎭	memory location
8	JUMP,Z	
9	0000 0000 ⎫	0013_{16}
A	0001 0011 ⎭	
B	LOADI,A	
C	0000 0000 ⎬	code to turn heater off
D	LOAD,M	
E	1111 1001 ⎫	heater control
F	0000 0000 ⎭	memory location
10	JUMP	
11	0000 0000 ⎫	jump back
12	0000 0101 ⎭	to 5
13	LOADI,A	
14	1111 1111 ⎬	code to turn heater on
15	LOAD,M	
16	1111 1001 ⎫	heater control
17	0000 0000 ⎭	memory location
18	JUMP	
19	0000 0000 ⎫	jump back
1A	0000 0101 ⎭	to 5
150	PUSH,A	interrupt routine starts, save A
151	LOAD,A	
152	1111 0010 ⎫	pressure transducer
153	0000 0000 ⎭	memory location
154	JUMP,Z	
155	0000 0001 ⎫	0162_{16}
156	0101 0010 ⎭	
157	LOAD,A	
158	0000 0000 ⎬	code to open safety vent
159	LOAD,M	

Memory Location (hexadecimal)	Instruction	Comments
15A	1111 1010 ⎱	safety vent control
15B	0000 0000 ⎰	memory location
15C	LOAD,M	
15D	1111 1001 ⎱	heater control
15E	0000 0000 ⎰	memory location
15F	JUMP	
160	0000 0001 ⎱	
161	0101 0001 ⎰	0151_{16}
162	LOAD,A	
163	1111 1111 ⎱	code to close safety vent
164	LOAD,M	
165	1111 1010 ⎱	safety vent control
166	0000 0000 ⎰	memory location
167	POP,A	
168	POP,PC	
169	RET	

Now let us discuss the program. We have used the following memory locations:

$$1111\ 1001\ 0000\ 0000 = F900_{16} \quad \text{heater control}$$
$$1111\ 1010\ 0000\ 0000 = FA00_{16} \quad \text{safety vent control}$$
$$1111\ 0001\ 0000\ 0000 = F100_{16} \quad \text{temperature transducer}$$
$$1111\ 0010\ 0000\ 0000 = F200_{16} \quad \text{pressure transducer}$$

The following control codes are used:

0000 0000	open safety vent or turn heater off
1111 1111	close safety vent or turn heater on

Transducer outputs are:

0000 0000	temperature or pressure low
1111 1111	temperature or pressure high

Now let us explain the program. In memory locations 0 to 5, the stack pointer's initial location is established as 0000 0000 1001 0110 = 0096_{16}. The control portion of the program commences with location 5. The A-register is loaded with the temperature transducer location. The next instruction (in memory location 8) causes a branch if the number now stored in the A-register is 0 (temperature too low). Assume that the number contained in A is 1111 1111 (temperature too high). Then the branching does not occur. A is then loaded with 0000 0000. Then, in locations D to F, this value is loaded into memory location 1111 1001 0000 0000 which is the heater control location. This results in the heater's being turned off. Memory locations 10 to 12 cause control to jump back to memory location 5 and the process is repeated.

If we are executing the instruction in memory location 8 and the A-register contains 0000 0000, then we jump to location 13_{16}. Now A is loaded with 1111 1111 which is the code to turn the heater on. This code is next put into the heater control location 1111 1001 0000 0000. This causes the heater to be turned on. After this we jump back to the instruction in memory location 5. The process keeps repeating itself, turning the heater on or off as required.

Now suppose that an interrupt is transmitted by the pressure transducer. The program jumps to the interrupt location, which we assume is 150_{16}. Remember that this instruction causes the (PC) to be saved and (SP) to be incremented by two. We also save A by pushing it onto the stack. Actually this is not necessary (or even desirable) for this particular control process. However, we do it for illustrative purposes. Now the interrupt routine takes care of the excess pressure. The instruction in memory locations 151-153 loads A with the contents of the pressure transducer memory location. The next instruction is a branch on zero. If the pressure is too high, then this branch command will be ignored and the instruction in locations 157 to 15E open the safety vent and turn off the heater. Control then jumps back to memory location 151_{16} and the process repeats until the pressure has returned to normal. Now the branch on zero in locations 154 to 156 causes control to jump to memory location 162. Now A is loaded with 0000 0000. The next instruction in memory locations 164-166 causes the safety vent to be closed. The need for the interrupt routine no longer exists. The old (A) is popped off the stack and put back in the A-register. Similarly, the old (PC) is popped off the stack and put in the PC. Normal operation of the program resumes.

Now let us consider a second example. Let us compute

$$\sum_{k=1}^{100} k \qquad (11\text{-}31)$$

Again, we shall not output the value but simply store it in a memory location. In the next section we shall discuss how such values can be output. The program is: $\qquad (11\text{-}32)$

Memory Location (hexadecimal)	Instruction		Comments
0	LOADI,A		
1	0110 0110	100_{10}	
2	LOADI,B		
3	0110 0100	100_{10}	
4	DEC,A		
5	LOAD,C,A		
6	ADDA,B		
7	LOAD,B,A		
8	LOAD,A,C		
9	JUMP,Z		
A	0000 0000 ⎫		
B	0000 1111 ⎬ $000F_{16}$		
C	JUMP		
D	0000 0000 ⎫		
E	0000 0100 ⎬ 0004_{16}		
F	LOAD,A,B		
10	LOAD,M		
11	0000 0000 ⎫		
12	0100 0000 ⎬ 40_{16}		
13	STOP		

Let us consider this program. We start by loading $0110\ 0100 = 100_{10}$ into A and B. Now A is decreased by 1. This new value of A is stored in C. Now the value of (B) is added to (A) and the result stored in A. $(A) = 100_{10} + 99_{10}$. The new value of A (199_{10}) is now put into B. Now the (decremented) value of 99, which has been stored in C, is put back into A. If the value stored in A is not zero, then control branches back to statement 4_{16} and the process is repeated. Now 98 is added to the sum stored in B. The process repeats until the decremented value of A becomes zero. Now the program branches to memory location $000F_{16}$. The final value of B is put into A. Now this value is put into memory location 40_{16}. This is the final answer and execution ceases.

11.7 INTERFACING—INPUT/OUTPUT

In the previous examples of machine language programs, we assumed that data could be input and output by writing to and reading from memory locations. We shall now discuss several procedures for the actual input and output of data.

Memory-Mapped Input/Output

We shall now consider how input and output devices can be treated as part of memory. We shall illustrate the procedure with the input/output for the chemical process control program (11-30). In that program, four words of memory were used, two to input data from the temperature and pressure transducers and two for output instructions to the heater and vent controls.

Let us assume that we construct a 4-word memory that is external to the main memory. Both the main memory and this external memory receive their data from the 8-bit data bus. When either circuit is not supplying data to the data bus, outputs must be isolated from the bus. If this were not the case, then the unused output circuits would "load down" the bus and reduce any signals on the data bus that originated elsewhere. The main memory and our external memory are fabricated with what is called a *tristate output*. Not only do they supply 0's and 1's to the data bus, but their output circuits are such that they can be effectively isolated from the data bus. This isolation is controlled by a tristate-enable lead. When there is a 1 on this lead, the outputs are connected to the bus. When there is a 0 on this lead, then the outputs are isolated from the bus.

For our program, we want to enable the main memory tristate outputs when memory is to be read and to enable the external memory tristate output when input/output is to be performed. Let us see how this can be accomplished. The MREQ signal from the microprocessor indicates whether the main memory or the external memory is required.

We have written the program so that the external memory locations all have 1's as their most significant bits while all the main memory locations that we have used have 0's as their most significant bits. Now consider the circuit of Fig. 11-7. If, when the MREQ signal is given, the most significant bit of the memory address is a 0, then the main memory will be enabled and the external memory will not be enabled. On the other hand, if the MREQ signal is given when the most significant bit of the address bus is a 1, then the main memory tristate output will not be enabled while the external memory tristate output will be enabled. Thus the desired result has been achieved. We have, thus far, simplified the discussion. Actually, neither tristate output should be enabled unless a READ signal is given.

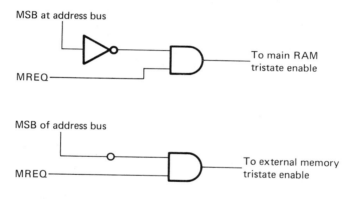

Figure 11-7 A procedure for enabling one of two memories

Now let us consider the transducers. memory locations. The transducer produces analog signals that must be converted to digital information using an analog-to-digital converter. Actually, for the simple on/off operation of this program, the A/D conversion could be accomplished by a temperature- or pressure-controlled switch and only one bit need be used for control. However, for generality, let us assume that the transducers' analog-to-digital converters supply all eight bits. Thus an 8-bit external memory word must be set up for each transducer. In Fig. 11-8 we illustrate one bit of such a transducer's memory. If the transducer's memory location is not addressed AND if MREQ AND READ are not high, then the decoder is such that its output is 0. In this case clock pulses are applied to the flip-flop and its state will constantly be updated by the transducer's analog-to-digital converter. In addition, the tristate output will be disabled.

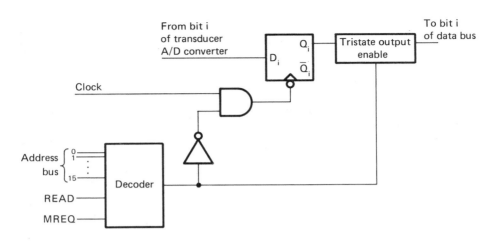

Figure 11-8 One bit of the transducer memory

Now consider that this memory location is addressed AND MREQ AND READ are high. Now the decoder output becomes 1. The clock pulses are removed from the flip-flop so that its state does not change. The tristate output is enabled. Hence, the appropriate data is put on the data bus. Remember that there is a flip-flop and tristate output for each data bus lead. However there is only one decoder. For program (11-30) we need two such memories (e.g. there must be two different decoders).

Now let us discuss the memory locations for the safety and pressure vent valve controls. The data, which is output to the external memory, and controls a heater (or vent) is stored in a local register composed of eight flip-flops. The output of this register is decoded and it controls the heater (or vent). The circuit for one bit of the control memory is shown in Fig. 11-9a. The only time that the decoder output is a 1 is when the correct address is put on the address bus, AND MREQ is 1 AND READ is 0. Now the flip-flop can change its state. The microprocessor puts the data on the appropriate data lines and this data is stored in the external flip-flop memory just as it would be in main memory. The output of these flip-flops is connected to another decoder called the output decoder. This outputs a 1 if $Q_0, Q_1, \ldots, Q_7$ are all 1, and a 0 otherwise. This signal is applied to the base of the transistor, see Fig. 11-9b. If the decoder output is a 0 then the transistor is turned off, while if the decoder output is a 1 the transistor is turned on. Thus, v_0 can be varied between a low and a high value. This voltage v_0 is then used to control the appropriate relays that turn the heater on or off or

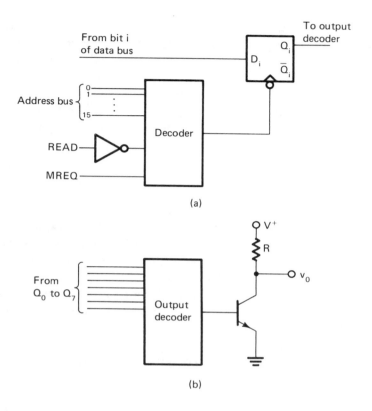

Figure 11-9 (a) one bit of the control memory; (b) the output decoder and control transistor

open or close the vent. For program (11-3) there must be two circuits of the type shown in Fig. 11-9. The decoders are different for each circuit.

We have made the input and output circuitry more complicated than necessary for program (11-30). There, only one bit is needed for either input or output. However, we have used all eight bits for illustrative purposes.

In this example we have assumed that the input and output are used in control applications. However, very similar ideas would be used if numerical information were input and output. For instance, in Fig. 11-8, the data might come, not from an analog-to-digital converter, but from a keyboard. In this case, the binary information that was supplied would represent a number that was typed in. We have oversimplified here in that the keyboard might not directly supply the eight bits of data in parallel form. Serial-to-parallel conversion might have to 1 performed. In addition, we assume that the keyboard has appropriate circuitry to convert signals from the switches on each key to the appropriate binary number.

If the output were to a printer, then the flip-flop of Fig. 11-9a could be a part of the printer's buffer register. Some printers require that the data be supplied in serial form and we would then require parallel-to-serial conversion.

Port Input/Output

The memory-mapped input/output that we have discussed has one disadvantage. Memory locations which are used for input/output can no longer be used for storage of programs or data. In many applications this is not a problem. However, if the program uses many memory locations, then it is undesirable to use up memory locations for input/ output. This is especially true if the microprocessor is used in a general-purpose computer. There are microprocessor instructions that provide the simplicity of memory-mapped input/output but which do not take up memory locations.

We shall illustrate this with two typical instructions. The first is the 2-word instruction:

OUTA   (11-33a)
1001 1000   (11-33b)

In this case, the following would take place. 1001 1000 would be placed on the least significant bits of the address bus, the READ signal would go low and the contents of A would be put on the data bus. This is just like a LOAD,M instruction with one important difference. The MREQ *signal is not made high.* Thus, the main memory is not affected by this instruction. However, this instruction can be used for output. For instance, the circuit of Fig. 11-9a can be used for this output if we simply invert the MREQ signal before it is input to the decoder. That is, in Fig. 11-9a, a NOT gate would be put in cascade with the MREQ lead. (Alternatively, this NOT gate could be omitted and the decoder suitably modified.) In addition, the decoder would be modified so that it only used the eight least significant bits of the address bus. The circuit functions in essentially the same way as the memory-mapped input/ output case. However, we have not affected any other main memory addresses. This is called *port input/output.* The number addressed by the least significant bits of the address bus is called the port number. We can address $2^8 = 256$ different ports with this procedure. Note that, in (11-33b), 1001 1000 is simply a typical number. Any 8-bit binary number can be used here.

Port input is accomplished in a similar way. There is the 2-word instruction

INA   (11-34a)
1111 0110   (11-34b)

In this case 1111 0110 is placed on the least significant bits of the memory address bus, the READ signal will be high. At the appropriate time, data will be read from the data bus and put in the A-register. Again MREQ is not made high so that the main memory is not affected. The circuit of Fig. 11-8 can now be used for data input with two modifications. The MREQ signal must be inverted (or the decoder suitably modified) and the decoder only inputs and decodes the eight least significant bits of the address bus.

11.8 ADDITIONAL ADDRESSING TECHNIQUES

There are other addressing techniques that are used at times. These provide additional versatility for the programmer.

Indexed Addressing

We often perform operations on several lists of data. Suppose that one list is stored in memory locations 500-600. A related list may be stored in memory locations 800-900. Now consider that we want to add the number stored in memory location 500 to the number stored in memory location 800. We also want to add the numbers stored in 501 and 801, etc. Indexed addressing provides a simple means of doing this. A number called a *displacement* can be specified; in this case the displacement would be 300. When the inedexed addressing mode is used, the address is incremented by the displacement. Thus, when items in two different memory lists are specified, we need only specify one memory location and a displacement. Even if the memory location has to be computed, the displacement is always constant. Thus, it becomes easier to specify the second address.

Indirect Addressing

The addressing modes that we have considered are called *direct addressing*. Here, a memory location is specified and that memory location contained an operand (e.g. a number that was added to the A-register). In indirect addressing, the memory location(s) contains, not an operand, but another address. The operand is stored in the second address. This provides the programmer with additional flexibility to perform more sophisticated operations.

11.9 ASSEMBLY LANGUAGE AND HIGHER-LEVEL LANGUAGES

Programming using machine language is a very tedious procedure. The programmer must remember numerical codes which correspond to each instruction. The job of remembering them can be very difficult. Another difficulty with machine languages is that the programmer must remember the numerical values of the storage locations of the variables. For instance, suppose that we want to obtain

$$b = x + y + z \qquad\qquad (11\text{-}35)$$

We must choose and remember the memory locations for x, y, z, and b. This is not much of a problem in a simple program. However, if we have many variables this becomes very tedious. Note that the memory locations specify the variable. For instance if, later in the program, we want to add the numerical value of b to the A-register, we would have to remember the memory location where b was stored.

To eliminate some of these difficulties, other languages have been developed. We shall start by considering *assembly languages*. Note that the objective here is not to teach programming but to discuss what these languages are.

When computation is performed, a machine language program always directs it. Thus, if another form of language is used, it must be translated into machine language. This translation can be done by the computer. When we work with an assembly language, a computer program called an *assembler* is written (in machine language). The assembler directs the computer, which translates the assembly language program into a machine language program which is then actually run by the computer.

In an assembly language, the numerical commands of the machine language are replaced by *mnemonics*. That is, instead of writing binary code, we simply write the mnemonic. The second major difference between assembly languages and machine languages is that, in assembly languages, variables are referred to by name instead of by memory location. Usually, a variable can consist of up to six alphanumeric characters (letters and numbers).

For instance, if we are writing a program in machine language, the instructions would be written in actual binary codes. On the other hand, if it were written in assembly language, then mnemonics would be used for the instructions. In addition, the memory locations would be referred to by name rather than by number. Assemblers are not standard and the actual form of the assembly language program might not exactly correspond to the machine language program. For instance, a 3-byte machine language instruction might be represented by a mnemonic and name which are both written on the same line.

Higher-Level Languages

Assembly language is considerably easier to use than machine language. However, it is still very tedious to use. For instance, suppose we wanted to execute the simple equation

$$b = (x+y)(z+x)/(y-z) \tag{11-36}$$

Very many machine or assembly language statements would have to be written to execute this. However, there are higher-level languages where this is not the case. In one such language, FORTRAN, a program which would evaluate this for $x=3$, $y=4$, and $z=5$ is

X=3	(11-37a)
Y=4	(11-37b)
Z=5	(11-37c)
B=(X+Y)*(Z+X)/(Y–Z)	(11-37d)
STOP	(11-37e)
END	(11-37f)

Note that, in theory at least, the greatest amount of versatility is obtainable in a machine language program since all other forms of programs must be translatable into machine language. However, most of the programming is done in higher-level languages because they are much easier to use and fewer mistakes are made by the programmers.

BIBLIOGRAPHY

Booth, T. L., *Digital Networks and Computer Systems*, Second Ed., Chaps. 10-15, John Wiley and Sons, Inc., New York 1978

Hill, F.J. and Peterson, G.R., *Digital Systems: Hardware Organization and Design*, Chap. 6, John Wiley and Sons, Inc., New York 1973

Mano, M.M., *Digital Logic and Computer Design*, Chaps. 9-12, Prentice Hall, Inc., Englewood Cliffs, NJ 1979

O'Malley, J., *Introduction to the Digital Computer*, Chaps. 10-12, Holt, Rinehart and Winston, Inc., New York 1972

Rhyne, V.T., *Fundamentals of Digital System Design*, Chap. 7, Prentice Hall, Inc., Englewood Cliffs, NJ 1973

Souček, B., *Minicomputers in Data Processing and Simulation*, Chap. 5, John Wiley and Sons, Inc., New York 1972

PROBLEMS

11-1. Discuss the basic components of the digital computer.

11-2. Describe the function of the program counter.

11-3. Describe the function of the instruction register.

11-4. Describe the data bus.

11-5. Describe the address bus.

11-6. Describe the function of the stack pointer.

11-7. Describe pushing and popping.

11-8. What is an interrupt? Describe its use.

11-9. Describe the operation of the external control signals.

11-10. Describe 1-word microprocessor instructions.

11-11. Describe 2-word microprocessor instructions.

11-12. Describe 3-word microprocessor instructions.

11-13. Draw a timing diagram that describes the operation of the microprocessor when it is executing a memory READ.

11-14. Describe the operation of program (11-30).

11-15. Describe the operation of program (11-31).

11-16. Write a program for a microprocessor that controls a chemical process. The temperature of a solution must be kept within specified limits. The level of the solution must also be held within specified limits (the level can be sensed by sensing pressure).

11-17. Assume that N_1, N_2, and N_3 are three numbers which are stored in $M(101)$, $M(102)$ and $M(103)$, respectively. Write a machine-language program which performs the following: It adds N_1 and N_2. If the sum is greater than 25, then N_3 is added. If the sum equals 25, then the operation stops. If the sum is less than 25, then $2N_3$ is added. Store the following values:

$N_1 = 25$
$N_2 = 50$
$N_3 = 150$

11-18. Write a machine-language program which adds $\displaystyle\sum_{k=51}^{250} 2k$.

11-19. Write a machine-language program which performs $\displaystyle\sum_{k=1}^{100} (-1)^k k$.

11-20. Draw the interface circuits for the program of Prob. 11-16. Use memory-mapped input/output.

11-21. Repeat Prob. 11-20 for port input/output.

11-22. Discuss the differences between machine language, assembly language, and higher-level languages.

11-23. What is the difference between an assembler and a compiler?

11-24. What is the most versatile language in which to program? What is the most convenient?

Appendix

Logic Families

We shall now discuss the implementation of logic gates using semiconductor circuits. Since virtually all computer circuits make use of integrated circuits, we shall discuss such circuits here. We start our discussion by briefly considering semiconductor devices and their switching. Those readers who desire a more thorough (or more basic) discussion should consult the references cited at the end of this appendix.

A-1. SEMICONDUCTOR DEVICES

In this section we shall briefly consider the various semiconductor devices that are used in switching circuits. We shall not consider the physics of these devices here but shall just consider them as circuit elements.

The p-n Junction Diode

This device is one which essentially conducts current in one direction but not in the other. A symbol for the diode and a typical voltage-current characteristic is shown in Figs. A-1a and b, respectively.

Let us consider Fig. A-1b. When v_b is positive, the diode conducts. This polarity of v_b is termed *forward bias*, and the diode is said to be conducting in the *forward* direction. When v_b is negative, we have *reverse bias*, and the current i_b is essentially zero. If a sufficiently large reverse bias voltage is applied, substantial current will result and this is termed *reverse breakdown*. Diodes are usually operated so that the reverse breakdown region is avoided. However, certain diodes, called *Zener diodes*, make use of reverse breakdown.

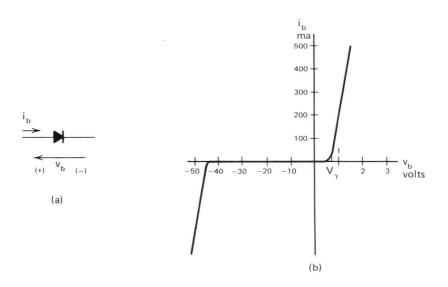

(a)

(b)

Figure A-1   The p-n junction diode. (a) its symbol; (b) a typical voltage-current characteristic

When the diode is forward biased, its current is small until the voltage V_γ is reached, see Fig. A-1b. This is called the *turn-on voltage*. We often approximate the diode's characteristics by considering that its current is zero if v_b is less than V_γ and that the diode conducts if v_b is greater than V_γ.

In switching circuits, the polarity of v_b is often switched from positive to negative (or vice versa). This represents switching from a 0 to a 1 or vice versa. If v_b is switched from a positive voltage to a negative voltage, we would expect that i_b would switch from a relatively large current to essentially zero. This is what does happen, but it does not occur instantaneously. The time that it takes to switch the diode is called its *switching time*. This is an important quantity since the speed at which a device switches determines the speed of its gate's response. Since a digital computation involves the switching of very many gates in succession, the switching speed of each device can have a profound effect on the time required by the computer to perform an operation.

In general, the time required to switch a diode off (i.e. v_b switches from positive to negative voltage) is greater than that required to switch it on. When the diode conducts in the forward direction, it is said to be turned on. The more heavily it conducts, the longer the switching time required to turn it off. When a diode is conducting, excess charge is said to be stored. This excess stored charge must be removed before it can be turned off.

The Junction Transistor

The element which is most commonly used in switching circuits is the junction transistor. In contrast with the diode, a small current can be used to control (i.e. to switch) a much larger current when the transistor is used. This is a very desirable property which can be used to advantage in logic circuits.

Transistors are either p-n-p or n-p-n type. Their respective symbols and terminology are shown in Fig. A-2. To describe the operation of the transistor, let us consider the simple n-p-n transistor circuit of Fig. A-3. The power supply voltage V_{CC} causes v_{CE} to be positive (i.e. the collector is positive with respect to the emitter). Under these conditions, if v_{BE} is positive as well as greater than a fraction of a volt, the collector current i_c will be large and essentially limited by the external resistance, R. In this case, the voltage v_{CE} will be small (a fraction of a volt), and the transistor is said to be *saturated*. The minimal value of v_{BE} that produces current is called the *base-emitter turn-on voltage; $v_{BE\gamma}$*. A typical value is 0.6 volt.

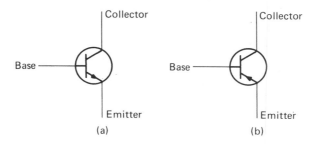

Figure A-2 Symbols for the junction transistor. (a) n-p-n; (b) p-n-p

If v_{BE} is less than $V_{BE\gamma}$, the current i_C will essentially be zero. In this case, the transistor is said to be *cut off*. We can often approximate a transistor's operation by assuming that it is cut off if v_{BE} is less than the base-emitter turn-on voltage and that it is saturated if v_{BE} is greater than the base-emitter turn-on voltage. (This assumes that V_{CC} is sufficiently large.)

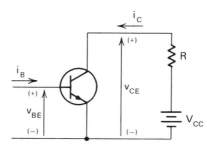

Figure A-3   A simple transistor circuit

We have considered the n-p-n transistor. The previous discussion would also apply to the p-n-p transistor if all potentials were reversed.

In switching circuits, a quantity that is often of use is the average ratio of the collector current to the base current. This is given by

$$\beta_F = i_C/i_B \tag{A-1}$$

Typical values of β_F are 100 or more.

If, in Fig. A-3, v_{BE} were switched from a saturation to a cut off value, then i_C would fall from its saturation value to essentially zero. However, this does not occur in zero time. In general, the greater the saturation value of i_C, the greater will be the time required to switch the current off (i.e. the greater will be the switching time). When specific logic circuits are considered, a compromise must often be made. For instance, increasing the saturation value of i_C increases the reliability of the circuit, which is an advantage. However, increasing the saturation value also increases the switching time, which is a disadvantage.

When the transistor is switched from cut off to saturation, there is also a finite switching time involved. However, in this case, the switching is faster and the saturation effects do not slow it down.

The Insulated Gate FET–IGFET–MOSFET

Another semiconductor device which finds extensive use in switching circuits is the insulated gate FET (IGFET). It is also called the metal-oxide-silicon FET (MOSFET). These devices can be switched on or off by the application of extremely small currents. These control currents are, in general, very much less than the base currents of junction transistors.

MOSFETs are either n-channel or p-channel types. Their symbols are shown in Fig. A-4. The MOSFET acts as a controlled resistance between the drain and source. A signal applied to the gate is capacitively coupled to the semiconductor material between the drain and source. The electric field due to this signal varies the resistance. There is only capacitive coupling between the gate and the bulk of the device, and the gate current is extremely small. MOSFETs used for digital computer applications are usually of a type called enhancement MOSFETs.

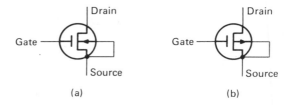

Figure A-4   Symbols for the MOSFET. (a) n-channel; (b) p-channel

Let us consider the circuit of Fig. A-5 to explain the operation of the MOSFET. When v_{GS} is zero or negative, the current i_D will be essentially zero and the MOSFET is said to be *cut off*. When v_{GS} is made positive, i_D increases. If v_{GS} is sufficiently large, the voltage drop v_{DS} becomes very small and the current is essentially limited by the resistance R. This is the *saturation* condition.

MOSFET devices, in general, can be used in circuits which consume very much less power than those circuits which use junction transistors. However, the switching speed of fast junction transistor circuits is considerably greater than that of fast MOSFET circuits.

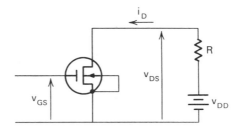

Figure A-5   A simple MOSFET circuit

Integrated Circuits

The modern digital system uses electronic components which are fabricated in such a way that many elements such as transistors, MOSFETs, resistors and capacitors can be incorporated in a single silicon chip. It would not be feasible to construct very large computers if it were not for integrated circuits. In addition, small minicomputers would not be practical without integrated circuits. Integrated circuits can now be fabricated containing thousands of elements on a single chip. This is called *large scale integration* (LSI). Small digital systems can be constructed on several chips using LSI. The entire CPU can be fabricated on a single chip. This is the *microprocessor*, which we have considered in detail in Chapter 11.

A-2 LOGIC FAMILIES

The various combinational logic gates can be constructed using many different circuit combinations. Some use transistors, other use FETs. Some use devices which are driven into saturation and others do not. It is usual to catagorize logic circuits into groups called *logic families*, such as those utilizing diode-transistor logic or those utilizing transistor-transistor logic. In the following sections we shall discuss the commonly used logic families.

Usually, each logic family consists of one or of a few gates which form a functionally complete set, see Sec. 3-3. These are used to constitute all the switching functions. For instance, all switching functions may be made up of NAND gates. Thus, the NAND gate can be the basic gate of the particular family.

Gate families are compared on the basis of several quantities. One of the most important of these, the propagation delay, is a function of the switching speed. The propagation delay is illustrated in Fig. 6-20. Another basis of comparison is the power dissipated by each gate. Let us consider some other standards of comparison.

All electrical devices are subject to extraneous electrical signals called *noise*. This can be caused by stray pick-up of other signals, which could even be generated within a computer, itself. For instance, a pulse generated in one part of the computer could be coupled by stray capacitance to another circuit. This is termed *crosstalk*. Noise may be generated externally by motors, etc. Finally, when very low level signals are encountered, the noise due to the random motion of electrons can become troublesome. Noise can obscure the signals and make 0's appear as 1's and vice versa.

Interference is not the only thing that can cause erroneous signals. Assume that a transistor saturates when its base ucrrent is greater than 10 ma. Because of variation in components, suppose that the driving circuit only supplies 9 ma. The transistor will not be fully saturated, but will be close to it. The succeeding stage will then be driven by a signal which is not as small as it should be. However, the departure from the desired signal will be small and an error will not result. However, if there are a great many circuits cascaded (i.e. one stage drives the next, etc.) then those small departures can eventually accumulate to produce an erroneous signal. This is also termed noise in a computer. Since the manufacturing tolerances on semiconductor devices are reasonably large, the designer should take such departures into account. For instance, cut-off voltage could be larger than necessary to insure that the device is really cut off. Different logic families have different immunities to all types of noise. This is another criterion used in picking logic families.

Often, one gate drives several other gates. Each gate requires some current to drive it. The driving gate must be capable of supplying the current to, and maintaining the voltage levels required by, the succeeding gates. This is a function of both the output capabilities of the output gate and of the input requirements (e.g. impedance) of the driven gates. Usually, for any given logic family, one gate will drive others of the same type. If the output impedance is low, and the input impedance high then, all other things being equal, many gates can be driven. Conversely, few gates can be driven if the output impedance is high and the input impedance is low. The maximum number of outputs that a gate can drive, and still function properly, is called the *fanout*. This is also a criterion used for logic families. Additional criteria which determine the choice of logic family are ease of fabrication and the stability of the device one it is fabricated. In the remainder of this appendix we shall discuss the commonly used logic families and see how the discussions of this section apply to them.

A-3. DIODE-TRANSISTOR LOGIC—DTL

One form of logic circuit uses both diodes and transistors, and is thus called *diode-transistor logic*, abbreviated DTL. The basic gates of DTL for positive logic are the NAND gate and the NOR gate. (Remember that each of these is functionally complete, see Sec. 3-3.)

In Fig. A-6a we have a positive logic DTL, NAND gate. The common lead is marked with a ground symbol. (This also means that this lead may be connected to a common chassis point.) The power supply is

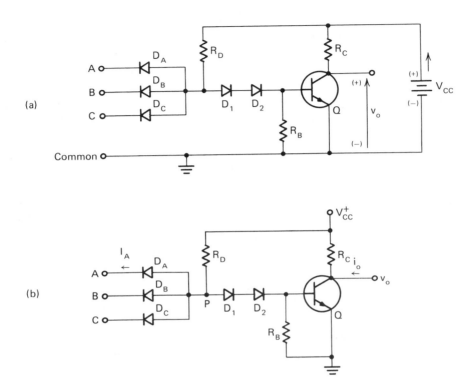

Figure A-6   (a) A positive logic DTL NAND gate; (b) a simplified schematic
representation of (a)

connected between the common lead and R_C as shown. The three
inputs are between A, B and C and the common lead. To conserve space
on the wiring diagram, the common lead is often omitted. Thus, Fig. A-
A-6b is equivalent to Fig. A-6a. We shall use this simplified form of
schematic diagram. Remember that the unidentified sides of a power
supply, or input voltage, are connected to the common lead. The voltage
of any point, unless otherwise marked, is referred to the common lead.

Let us discuss this gate qualitatively. We assume positive logic here.
Suppose that a 0 is represented by 0 volts and a 1 by 5 volts. Assume
that $V_{CC} = 5$ volts. In an actual circuit, these voltages are slightly
different. Suppose that any of A, B or C represents a 0. Then, V_{CC} will
cause the corresponding input diode to conduct. Point P will then be at
a low voltage. We say that point P is low. The diodes D_1 and D_2 will be
reverse biased since the voltage at point P will be less than their cut-in

voltages. There will be no base current and the transistor will be essentially cut off. Thus, if any of A, B or C is a 0, the output v_o will be a 1.

Now suppose that A, B and C are 1's. Their diodes will be cut off. Now, consider the series circuit consisting of V_{CC}, R_D, D_1, D_2, and R_B and the emitter-base junction in parallel. The voltage V_{CC} will now forward bias D_1 and D_2, and there will be base current i_B in the transistor Q_1. The values of R_D and R_B are chosen so that i_B will become large enough to produce saturation. Thus, v_o will be a small voltage which represents a 0. Hence, if A AND B AND C represent 1's, then v_o will represent a 0. Thus, this circuit is a NAND gate. That is

$$v_o = \overline{ABC} \tag{A-2}$$

The fanout of this circuit is a function of the amount of current which can be supplied by it to an external element (i.e. the maximum value of i_o, see Fig. A-6b). This is, in turn, a function of the base current of the transistor and of its β. The details of these calculations are given in the electronics circuits texts cited at the end of the appendix.

The fanout of this circuit can be greatly increased by replacing the diode D_1 by a transistor. This is shown in Fig. A-7. The emitter-base junction of transistor Q_{D1} takes the place of D_1. Now the current i, see Fig. A-7, is increased because of the presence of Q_{D1}. This increases the fanout.

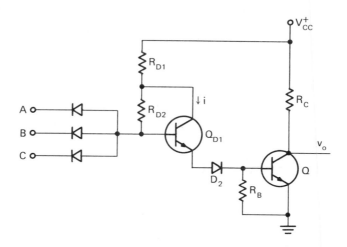

Figure A-7 A modification of Fig. A-6b which has increased fanout

A.4 HIGH-THRESHOLD LOGIC—HTL

If the computer is to be used in a very noisy atmosphere, for instance, near heavy electrical machinery, then the resulting electrical noise can cause errors. There are circuits with lower noise immunity than DTL circuits. These are called *high-threshold logic* circuits, HTL. For these circuits, the power supply voltage is increased to about 15 volts. (Note that by increasing the signal levels and, thus, the difference between a 0 and a 1, we increase the noise immunity.) However, the power dissipated is increased and the switching speed reduced. Thus, this circuit is only used when very high noise requires it.

A.5 TRANSISTOR-TRANSISTOR LOGIC—TTL—T^2L

The most commonly used logic circuit at the present time is called *transistor-transistor logic* (TTL). This is because it has the fastest switching of all the logic circuits that use saturated devices. In many ways, it is closely related to DTL. The basic (positive logic) gate is the NAND gate.

The TTL circuits usually make use of a transistor which has several emitters, as shown in Fig. A-8a. This drawing is not to scale and the base region between the emitter and the collector is really very narrow.

Figure A-8b is an oversimplified model for this transistor, showing only the emitter-base and collector-base junction diodes. We have neglected transistor action which we shall see is important to the switching speed of the circuit. However, we shall use the simplified model to present the basic ideas of the logic circuit.

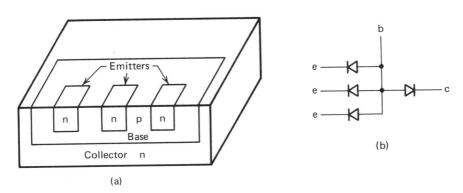

Figure A-8   (a) A 3-emitter transistor; (b) an oversimplified model for this transistor

A basic TTL circuit is shown in Fig. A-9. If we replace the multi-emitter transistor by the simplified model of Fig. A-8b, then this circuit becomes essentially the same as that of Fig. A-7. The positions of the diode D_2 and the transistor Q_{D1} are interchanged but the operation is essentially the same.

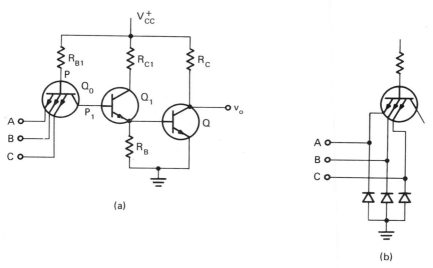

(a)

(b)

Figure A-9 (a) A basic positive logic TTL NAND circuit; (b) an actual TTL input circuit

Let us now consider the operation of this circuit. The following are typical values.

V_{CC} = 5 volts

R_C = 2000 ohms

R_{B1} = 4000 ohms

R_{C1} = 1000 ohms

R_B = 4000 ohms

For the transistor, we assume the following typical values: $V_\gamma = 0.6$ volt and, under saturation conditions, $V_{BE,sat} = 0.7$ volt and $V_{CE,sat} = 0.3$ volt. If all the inputs are 1's, then the "input diodes" will be cut off. Thus, there will be no emitter current in the transistor Q_0. However, we can consider that there are three diodes in series, the base-collector junction of the input transistor Q_0 and the base-emitter junctions of transistors

Q_1 and Q. The voltage drop across this junction under saturation conditions is

$0.7 + 0.7 + 0.3 = 1.7$ volts

Since $V_{CC} = 5$ volts > 1.7 volts, these junctions will be forward biased. Thus, Q_1 and Q will be saturated. Actually, the emitter current of Q_1 will be higher than its base current. This certainly causes transistor Q to saturate. As in the case of the DTL circuit, the transistor action of Q_1 increases the fanout.

Now suppose that one (or more) of the inputs is 0. The input voltage is then 0.3 volt. That is, the voltage level representing a 0 is $V_{CE,sat}$. That (Those) emitter-base junction(s) of Q_0 will be forward biased. Thus,

$$v_p = 0.3 + 0.7 = 1.0 \text{ volt} \tag{A-3}$$

Note that $0.7 = V_{BE,sat}$ and is the voltage drop which is assumed to exist across a forward-biased emitter-base junction. This is less than $V_{\gamma 1} + V_{\gamma 2} = 1.2$ volts, so the collector-base junction of the input transistor will not be turned on. Thus, Q_2 will be off and the v_o rises to 5 volts.

We have thus far ignored the transistor action of the input transistor. Actually, it is important because the emitter currents are much larger than the base currents. These larger currents tend to drive excess stored charge from Q_1 to Q rapidly. This excess stored charge is one factor that reduces the switching speed. The increased currents greatly increase the switching speed. For this reason, the TTL circuit has the fastest speed of any logic that uses saturated transistors.

In an actual integrated circuit, extra diodes are added to the input circuit. These are shown in Fig. A-9b. These diodes are reverse biased and normally do not affect the operation of the circuit. At times, because of inductance in the transmission lines that interconnect the various circuits, damped sinusoid oscillations are set up. These may lead to the occurrence of relatively large negative voltages in a positive logic circuit. The added diodes in Fig. A-9b keep these negative voltages at a very low level preventing them from causing damage.

A.6 SCHOTTKY DIODE—TRANSISTOR-TRANSISTOR LOGIC —STTL—ST²L

The switching speed of TTL circuits is fast because the current due to transistor action helps to rapidly remove the excess stored charge.

The switching speed could be greatly increased if we could prevent the transistor from becoming saturated, in which case, stored charge would not have to be removed. If we could incorporate a device in the circuit that would automatically hold the transistor at the *edge* of saturation, then we would obtain all the advantages of saturated operation (e.g. reliability, low power compensation) without sacrificing the switching time required to remove the excess stored charge that results when the transistor is heavily saturated.

Let us consider the circuit of Fig. A-10a. Here we illustrate a single transistor used in a logic gate. A diode is added between the base and collector of the transistor. (We shall explain the special symbol for the diode subsequently.) If we attempt to forward bias the collector-base junction of the transistor, the diode will also be forward biased. This will divert current from the transistor. Thus, the excess stored charge will be reduced. Note that the diode will not conduct until V_γ is exceeded, so that the diode does not substantially affect the circuit until the "edge of saturation" has been reached. Therefore, this circuit appears to have achieved the desired results.

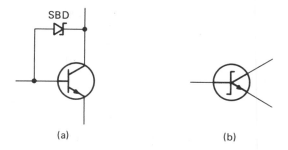

(a) (b)

Figure A-10 (a) A transistor-Schottky barrier diode circuit that improves switching speed; (b) a symbol for the composite transistor

If the diode is an ordinary p-n junction diode, then the performance of this circuit will not be satisfactory. There will be excess charge stored in the diode when it draws a large forward current. When the transistor is to be cut off, the diode must act as an open circuit. However, until its excess stored charge is eliminated, the diode will not switch and acts as a low impedance. Thus, when switching times are considered, the reduced storage time of the transistor will be offset by the increased storage time of the diode and great improvements in switching speed will not be obtained. However, there is a diode called a *Schottky-barrier diode* which can be used to remove this difficulty.

This diode consists of a metal-to-doped-semiconductor junction, rather than a p-n junction. Such a device can exhibit all the rectifying properties of the p-n junction diode. However, when Schottky diodes are forward biased, they have essentially zero charge storage. Hence, they do not exhibit storage time effects and their switching is very rapid. The diode symbol of Fig. A-10a is that of the Schottky diode. In addition, Schottky diodes can be constructed so that their forward bias cut-in voltage can be made slightly less than that of the forward bias collector-base cut-in voltage of a transistor. Hence, in the configuration of Fig. A-10a, most of the collector-base current will be diverted to the diode when saturation conditions exist. Thus, during "saturation," the excess stored charge of the transistor will be very low, and the addition of the Schottky barrier diode can significantly improve the switching speed.

When Schottky barrier diodes are incorporated in the already fast TTL circuits, a very rapid circuit results. These are called Schottky diode transistor-transistor logic circuits, which are written as STTL or ST^2L. All of the transistors of the gate, including the multiple-emitter input transistor, incorporate Schottky diodes. The transistor-Schottky diode combination is represented by a single symbol, see Fig. A-10b.

Another advantage of these circuits is that they are not very difficult to fabricate as integrated circuits. In fact, they are only slightly more difficult to fabricate than are ordinary TTL integrated circuits.

A.7 ACTIVE PULL-UP CIRCUITS

The output stage of the logic circuits we have considered consists of the resistor R_C in series with the transistor Q, see Fig. A-6. This is called a *passive pull-up* circuit, since the change in voltage v_o can be considered to be brought about by the change in voltage drop across the passive R_C. Now consider that Q is switched from saturation to cut-off. The output voltage across Q rises from its saturation value to its cut-off value. There is an effective capacitance C_o between the collector and emitter of the transistor. As the voltage rises, this capacitance must be charged. The time constant of the RC circuit is

$$T = R_C C_o \tag{A-4}$$

The larger T is, the longer it takes for the output voltage to reach its "final" value. This limits switching speed. In general, C_o is fixed. We can reduce R_C. However, this will increase the dissipated power under saturation conditions, which is undesirable. Note that this power, dissipated in R_C is

$$P = (V_{CC} - V_{CE,sat})/R_C \tag{A-5}$$

and V_{CC} and $V_{CE,sat}$ are independent of R_C.

An alternative circuit can eliminate the difficulty of increased power. We desire that R_C be small under cut-off conditions when the current through it and, hence, the power dissipated by it, is small. Then, C_o can be charged quickly. However, we want R_C to be large when Q is saturated. (Note that C can be discharged quickly through the saturated transistor.) To obtain such a variable R_C, we replace it by a transistor, see Fig. A-11. Assume that the voltages v_1 and v_2 are "out of phase" so that when v_1 produces saturation of Q, v_2 results in the cut-off of Q_P and vice versa. Now, when Q is switched from saturation to cut-off, Q_P is saturated. Thus, it acts as a small resistance and the switching can take place rapidly. On the other hand, when Q is saturated, then Q_P is cut off. This causes its collector current to be extremely small. This, in turn, causes the dissipation to be limited. This circuit is called an *active pull-up* because the passive R is replaced by the active transistor Q_P. It is also called a *totem pole* circuit because one transistor is above the other one. This circuit can switch faster, with lower power dissipation, than the passive pull-up circuit.

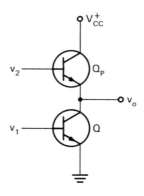

Figure A-11 A basic active pull-up output circuit

Now let us discuss the practical circuit of Fig. A-12. When Q_1 is cut off, then V_a will be high. Thus, Q_P will be saturated. Since Q_1 is cut off, so is Q. When Q_1 is saturated, V_a will be low and Q_P will be

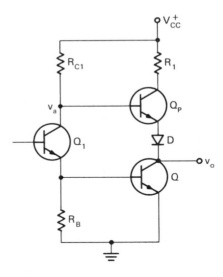

Figure A-12 A practical active pull-up output circuit

cut off and Q will be saturated. Thus, we have achieved the desired operation. The resistor R_1 has a small value and is used to limit current during switching. A transistor switches from cut-off to saturation faster than in the reverse direction. Thus, there will be a short transient period of time, near the switching time, when both transistors are saturated. If R_1 were not present, then the power supply would be "short circuited" during this period. R_1 limits the current, reducing both power dissipation and large power supply current pulses which could act as noise in other parts of the computer. A typical value for R_1 is 100 ohms. The diode D is included to obtain proper bias for the transistor Q_P.

A.8 RESISTOR-TRANSISTOR LOGIC—RTL

Another form of logic circuit which makes use of resistors and transistors is shown in Fig. A-13. A NOR gate is illustrated there. This is the basic gate of this family for positive logic. If any input is a 1, then its transistor saturates and v_o becomes low. If all inputs are 0's, then the transistors are cut off and v_o is high. Thus, the desired logic function is obtained.

This circuit uses very little space on an integrated circuit chip. It also can operate at low voltage levels. For instance, typical values are V_{CC} =3.5 volts, R_B=450 ohms, and R_C=700 ohms.

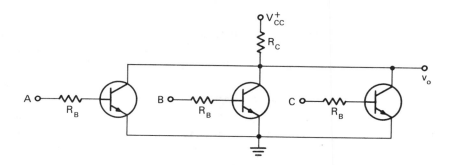

Figure A-13 A positive logic RTL NOR gate

Let us consider what happens when this circuit drives similar ones. Suppose that the output wire is connected to two inputs. This would be as shown in Fig. A-14. In general, transistors do not have identical characteristics. Thus, if Q_1 has a lower v_{BE} than Q_2, it would take more than its share of base current were it not for R_B. In saturation, i_B is essentially a function of R_B and not of v_{BE}, provided that R_B is large enough (i.e. greater than several hundred ohms). This assumes typical values of V_{CC}. Thus, keeping the two values of R_B nearly equal will insure that a proper division of current results even if the transistors are different.

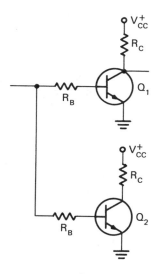

Figure A-14 The inputs of two RTL circuits that are driven by a single output

A.9 DIRECT-COUPLED TRANSISTOR LOGIC—DCTL

If we set the value of $R_B = 0$ in Figs. A-13 and A-14, the logic circuit obtained is called *direct-coupled transistor logic*. Its advantage over RTL is that the switching speed is increased since the larger base current can remove excess charge faster.

However, DCTL has several disadvantages. Under saturation conditions, there is not any R_B to regulate the base current. Thus, great differences in base current can result. Actually, one transistor can take so much current that the other is not driven into saturation. This is called "*current hogging.*" Thus, great care must be taken to make the transistors nearly identical. In addition, the very large values of base currents can result in large amounts of excess charge which tends to offset the previously discussed advantage of this circuit. This circuit can operate at very low voltage levels, which is an advantage. Other logic circuits are now much more commonly used than is this one. In the next section we shall discuss a circuit that incorporates many of DCTL's advantages without its disadvantages.

A.10 INTEGRATED-INJECTION LOGIC—I²L

We shall now consider a circuit that can be considered to be based upon DCTL but which does not have the current-hogging problem. Like DCTL, it does not have any resistances so it can be easily fabricated using integrated circuit techniques.

Consider the circuit of Fig. A-15a. This is a direct-coupled circuit wherein one transistor Q drives two others, Q_A and Q_B. This circuit is prone to current hogging. If $v_{BEA} < v_{BEB}$, then i_A may be very much greater than i_B and the circuit may not function properly. Now consider Fig. A-15b. The two transistors Q_A and Q_B have been combined into one transistor with two collectors. Thus, we still effectively have one transistor Q, driving two stages, but now there is only one base that is being driven by Q. Hence, the current-hogging problem does not exist, since the output current of Q does not have to divide.

In integrated circuits, transistors can be fabricated in less space than can resistances. Thus, it would be desirable to eliminate the resistors from the circuit. This is accomplished in Fig. A-16. The transistor Q_I acts as a current generator, *injecting* a fixed amount of current. Hence, it acts as a large resistance. This injecting action gives the logic circuit its name.

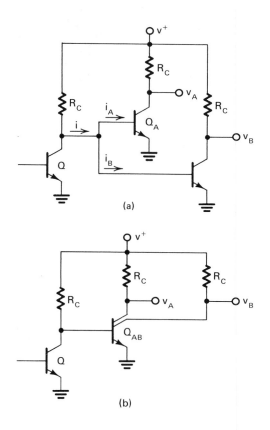

(a)

(b)

Figure A-15 (a) A direct-coupled circuit that may suffer from current hogging; (b) a circuit that does not have that problem

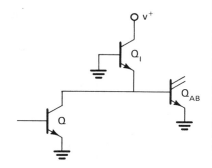

Figure A-16 A modification of Fig. A-15b that does not require resistances

An I^2L NOR gate is illustrated in Fig. A-17. If any of v_A, v_B, or v_C is high, then its corresponding transistor will be turned on. Hence, if any of v_A, v_B, or v_C is high, v_o will be low. The only time that v_o will be high is if all of v_A, v_B, and v_C are low. In terms of logical values, the operation can be described by

$$v_o = \overline{v_A + v_B + v_C} \tag{A-6}$$

Hence, we have achieved a NOR gate.

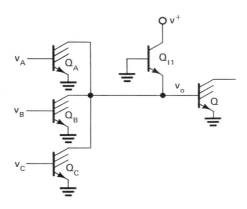

Figure A-17 A positive logic I^2L NOR gate

A.11 EMITTER–COUPLED LOGIC—ECL

The previously discussed logic circuits all utilized saturated transistors. This makes for reliable operation and low power dissipation since the voltage across the transistor is low when the current through it is high and vice versa. However, saturation slows switching speed. Now let us consider a nonsaturated logic circuit which has the fastest switching speed. However, its power dissipation is relatively high. The basic idea of this circuit is illustrated in Fig. A-18a. Here a current generator drives the parallel combination of two resistors. For this circuit we have

$$I_1 = IR_2/(R_1+R_2) \tag{A-7a}$$

and

$$I_2 = IR_1/(R_1+R_2) \tag{A-7b}$$

Now suppose that R_1 is a variable resistor. If $R_1 \gg R_2$, then

$$I_1 = IR_2/R_1 \approx 0$$

$$I_2 = I$$

Thus, essentially all the current passes through R_2. Similarly, if $R_1 \ll R_2$, then essentially all the current will pass through R_1. Again, we can implement the "variable resistance" using a transistor. This circuit is shown in Fig. A-18b. The voltage V_{BB} is a fixed voltage while v_1 can be varied. If v_1 is such that Q_1 is cut off, then the current I will be through Q_2. Similarly, if we make v_1 much larger than V_{BB}, then I will be through Q_1 while the current in Q_2 will be very small. Since I is limited, the transistors can be kept unsaturated even though heavily conducting. Since it is the current that is switched in this circuit, the circuit is called *current-mode logic, CML*. Also, because the two transistors are coupled through their emitter circuits, this is also called *emitter-coupled logic, ECL*. The basic gate of this family for positive logic is the OR or NOR gate.

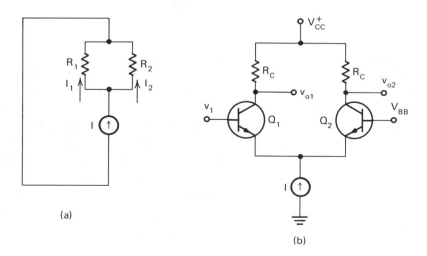

(a)

(b)

Figure A-18 (a) A simple current-mode model; (b) a transistor implementation of this model

Now let us consider the more practical circuit of Fig. A-19. Here the one "variable" transistor has been replaced by three, so that we have a 3-input gate. The constant current generator has been replaced by a large resistor in series with a voltage source. If R_E is large enough, the current through it will be essentially a constant.

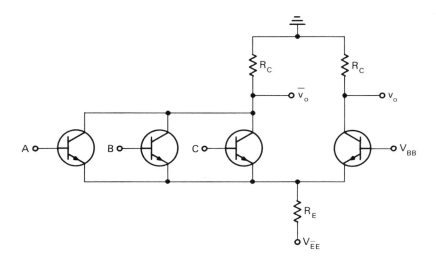

Figure A-19 An ECL NOR/OR positive logic gate

Typically, $V_{EE}=-5$ volts. Now, if A or B or C represents a 1, then the corresponding transistor will conduct heavily. Then, $\bar{v}_o$ will represent a 1 and v_o a 0. If all of A, B and C are 0's, then $\bar{v}_o=0$ and $v_o=1$. Thus, the output v_o gives a NOR gate while $\bar{v}_o$ is its complement, an OR gate. One problem with this circuit is that the voltage at the collector of the transistor differs from the base voltage. Thus, the voltage levels at the output will differ from those at the input. It is desirable to keep the voltages corresponding to 0's and 1's constant throughout the system. This can be accomplished by adding the common-collector amplifiers shown in Fig. A-20. This common-collector amplifier has a very low output impedance and, thus, increase the fanout.

The emitter-coupled logic is faster, but dissipates more power and has worse noise immunity than TTL. Thus, each has its advantages.

A.12 MOSFET LOGIC FAMILIES—MOS LOGIC FAMILIES

We shall now consider logic families that use MOSFET devices, specifically those using enhancement MOSFETs. In general, MOS

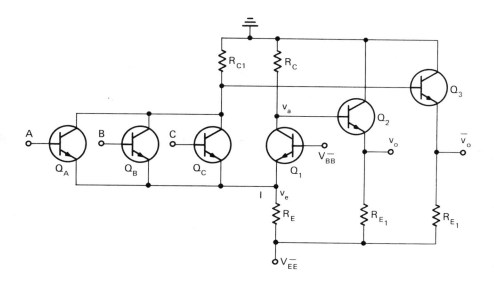

Figure A-20 An ECL NOR/OR gate with no level shift between input and output

integrated circuits can be fabricated using fewer steps than those using junction transistors. In addition, power dissipation can be much smaller with MOSFET logic. However, the switching speed is much faster using junction transistors.

Basic MOSFET Logic—MOS Logic

A basic MOSFET logic element consists of an FET in series with a resistive pull-up element, see Fig. A-5. However, it is easier to fabricate MOSFETs than resistors in integrated circuits. Thus, an active pull-up is used, see Fig. A-21. Assume that the upper FET acts as a constant resistance. If A represents a 1 which is a positive signal, then FET F_1 will saturate. This results in v_o's being a small positive voltage output. Typically, $V_{DD} = 10$ volts and the level for a 0 is a fraction of a volt. If A represents a 0, then the input bias will be small and the transistor F_1 will be almost cut off. In this case, v_o will be close to V_{DD}. Thus, we have achieved inversion.

A positive logic NOR gate is shown in Fig. A-22a. Note that if any input is a 1, the corresponding FET will saturate and v_o will represent a 0. If all inputs are 0's then the FETs F_1, F_2 and F_3 will be cut off and v_o will represent a 1. Similarly, Fig. A-22b is a positive logic NAND gate. The NOR and NAND are the basic MOS logic gates.

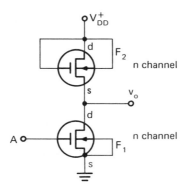

Figure A-21 A basic positive logic MOSFET inverter

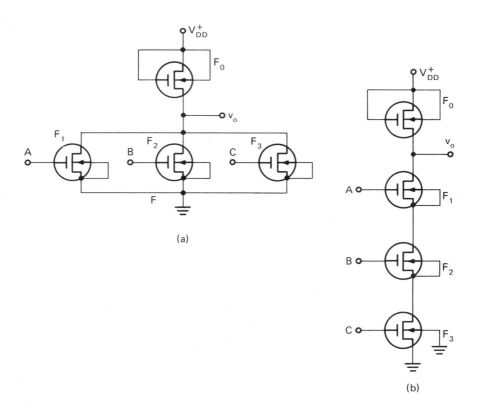

Figure A-22 MOSFET positive logic gates. (a) NOR; (b) NAND

A.13 COMPLEMENTARY SYMMETRY MOSFET LOGIC— MOS LOGIC—COSMOS LOGIC

If we use both n-channel and p-channel enhancement MOSFETs in the logic circuit, circuits which draw very little current can be obtained. In Fig. A-23 we illustrate a basic inverter circuit using this complementary symmetry. Let us assume that a 0 is represented by a small positive voltage essentially equal to zero, and a 1 by a positive voltage essentially equal to V_{DD}, typically 10 volts. Suppose that input A is a 1. Then, F_1 will be saturated while F_2 will be cut off. Note that F_1 is an n-channel device while F_2 is a p-channel device. Thus, the effective resistance of F_2 becomes extremely large while that of F_1 is very small. Thus, $v_o \approx$ 0 volts. Now suppose that A represents a 0. Then, F_1 will be cut off. It may appear as though F_2 would also be cut off. However, the gate-to-source voltage of F_2 is

$$v_{GS} = v_1 - v_o \qquad \text{(A-8)}$$

Since v_o is positive and $v_1 = 0$, then v_{GS} will be negative and equal in magnitude to V_{DD}. Then, F_2 will be saturated. Hence, $v_o = V_{DD}$.

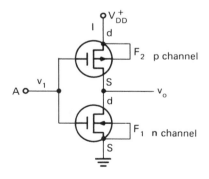

Figure A-23 A COSMOS positive logic inverter

Note that either FET F_1 or F_2 will always be cut off. Thus, the current through the circuit will be extremely small. Hence, this circuit dissipates very little power and, consequently, many circuits can be put on a single chip. The very low power and high packaging density are prime advantages of this circuit.

For a transient period after switching, there will be some current. This is because the various capacitances of the circuit must be charged or discharged. If the gates are switched very rapidly, the power dissipation will increase.

Since these are insulated gate FETs, the input impedances of the gates are very high. Thus, the fanout of these circuits is high. Note that there is a capacitance across the input. Thus, if very short pulses are considered, their capacitances will act as a load. Hence, the very high fanout is obtained only when very short pulses are not used.

A positive logic NOR gate is shown in Fig. A-24. We can consider pairs of FETs, one in the lower "parallel" section and the other in the upper "series" section. The behavior of each pair is similar to that of the pair of Fig. A-23. If any input is a 1, the corresponding parallel lower FET will saturate, while the corresponding series upper FET will cut off. Thus, the output will be a 0. If all the inputs are 0, then all the lower FETs will be cut off and all the upper ones will be saturated and the output will be a 1. Thus, this device functions as a NOR gate.

A positive logic NAND gate is illustrated in Fig. A-25. The operation of this gate is the converse of that for the gate of Fig. A-24.

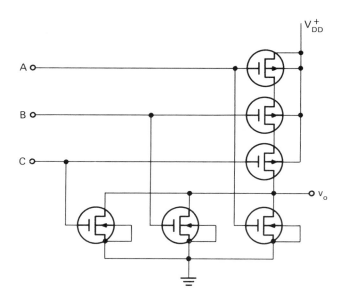

Figure A-24 A COSMOS positive logic NOR gate

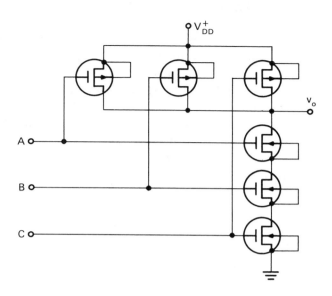

Figure A-25 A COSMOS positive logic NAND gate

A.14 SOME GENERAL COMMENTS ABOUT LOGIC FAMILIES

We have considered the basic logic gates of various families. In LSI, many interconnected gates are fabricated on one chip. The user must then connect the proper power supply voltage(s) and provide for the appropriate input and output signals. Care should be taken to see that the signals are of the proper polarities and are at the correct levels. The impedance of the input and output circuits should also be compatible. Appropriate values can usually be obtained from the manufacturer's specifications.

When small-scale integration is used, gates of one logic family are usually interconnected by the user to obtain a switching function. Again, the correct power supply voltages must be obtained. If all the gates are from the same family and are produced by the same manufacturer, then there will probably be no incompatibility of levels. However, care should be exercised. (Note that gates other than the basic ones are produced.) The previous comments about input and output circuits apply. In addition, the rated fanin and fanout should not be exceeded.

Comparison of Logic Families

We have indicated the various advantages and disadvantages of logic families. Let us now summarize the most important points.

At the present time, TTL is probably the most widely used logic family. Its switching speed is the fastest of any saturated logic circuit. The use of STTL increases the speed even more.

ECL gates are the fastest available. However, the power dissipated is relatively high and the noise immunity is much poorer than in TTL.

DTL is comparable to TTL except that its switching speed is slower and its fanout slightly less. The noise immunity is somewhat less than with TTL.

RTL has a switching speed slightly slower than TTL. However, its fanout and noise immunity are poorer.

I^2L and MOS have very low power dissipations when operated at slow switching speeds. They are very easy to fabricate as integrated circuits. CMOS is widely used in microprocessors and in special-purpose computers.

Table A-1 summarizes the pertinent characteristics of logic families. The delay is the amount of time required for a signal to propagate through the gate. High speed means small delay times.

Table A-1 Typical Characteristics of Logic Families

	RTL	*DTL*	*TTL*	*STTL*	*ECL*	*I^2L*	*MOS*	*CMOS*
Delay	15ns	30ns	10ns	3ns	2ns	50ns	100ns	50ns
Fanout	6	8	12	12	16	12	12	12
Power dissipation	20mW	9mW	10mW	20mW	25mW	100nW	1mW	10nW
Noise immunity	0.3v	0.9v	1.0v	1.0v	0.05v	0.2v	0.9v	1.7v

Note: n = nano = 10^{-9} ; hence, 1ns = 1×10^{-9} sec.
      m = milli = 10^{-3}

BIBLIOGRAPHY

Belove, C., Schachter, H., Schilling, D.L., *Digital and Analog Systems, Circuits and Devices*, Chap. 5, McGraw-Hill Book Co., New York 1973

Chirlian, P.M., *Analysis and Design of Integrated Electronic Circuits*, Chap. 9, Harper & Row, New York 1981

Gray, P.E., and Searle, C.L., *Electronic Principle Physics, Models and Circuits*, Chaps. 4-10, 22, John Wiley and Sons, New York 1969

Kohonen, T., *Digital Circuits and Devices*, Chaps. 6 and 7, Prentice Hall, Inc., Englewood Cliffs, NJ 1972

Millman, J., *Microelectronics*, Chap. 5, McGraw-Hill Book Co., Inc., New York 1979

O'Malley, J., *Introduction to the Digital Computer*, Chap. 3, Holt, Rinehart and Winston, New York 1972

PROBLEMS

A-1. Describe the operation of the DTL circuit of Fig. A-6.

A-2. Why is Fig. A-7 an improvement over Fig. A-6?

A-3. Describe the operation of the TTL circuit of Fig. A-9.

A-4. What are the advantages of STTL over TTL?

A-5. What are the advantages of active pull-up?

A-6. Describe the operation of the RTL circuit of Fig. A-13.

A-7. What is the major disadvantage of DCTL?

A-8. Compare DCTL and I^2L.

A-9. Describe the operation of the I^2L NAND gate of Fig. A-17.

A-10. Describe the opeation of the ECL gate of Fig. A-20.

A-11. Describe the operation of the MOSFET gates of Fig. A-22.

A-12. Describe the operation of the CMOS NOR gate of Fig. A-23.

A-13. Repeat Prob. A-12 for the NAND gate of Fig. A-24.

A-14. Describe why ECL can be faster than TTL.

A-15. What are the advantages of TTL over ECL?

A-16. Compare the various logic gates.

A-17. If speed is the most important factor, which logic gate family would be used?

A-18. If power dissipation is the most important factor, which logic gate family would be used?

A-19. If noise immunity is the most important factor, which logic gate family would be used?

A-20. Which gate family would be used at the present time for most digital computer applications? Is it one of those picked in Probs. A-17 to A-19? Why?

Index

1 3 57

Engin.

TK7868.D5C45
    Chirlian, Paul M.
      Digital circuits with
microprocessor applications.

9/26/90